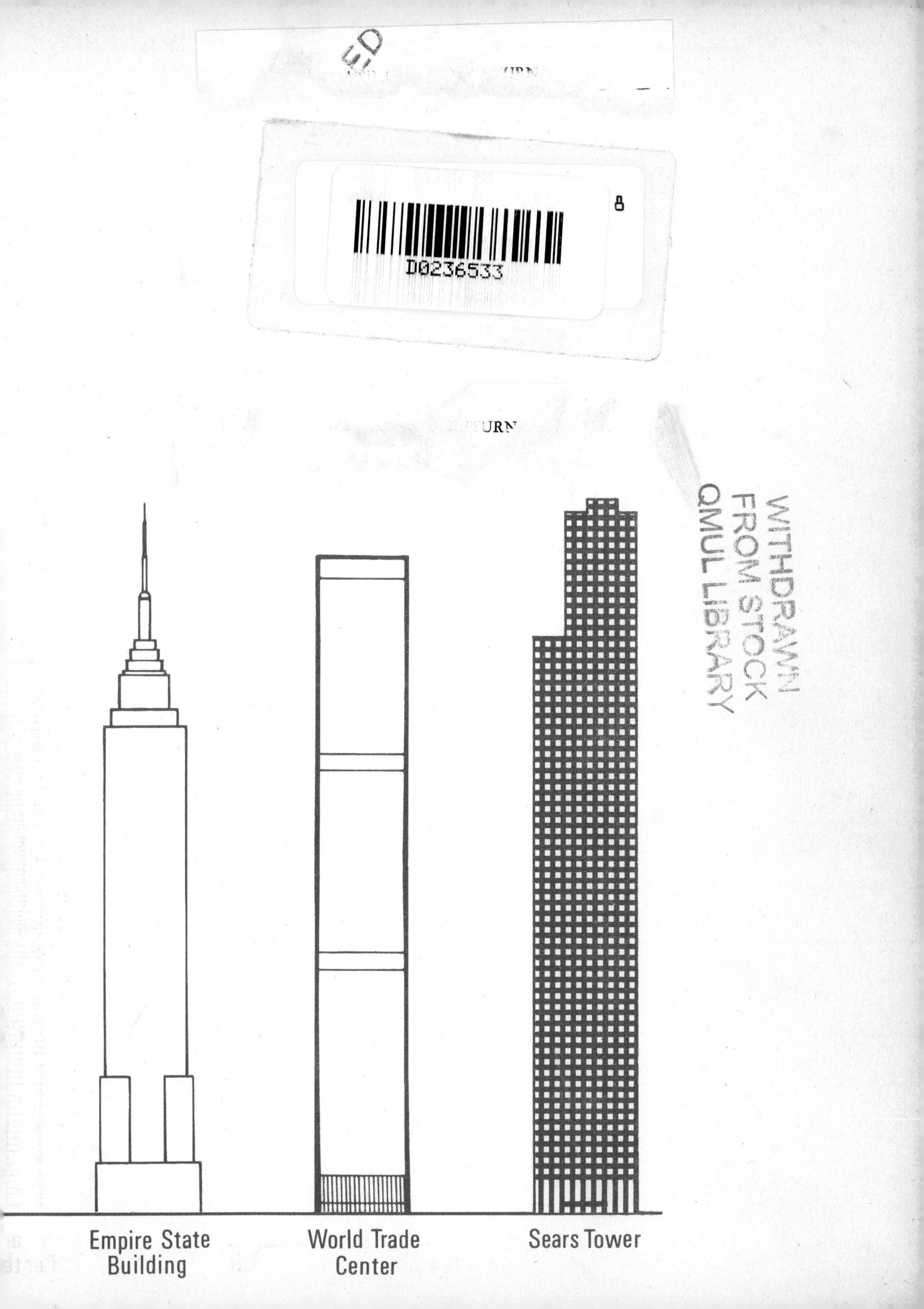
Empire State Building
World Trade Center
Sears Tower

THE GUINNESS BOOK OF STRUCTURES

bridges, towers, tunnels, dams

John Lucas's painting of eminent Victorian civil engineers at Britannia Bridge, often called *The committee meeting*, was the idea of Samuel Morton Peto, the railway contractor. Personalities shown are, left to right, sitting: Robert Stephenson, MP, Chas H Wild, Joseph Locke, MP, I K Brunel. Left to right standing are: Admiral Moorson, Latimer Clark, Edwin Clark, Frank Forster, G P Bidder ('the calculating boy'), Mr Hemmingway (initial unknown, but he was the master mason who was contractor for the piers of the bridge), Capt Claxton RN, Alex Ross. Posterity has not recorded the identity of the two workmen.
The painting hangs in the building of the Institution of Civil Engineers at Westminster. *This reproduction is from a photograph by British Transport Films*

THE GUINNESS BOOK OF STRUCTURES

bridges, towers, tunnels, dams

John H Stephens

GUINNESS SUPERLATIVES LIMITED
2 CECIL COURT, LONDON ROAD, ENFIELD, MIDDLESEX

Published in Great Britain by

Guinness Superlatives Ltd, 2 Cecil Court, London Road,
Enfield, Middlesex

ISBN 0 900424 28 1

Set in 'Monophoto' Baskerville Series 169
printed and bound in Great Britain by
Jarrold and Sons Ltd, Norwich

OTHER GUINNESS SUPERLATIVES TITLES

Facts and Feats Series:

Air Facts and Feats, 2nd ed.
John W R Taylor, Michael J H Taylor,
David Mondey

Rail Facts and Feats, 2nd ed.
John Marshall

Tank Facts and Feats, 2nd ed.
Kenneth Macksey

Yachting Facts and Feats
Peter Johnson

Plant Facts and Feats
William G Duncalf

Car Facts and Feats, 2nd ed.
edited by Anthony Harding

Business World
Henry Button and Andrew Lampert

Music Facts and Feats
Bob and Ceila Dearling

Animal Facts and Feats, 2nd ed.
Gerald L Wood

Guide Series:

Guide to Fresh Water Angling
Brian Harris and Paul Boyer

Guide to Mountain Animals
R P Bille

Guide to Underwater Life
C Petron and J B Lozet

Guide to Formula 1 Motor Racing
Jose Rosinski

Guide to Motorcycling, 2nd ed.
Christian Lacombe

Other titles:

English Furniture 1550–1760
Geoffrey Wills

English Furniture 1760–1900
Geoffrey Wills

The Guinness Guide to Feminine Achievements
Joan and Kenneth Macksey

The Guinness Book of Names
Leslie Dunkling

Battle Dress
Frederick Wilkinson

Universal Soldier
edited by Martin Windrow
and Frederick Wilkinson

History of Land Warfare
Kenneth Macksey

History of Sea Warfare
Lt Cmdr Gervis Frere-Cook
and Kenneth Macksey

History of Air Warfare
David Brown, Christopher Shores
and Kenneth Macksey

The Guinness Book of Facts
edited by Norris D McWhirter

The Guinness Book of Records, 23rd.
edited by Norris D McWhirter

CONTENTS

	List of tables	6
	Some engineering superlatives	8
	Introduction	9
Chapter 1	*ABOVE GROUND*	
	bridges	11
	dams	61
	towers, masts and tall buildings	85
	clear spans within buildings	101
Chapter 2	*BELOW GROUND*	
	tunnels and underground workings	105
	excavation and dredging	141
	foundations	155
	ground engineering	165
	HYDRAULIC WORKS	
	harbours and canals	174
	barrages and spillways	193
	flood control and sea defence	197
	hydro-electric power	200
	pumping stations	202
	irrigation	204
	aqueducts	209
	pipelines	212
	water supply and treatment	216
	main drainage and sewage treatment	219
Chapter 4	*ENGINEERS AND CONSTRUCTION*	
	great civil engineers	224
	great schemes	232
	constructional miscellany	262
	nicknames	274
	Index	277

list of tables

The longest span for a stressed-ribbon bridge is that of the bridge carrying a belt conveyor at the cement works of Holderbank–Wildegg in Switzerland. This bridge has a span of 216·4 m (710 ft) and the span sags 14·75 m (48·4 ft). It consists of steel cables with precast concrete deck elements clipped to them. The extreme depth of these precast elements is 0·26 m (10·25 in) giving the bridge a span : depth ratio of 832. This is the world's slenderest bridge. It is also unique in being the only bridge that passes in free span through a tunnel. The box-like shape on the bridge is the conveyor. *Photo: Ed. Züblin AG*

TABLE		PAGE
1	The world's longest spans	27
2	Chronological list of the world's longest span	29
3	The world's longest arch spans	32
4	The world's longest cantilever spans	39
5	The world's longest steel plate and box girder spans	44
6	The world's longest spans of concrete (except arches)	45
7	Chronological list of the world's longest simply supported truss span	47
8	The world's longest spans of opening bridges of various types	53
9	The longest bridges (as opposed to bridge spans) in the world	56
10	The slenderest bridges; the shallowest arches; structural continuity	59
11	The world's largest reservoirs	67
12	The world's largest dams; embankments of earth or rock	70
13	The world's largest concrete dams	80
14	The world's longest dams	81
15	Major sea dams	82
16	Chronological list of the world's highest dam	83
17	The world's tallest structures	87
18	Chronological list of the tallest structure in the world	89
19	The world's tallest structures by material of construction	92
20	Comparative weights of steel	94
21	The world's roofs of longest span	96
22	Roofs spanning more than 100 m (328 ft)	97
23	The world's largest buildings	99
24	Examples from Moh's scale of hardness	106
25	The world's longest tunnels	109
26	The world's longest hydro-electric tunnels	113
27	Hydro-electric tunnels of largest cross-section	115

The manifold of Chute-des-Passes Tunnel where five branches each lead to one of the turbines in the power station. The main tunnel is 10·5 m (34·3 ft) in diameter and transmits one million horse power (750 MW) at a head of 196 m (640 ft). *Photo: Aluminium Company of Canada, Ltd (Principal fully-owned subsidiary of Alcan Aluminium Ltd)*

TABLE		PAGE
28	Chronological list of the world's longest tunnel	117
29	Chronological list of the world's longest subaqueous tunnel	119
30	Chronological list of the largest immersed-tube tunnels	120
31	Weekly rates of advance for tunnels in rock	132
32	Tunnels and shafts under the highest pressures	140
33	Feats of excavation in construction	154
34	Chronological list of the deepest caisson or monolith	157
35	Cofferdams	166
36	The world's longest breakwaters and jetties	180
37	The world's largest dry-docks	184
38	Chronological list of the world's largest dry-dock	185
39	The world's largest locks	186
40	Inland navigation – the longest routes	189
41	Ship canals	190
42	The longest canals	191
43	The world's largest high-head spillways	193
44	The world's largest barrages	194
45	The world's largest hydro-electric power stations	201
46	The world's largest pumping stations	203
47	Areas of land irrigated	205
48	The world's longest aqueducts	210
49	The longest sea outfalls and pipeline crossings	212
50	The world's largest sewage treatment works	223
51	Statistics for the Kuanhsien irrigation scheme	238
52	Statistics for the Mississippi River	242
53	Suez and Panama Canals	258
54	The world's heaviest lifts	264

some engineering superlatives

The greatest span is 4888 m (16037 ft), the distance spanned by a high-voltage transmission line across the Sogne Fiord, Norway

The tallest structure is the Warszawa Radio Mast at Płock, Poland. Its height is 640 m (2099·7 ft)

The longest structure is the Great Wall of China. The best estimate of the length of the main line only of the wall is 3460 km (2150 miles)

The deepest foundations are those of the Forties North Sea oil production platforms. Piles penetrate up to 100 m (330 ft) into the sea bed, that is about 230 m (750 ft) below mean sea level

The greatest load carried by any structure will be 18 million tonnes, the load imposed by the full reservoir on Sayany Dam in the USSR. The dam is at present under construction

The largest flood passed by any structure is 70510 m^3/sec (2·29 million ft^3/sec), the flood of the Ganga River in 1971 at Farakka Barrage, India

The fastest rate of excavation on record is 69·9 million m^3 (91·4 million yd^3) at the North Kiangsu Canal, China, achieved in 80 days entirely with manual labour

The largest excavation is Bingham Canyon copper mine, USA, from which about 1200 million m^3 (1570 million yd^3) had been removed by the end of 1974

The largest man-made mound is New Cornelia tailings 'dam' near Ajo, Arizona, USA. Its volume is 209515000 m^3 (274026000 yd^3)

The deepest drill hole is in Washita County, Oklahoma, USA. It was drilled to a depth of 9853 m (31441 ft)

The biggest sewer is Mexico City's Central Outfall which has a maximum diameter of 6·5 m (21·3 ft)

The biggest pipes are the penstocks serving Grand Coulee number 3 power station, USA. Each one is 21·3 m (40 ft) in diameter

The heaviest weight ever lifted is 41 000 short tons (37 190 t), the weight of the roof of the Velodrome in Montreal, Canada, which was raised by jacks to strike its centering. This record excludes direct flotation, i.e. the displacement of a ship (maximum about 500000 t) or the weight of a ship raised by pumping out the tanks of a floating dock (maximum about 100000 t)

The heaviest mobile object will be the Ninian Field oil production platform, under construction at Loch Kishorn and due to be floated to site in 1977. This concrete structure will weigh about 600000 t when fully ballasted

introduction

Some statistics have a futility that is absolute. Others, when related to their subject with understanding, give a sense of scale to, and a quantitative judgment of, achievement. The author has tried to emphasise the latter kind, and to interpret, if briefly. The quality of human imagination, intellect and endeavour that had to be contributed to the great works of construction listed in this book cannot be conveyed by figures alone. A minimum of explanation, we hope, will set the reader facing the road to understanding.

In some ways the book records a threshold. A great expansion in competence and quantity of civil engineering construction is a characteristic of our time. Tunnelling is at the start of a golden age. There is greatly increased confidence in tackling all types of problems below ground level. The age of dam-building, now at its peak, will probably be as well defined as that of the Gothic cathedrals; sea dams only, remain in their infancy. In building construction, the structural engineer can meet virtually any demand – the greatest clear spans perhaps excepted – made upon him. Bridge-building matches other subjects, but the limit of span of the suspension bridge – generally considered to be about twice the present record – has not been reached. Revolution is still possible if some new material – perhaps carbon fibre for example – of very high strength could be employed. Revolution is taking place in the exploitation of oil far out in the open sea.

The other side of the coin is that failure and collapse are still too common; they are the penalty for carelessness, greed, or over-ambitious boldness in design. The standard of humdrum everyday buildings, like houses, is patchy through economic pressures, and was better 50 years ago than it is today. There is also a great pressure to prevent the construction of many big projects by those who see them as a threat to the natural environment. Clearly a balance has to be drawn, but injurious side effects are by no means inevitable.

In spite of all the achievement and promise, the world's greatest need in construction is rather simple. It is for houses for the vastly expanding cities of the less-developed countries. They must be cheap and simple – simple enough to be built, with guidance and advice, by the families that need them to live in. A kit of parts for a 'self-help' house, suitable in its locality to be built on a large scale, is world priority number one.

SCOPE OF THE SUBJECT

The term 'civil engineer' was first used in the English language in the 18th century by Smeaton, to distinguish his activities from military engineering. No other distinction was necessary at that time, but later mechanical and electrical engineering hived themselves off, and today civil engineers serve the military. In a word, the civil engineer constructs. His discipline carries him on occasion into the higher flights of pure science, but equally at other times he needs gum boots to wade through the stickiest empirical mud. He must understand and cross the borderlines of the disciplines of his other engineering colleagues and of architects on matters that are primarily technical, and equally of those of cost accountants, economists, financial specialists and management consultants in matters of the organisation and operation of projects. He may be purely professional – an independent consultant; he may run a contracting business; or he may serve the central or local government or some overseas arm of government or an international organisation. He may construct for, or operate, a State-run organisation such as a railway or harbour board. Each of these fields requires its own superstructure of bureaucratic specialisation; the technical knowledge cannot function in a vacuum. On the technical side, however, the way that civil engineering overlaps and mingles with other disciplines can be shown thus:

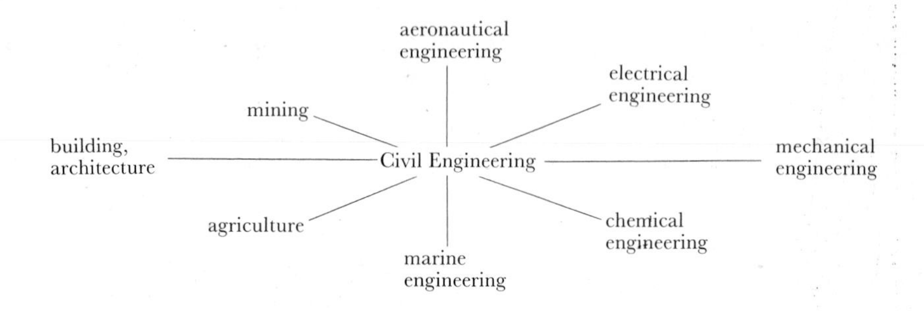

AUTHOR'S APOLOGY

One decision has been made with regret. The book does not give the names of the engineers responsible for each entry. To do this reasonably, it would have been necessary to state (1) the owner, (2) the engineer responsible for design, and (3) the contractor who built the job. Sometimes there is more than one name in each category. It has been difficult to concentrate the information into tables and text of reasonable size and inclusion of the names would have made this book too cumbersome for easy and pleasant use. Names have, however, been given wherever possible. For example most of the acknowledgments of photographs are to specialist firms concerned with the job illustrated.

In some of the other Facts and Feats books in this series, the subject is in the public eye and the record is pre-eminently important to it. Civil engineers tend to get on with the job rather than publicise the longest and greatest. Many of the subjects in the following pages have thus been poorly recorded, and it has not always been possible to explore every superlative with exhaustive authority. What follows is the best that could be done; new information and new claims will always be welcomed.

Accuracy can be excessive to the point of becoming spurious. Many of the figures quoted have been slightly rounded in metric conversion or if common sense so indicated. Nowhere, however, has a record been 'rounded' upwards.

chapter 1

ABOVE GROUND

bridges

A material that is strong in tension is invaluable to the bridge-builder. The arch is the only structural form in which there is no tension. Steel, either on its own or as reinforcement – that is resisting the tensions – in concrete, is that invaluable material in modern bridge-building. It is used in its most daring form in the suspension bridge, so called because the bridge hangs on steel cables. The cables and the suspenders are in tension, and a girder of some kind stiffens the bridge at deck level so that its geometry is not distorted when the deck is loaded. The world's longest spans are all suspension bridges (table 1) and the suspension bridge has been the dominant structural type for the longest spans for over a century (table 2).

Records and authentic facts about the earliest bridges are rare. The earliest reference is of a bridge across the Nile built in about 2650 BC. Sadd-el-Kafara Dam (page 75) is dated earlier than that, and the masonry still standing in the remains of the dam offers supporting evidence of the practicability of building a bridge in those days. The oldest surviving bridge is an arch over the River Meles at Izmir, Turkey, built in about 850 BC excepting perhaps the primitive clapper bridges of Dartmoor and Exmoor.

The arch, built of masonry, reached maturity of technique and architectural form in Roman and medieval times, but the longest spans were built in the modern era. The longest span of stone, however, was not contrived by man; it is the natural arch in Utah, USA.

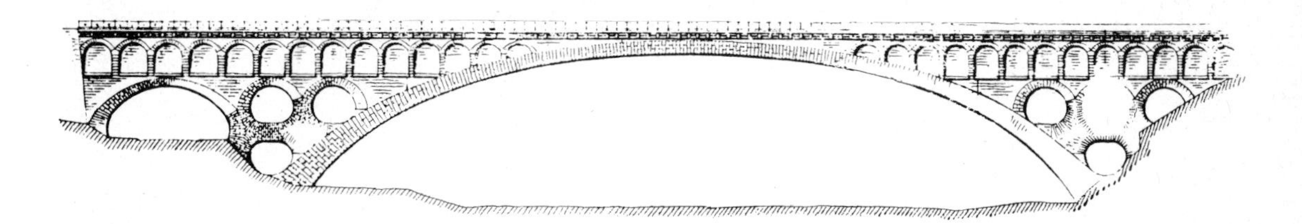

Plauen masonry arch in East Germany. *From Tyrrell's* History of Bridge Engineering

Timber and masonry (i.e. stone and brick) were substantially the only structural materials until about 200 years ago. The use of iron, then of steel and concrete, were alone sufficient to revolutionise bridge-building.

Grand Maître Aqueduct. The concrete of these arches is according to the Coignet system, and was made of sand, hydraulic lime and Portland cement in the proportion 800:167:33. The aqueduct has been in use for a century and the concrete is now somewhat vulnerable to bad weather. Maintenance has been necessary in the last twenty years or so, to ensure that the lining of sheet lead remains watertight, and to strengthen the exterior with reinforced concrete ribs in the soffits of the arches. *Photo: Ingénieur Général des Services Techniques de la ville de Paris*

But their advent has been accompanied by discoveries about the behaviour and analysis of materials and structures, and knowledge of this kind has been steadily growing over the two centuries since cast iron was first applied to bridge-building. Understanding has been perhaps more important than materials in accounting for the tremendous progress of that period.

STRUCTURAL TYPES

As materials improve or techniques become more proficient and cheaper so the relative attractiveness in cost of various structural types may change, within the limits of span over which they compete. Basically, there are just four forms: the beam, the arch, the suspension bridge, and the cantilever. The arch is virtually the reverse in shape and stress of the suspension bridge.

The beam is the simplest form of bridge. Its range can be extended by triangulating structural members into a truss – a framework that is a beam – or by using a shaped beam such as an I-section or a hollow box. The shape of the beam can also be varied along the length of the span, by increasing its

Winningen Bridge over the Mosel River exemplifies the clean line of the modern box girder bridge. This particular one is noteworthy for the height of the piers, the tallest of which is 124·3 m (408 ft) high. The record height for a bridge pier is 146·5 m (480·6 ft) held by the Europabrücke on the Brenner Motorway near Schönberg, Austria. Second is Viadotta Italia 145 m (476 ft). The Winningen pier is 19·02 × 7·33 m (62·4 × 24 ft) in cross-section at the base, and consists simply of a thin outer wall 30 cm (12 in) thick with two cross walls 25 cm (10 in) thick. *Photo: Leonhardt und Andra*

In contrast with the precasting techniques developed in France, German engineers developed a 'bridge factory' – a falsework bridge that spans between the columns of the permanent bridge, and is comprehensively equipped. The permanent span is constructed resting on this structure, casting the concrete in place. When the concrete has matured the prestressing tendons are tensioned. Then the 'factory' is moved forward to the next span. *Photo: Dyckerhoff und Widmann K.G.*

Concreting the bridge deck inside the 'factory'. These two pictures show construction of the Paschberg Bridge on the Brenner Motorway in Austria

depth gradually close to the supports for example. It can be held up by one or more stays attached to a tower or pylon on the bridge. It can be 'continuous' as opposed to 'simply supported'; a continuous beam extends over a number of spans, and the piers become props, holding up one long beam. All of these apparent complications permit material to be used more economically – that is to be distributed so that it is stressed more uniformly at around the maximum stress that it can safely bear – and so also increase the potential length of the span; or else they make erection simpler.

Each arm of a cantilever bridge is, or could be, free-standing. Cantilever bridges are the closest rivals of suspension bridges in length of span, still just having a slight edge on arches, even though the longest cantilever spans were built over 50 years ago. They are listed in table 4.

Today the cable-stayed box girder is more popular with designers, and this type has not yet reached its limit. Equally, neither has the plate or box girder bridge without stays, which in its latest example has a span only 50 m (164 ft) less than its contemporary cable-stayed rival. Cable-stayed girders are favourable with spans up to the range 1000–1500 m (3000–5000 ft) according to current opinion.

The arched viaduct in Britain and the timber trestle in the USA were the characteristic long, multi-span structures of the early days of railways. The steel trussed bridge was the typical bridge of great length in the railway-building days of the early 20th-century. Today such construction is typically a prestressed concrete road bridge (table 9).

CONCRETE BRIDGES

Concrete is not evident among the world's longest spans; it is nevertheless the second important material for bridges today. Prestressed concrete has potential for long spans, and is very competitive for the shorter spans that are commonly required in bridge-building. The structural types encountered with steel are found again, but with different shapes and characteristics (tables 3 and 6).

Variants of the suspension bridge have been built in concrete. Perhaps the most interesting is the stressed ribbon bridge, which consists of a flat ribbon of steel cables – analogous to the steel cables of the conventional

'Stitching on'; erection by cantilevering, casting the concrete in place. Four falsework cantilevers shown in use at Kingston Bridge, Glasgow, in 1959 *and right* Esbly Bridge over the River Marne. *Photo: C & CA*

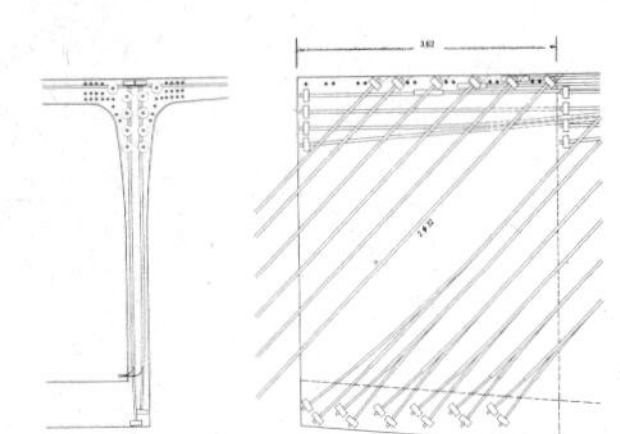

'Stitching on' to a prestressed concrete cantilever. These sketches are a detail from Bendorf Bridge. One side of the box-shaped cross-section of the bridge is shown on the left and a longitudinal section through the wall of the box on the right. The length of 3·62 m (12 ft) is cast on at one set-up of the falsework carriage (like the carriages shown at Kingston Bridge on page 13) and the prestressing cables, which extend through into the part already built, are tensioned when the new concrete has matured, to join the new piece, structurally, with the old. *Courtesy of Dyckerhoff und Widmann*

Salgina Gorge Bridge, an example of Maillart's work

suspension bridge – pulled very tight, so that they hardly sag at all; a rectangle of concrete is placed round them to form the deck of the bridge. This modern 'invention' is a rebirth of the traditional Chinese suspension bridge in which the bamboo ropes or iron chains were stretched across the span and planks laid across them to form the deck of the bridge. The longest span of this type yet built is the world's slenderest bridge (see page 6).

The first concrete bridge (i.e. concrete made with Portland cement and, therefore, ignoring Roman cement concrete) in France, and the first large-scale use of the material in modern bridge-building, was the Grand Maître Aqueduct (about 1870, as part of the River Vanne aqueduct (1867–74), supplying water to Paris), a long series of concrete arches through the Forest of Fontainebleau. The first in Britain was an arch of 23 m (75 ft) span built by John Fowler at Earls Court, London in 1867. An early example of a reinforced concrete beam bridge is Homersfield Bridge in Suffolk (15 m (50 ft) span, 1870). A notable early arch in Britain is Borrodale Bridge on the Fort William–Mallaig Railway (the West Highland Line) with a span of 39 m (127·5 ft) completed in 1898.

Rafael Urdaneta Bridge, over Lake Maracaibo, Venezuela. This bridge is the prototype of several similar ones, including Wadi Kuf Bridge, designed by Morandi. The ties are of prestressed concrete. The prestressed tie is more subtle in its behaviour than a plain cable, and the variation in stress in the tie is very much less when the bridge is in use

Arches apart, concrete and specifically reinforced concrete were used only timidly in bridge-building until the Swiss engineer Maillart transformed the scene by building some fine and daring bridges with this material. His first bridge, Chatellerault, was built in 1898, and his last in 1940.

Reinforced concrete then took its place in bridge-building, and elegant bridges like Waterloo Bridge over the Thames in London can be cited to exemplify its mature application in fine bridges, but it was sparingly used. The post-war revolution in prestressed concrete again radically changed the scene. Freyssinet's Marne bridges were as epoch-making as those of Maillart, but were followed by a world-wide adoption of prestressed concrete for many structural purposes, including its widepsread use for bridges.

TIMBER BRIDGES

Timber has played a major role in the history of bridge engineering, but as used historically proved to be not a permanent material and there are no early examples of timber bridges left. The beam of natural timber

The sculptural quality of structural concrete has been elegantly captured in the so-called Picasso Bridge over a motorway at Wuppertal, Germany. *Photo: C & CA*

was virtually the only alternative to a masonry arch (stone beams, which are cumbersome and strictly limited in span excepted) until the truss was invented, reputedly by Leonardo and Palladio. Trussed bridges of timber then began to be built when long spans were needed, and combined arch and truss timber bridges reached their peak in the 18th and early 19th centuries. Table 2 lists a few outstanding examples, but about a dozen spans of 60 m (200 ft) or over are on record. The invention of the truss was perhaps the first fundamental advance since the arch, but the truss did not become a fully explicit engineering form until it was used so widely and adventurously in railway engineering, in timber with some iron members, in iron and eventually in steel.

Langelinie, a footbridge of reinforced concrete built in Copenhagen in 1894 and designed by A Ostenfeld. The span is 20 m (66 ft) and the width 6·3 m (21 ft), and the bridge was designed for a live load of 500 kg/m^2 (about 100 lb/ft^2). The bridge is only 0·2 m (8 in) deep at the centre yet the stresses in it are low – 25 kg/cm^2 (356 lb/in^2) maximum in the concrete according to the designer's calculations (using influence lines). The bridge is a fixed arch, lightly reinforced with a mesh top and bottom, with 25 mm (1 in) of cover. The massive abutment walls are not reinforced, and are carried by timber piles. *Photo: Concrete Research Laboratory, Karlstrup, Denmark*

Bascule bridge at Cadiz. *Photo: Rheinische Stahlwerke*

Timber was extensively used in the early and middle days of railway engineering, particularly in the USA, in viaducts and trestles of great height and great length. However, timber had to yield to more permanent materials. According to Tyrrell a covered timber bridge could be expected to last about 30 or 40 years or roughly three times as long as one exposed to the weather. Today timber is used ambitiously in buildings but hardly at all in bridges.

A notable early trestle was the Portage Railway Viaduct over the Genesee River in New York State, USA, built in 1852, which carried a single-track railway (Erie Railroad) 71 m (234 ft) above the river, and was 267 m (876 ft) long. The towers were 15 m (50 ft) apart, and up to 67 m (220 ft) high. This was the boldest timber trestle ever built according to Tyrrell. It caught fire in 1875 and was replaced by a metal structure.

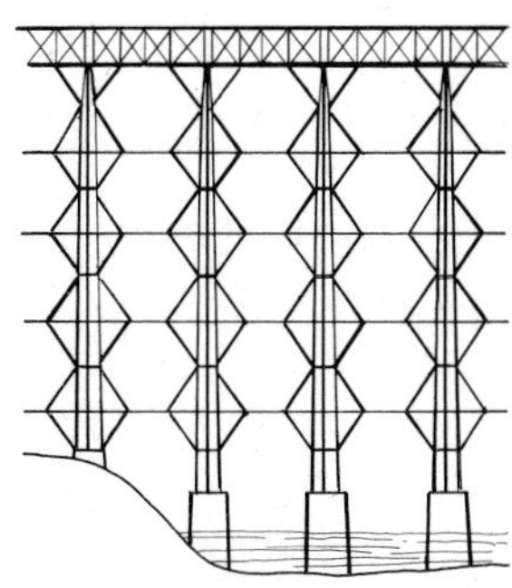

Timber trestle of Portage Viaduct, over the Genesee River, New York State, built in 1862. *From Tyrrell's* History of Bridge Engineering, *1911*

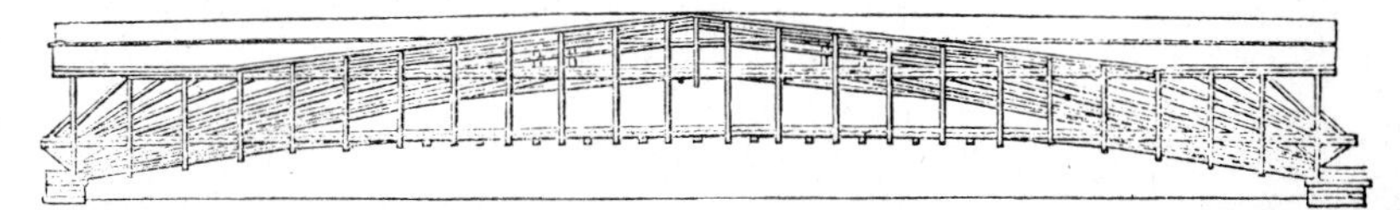

Wittengen timber bridge over the River Limmat – the longest span ever in timber. *From Cresy's* Encyclopedia of Civil Engineering, *1861.* (See table 2)

Its claim to be the longest span ever in timber is challenged by a claim that, in 1850, a bridge of timber with a span of 132·9 m (436 ft) was built over a ravine at Montgomery, Alabama, USA. The wood was used in tension, as in a suspension bridge, and consisted of relatively thin laminations, glued together. The engineer was John R Remington

IRON BRIDGES

The Industrial Revolution brought canal- and road-building. It changed the pace of construction, and added a new material – iron – to its resources. Wrought iron had been used in bridge-building, as chains, centuries before in China and Japan (see chapter 4). The earliest recorded example in Europe – apparently an authentic record of a bridge actually built – is Glorywitz Bridge, a military iron-chain bridge built over the Oder River in Prussia in 1734. The first in Britain was Winch Bridge, a footbridge over the Tees, about 2 miles (3 km) above Middleton, completed in 1741. The bridge had a span of 21 m (70 ft), a width of 0·6 m (2 ft), two wrought-iron chains and no stiffening except apparently two stay chains. It fell in 1802. Ironbridge, built in 1779 over the Severn at Coalbrookdale, is always cited as the world's first cast-iron bridge, but although an important prototype, apparently it was not the first. There is evidence of an iron arch, a footbridge of 22 m (72 ft) span in Kirklees Park near Leeds, reported in 1770, which was probably the first one.

Cast iron now became available in comparatively massive quantities and was the predominant metal used in bridge construction until the middle of the 19th century, competing with timber (still the main material) and masonry. Telford was the great master of the cast-iron arch and built shallow, graceful arches with this material. The longest spans of cast iron are in table 3. They are not very long (in spite of Telford's proposal for an arch of 183 m (600 ft) span of cast iron for London Bridge) because cast

Ironbridge over the Severn at Coalbrookdale in Shropshire was completed in 1779. It has a span of 30 m (100 ft) and is in use today as a footbridge. The bridge contains 378 tonnes of metal, and has no bolted joints but is slotted and dovetailed together. It is excellently detailed, but not in the optimum manner structurally. Some of the ribs do not continue right round the arch, and therefore cannot resist its thrust, though they do contribute to its stiffness. The bridge has been threatened by instability of the slopes of the Severn Gorge which it bridges, and an inverted arch of concrete has been built in the river bed to hold the abutments apart and so preserve the bridge. An odd repair indeed, for an arch thrusts outwards, at its abutments, and fails if they move apart. *Photo: Staifix*

The Lewiston–Queenston (New York–Ontario) Bridge about 10 km (6 miles) down the Niagara River from the falls, is the world's longest span arch with fixed (as opposed to hinged) abutments. The bridge was completed in 1962, and (table 3) replaced a light suspension bridge. After a chequered history of bridge building, came the Niagara Falls Bridge Commission which now manages three steel arch bridges below the falls – Rainbow Bridge (used by 35 per cent of US visitors to Canada), Whirlpool Rapids Bridge (the site of the railway suspension bridge from 1855 to 1897) and the one shown here. None of the early suspension bridges survive.

The picture also shows the Sir Adam Beck–Niagara generating stations on the right and the US Niagara power stations on the left, of around 4000 MW total installed capacity. Both stations are linked with pumped storage schemes; abstraction of water for power is controlled by treaty, so that the scenic spectacle of the falls is not vitiated – a policy influenced by the fact that the tourist revenue is much bigger than the power revenue. *Photo: Niagara Falls Bridge Commission*

Artist's impression of the Britannia Tubular Bridge with the Menai Suspension Bridge in the background. *A print from the* Illustrated London News, *1849*

Artist's impression of the Britannia Bridge reconstructed after damage by fire in 1970. *Courtesy of Husband and Company.* The bridge has been repaired in the form shown above, with arches replacing the tubes. This reconstruction has aroused controversy, and it has been suggested that a magnificent bridge has been ruined; on the other hand, the interests of navigation prevented Stephenson from using a very similar solution in the 1850s, and navigation is no longer of such importance. One wonders if the critics could have proposed a better solution to the problem

The only surviving Bollman truss bridge, illustrating precise assembly from pieces accurately prefabricated in a factory. An 80 ft (24 m) double-span truss of iron built by Wendel Bollman at Savage, Maryland in 1869. The only remaining example of the design which facilitated rapid expansion of early American railways, and of which the Baltimore and Ohio Railroad constructed 100 on its main line. *Photo ASCE*

iron was not trustworthy – it was brittle and fractured suddenly in tension. Typical tensile strengths are: cast iron 77–123 MN/m² (11 000–15 500 lb/in²); wrought iron 308–540 MN/m² (45 000–78 000 lb/in²). Wrought iron was conclusively proved superior in a famous series of tests carried out by Fairbairn and Hodgkinson in the 1840s. It reached its zenith with bridges such as Brunel's Saltash Bridge near Plymouth and Stephenson's tubular bridges at Conway and the Menai Strait joining Anglesey to the Welsh mainland. These tubular bridges have been claimed as the first bridges built entirely of wrought iron, but the claim should be interpreted in the sense that the tubular bridge was a girder in the modern engineering manner, and contained massive quantites of metal. Before 1784, which saw

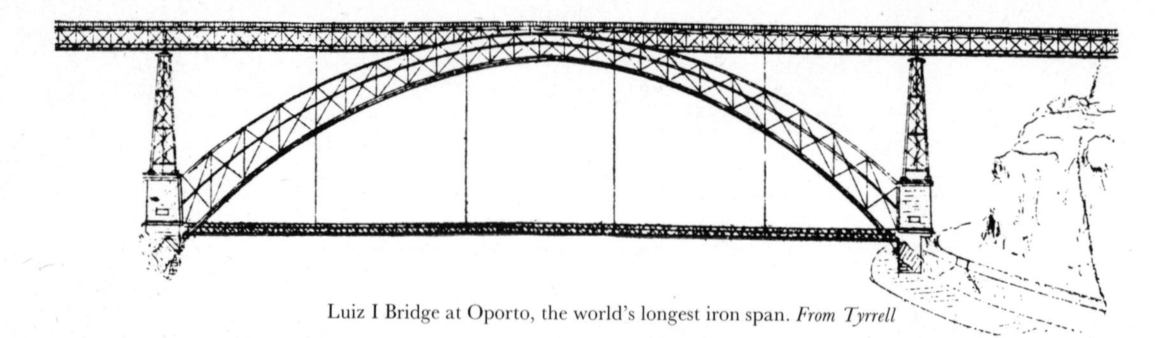

Luiz I Bridge at Oporto, the world's longest iron span. *From Tyrrell*

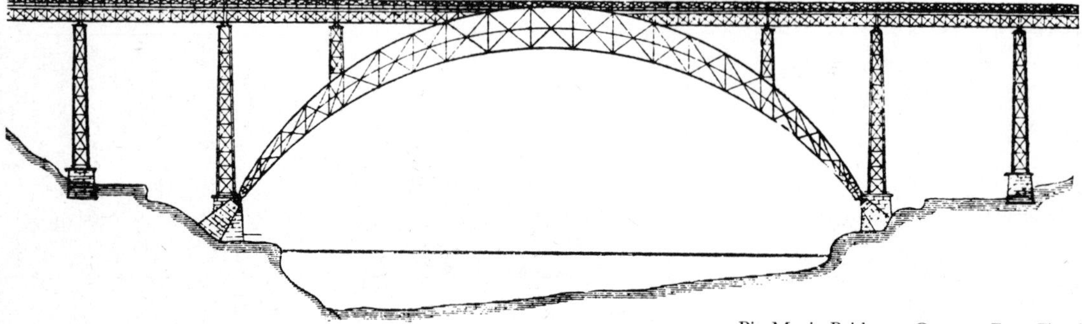

Pia Maria Bridge at Oporto. *From Tyrrell*

the introduction of Cort's puddling process, wrought iron was made a few kilograms at a time, and the links of a big chain were about the practical limit in terms of quantity. A bridge was built entirely of wrought iron in 1808 over the River Crou at Saint-Denis in France – a braced arch with straight upper chords and crossed diagonals, spanning 12 m (39 ft). Being an arch, it did not exploit the superiority of wrought iron in tension, but, chains excepted, it appears to be the world's first wrought-iron bridge. In typical bridges of the early railway period, cast and wrought iron were used in combination, the wrought iron taking the tensile loads. Trussed bridges of considerable size were built in this way. An early example is the Gaunless Bridge, now preserved in the York Railway Museum because it is claimed* to be the first iron railway bridge – it was built in 1825 on the Stockton to Darlington Railway and used until 1901. This bridge, which is a little tiddler, has curved wrought-iron bars, forming a bowstring-cum-truss shape, connected at each end by bosses of cast iron and by cast-iron verticals, a type of construction described as a 'Brummagen bedstead'. Later in the 19th century, wrought iron was exclusively used for trussed bridges. It was in general use by 1880, but it had a short life in bridge-building. The Bessemer process of steel-making was introduced in 1856, and the open-hearth process from about 1870. There was some hesitancy at first in using steel, but by 1890 or so it had become established, and superseded iron completely a few years later.

The world's longest span of iron is the wrought-iron arch of Luiz I Bridge at Oporto (table 3). The only iron span over 500 ft (152 m), except arches, was the truss span at Cincinnati listed in table 7.

*Also a claim that is challenged, in this case by a bridge dated 1811 at Robertstown, Aberdare.

THE INTRODUCTION OF STEEL

The first use of steel in bridge-building was in 1828, when a chain suspension bridge with a single span, generally given as 102 m (334 ft) and a width of 3·1 m (about 12 ft) was built over the Danube Canal in Vienna. The bridge was designed by von Mitis; and one report says that, compared with a corresponding bridge of iron, about half the weight was saved. The bridge had two eyebar chains, each comprising four lines of steel-bar links about 2 m (6·4 ft) long. It was replaced in 1860. Bessemer steel was used for three road bridges in Holland in 1862. They were single-span trusses of 30–37 m (98–121 ft) span; steel was then used on Dutch railway bridges but the results were not successful and the new material was distrusted.

Eads Bridge, St Louis, over the Mississippi River claims a whole family of superlatives, including the first major use of steel in bridge-building (an alloy steel comparable with present-day high-strength steels) and one of the first significant steel structures of any kind. It is named after its designer and builder, Capt James Buchanan Eads, and was opened in 1874. It was the first bridge to use hollow tubular members; its three spans were substantially longer than any other bridge spans at that time except suspension bridges; it is claimed to be the first bridge to use cantilever methods of construction entirely, thus eliminating falsework; it was the first bridge in the USA employing compressed-air caissons and was the deepest construction of that kind; and it was designed so that any part could easily be removed for repair or replacement. The bridge is in use today. *Photo: ASCE*

Astoria Bridge, Oregon. The three main spans exemplify the steel truss at a peak of lightness and elegance, and at a record length for a continuous three-span group. The bridge is 6·6 km (4·1 miles) long and it has about 300 spans altogether. The piers of the main span extend 21 m (70 ft) below mean water level, and piles another 58 m (190 ft) below that. *Photo: Oregon State Highway Department*

The first major use of steel in Europe, and its introduction as the basic bridge-building material, came about with its use entirely for the structure of the Forth Bridge (table 2) which was completed in 1889. The Board of Trade had prohibited the use of steel in bridge-building in Britain until 1877, and the contract for the steelwork of the Forth Bridge was let in 1882.

However, this use was anticipated by Eads in his steel arches over the Mississippi River completed in 1874 at St Louis. The bridge has three spans of 153+159+153 m (502+520+502 ft). The arch ribs are tubes, braced together, of corrosion-resistant chromium steel. The bridge is still in use and has required very little maintenance. It would appear that this early use in a big bridge was dependent upon the fact that the steel was in an arch and, therefore, in compression and there was not at that date confidence to use steel in tension.

What is claimed as the first all-steel bridge was built at Glasgow (USA) over the Missouri River and completed in 1879. It consisted of five spans, the longest of 96 m (314 ft) all Whipple trusses plus 427 m (1400 ft) of approach spans, apparently not of steel, and 264 m (867 ft) of wooden trestles. The engineer who specified steel was William Sooy Smith and here, for the first time, steel was seen in its modern structural role. But it is a tenuous 'first'. Steel in the form of hard-drawn wire was first used in the cables of suspension bridges in the Brooklyn Bridge. The bridge was completed in 1883; it took fifteen years to build, and the Missouri River Bridge took about two years.

The cables of Brooklyn Bridge consist of strands of parallel cold-drawn steel wires. Each wire is about 5 mm (0·2 in) in diameter and is galvanised for protection from corrosion. This cold-drawn steel will break in tension at an ultimate stress of about 70 t/in² (increased to 100 t or a little more in modern bridges), that is one wire will break under a load of a little over 2 t. The working load that the wire carries in the bridge is about a third of the breaking load, say about 0·75 t.

St James's Park Bridge. *Photo: Crown Copyright, Department of the Environment*

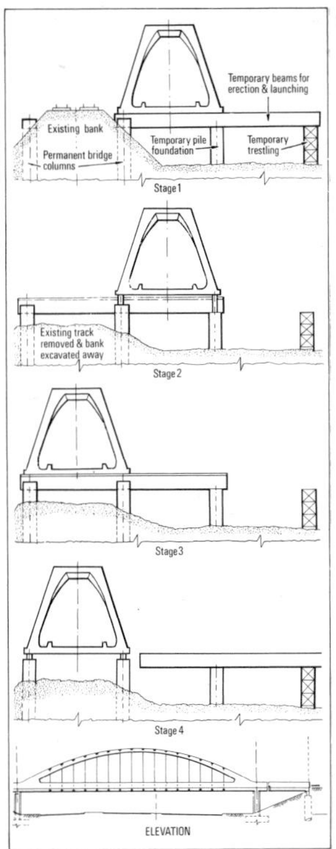

Frodsham bridges (two identical bridges), were both built beside the railway and rolled sideways into position. The span of each bridge is 63 m (207 ft) and the weight rolled into position was 4500 t.
Stage 1 Transverse sliding-in beams placed. The bridge is built, resting on these beams, with the railway in use.
Stage 2 Railway track removed and embankment excavated. The bridge is raised clear of the sliding-in beams by jacks. It is supported by the temporary and permanent columns at this stage, as the sketch shows. The sliding-in beams are pulled over to the position shown.
Stage 3 Sliding-in gear is inserted under the bridge, the weight of which is then transferred to it. The bridge is moved laterally into position, and then lowered on to its bearings.
Stage 4 The sliding-in beams are pulled back, and the tracks laid on the bridge, which is then ready to accept traffic. Interference with railway traffic was limited to one week-end.
Courtesy of Husband and Company and Chief Civil Engineer, London Midland Region, British Rail

BEAUTY OF LINE

Footbridges and short-span bridges, such as motorways need in large numbers, can bring out great talent. St James's Park in central London has a slender, chic footbridge of prestressed concrete that contributes with perfect manners to the serenity of the park. Another example of the slender elegance of prestressed concrete is Garrett Hostel Bridge over the Cam at Cambridge. The steepness of the Valley of the Wear at the City of Durham where the Norman Cathedral seems to have grown out of the rock, evoked a different mood – a 'thin, taut, white band' across the valley resting on a pair of V-shapes of slender, tapered fingers. This bridge – Kingsgate Footbridge – was built in two parts, one on each side of the valley. The halves were each constructed on temporary bearings, with each half-span aligned up and down the valley, so that elaborate falsework across the valley was not necessary. When completed, the two halves were rotated on their bearings through 90 degrees to their final alignment, jacked up and relevelled on their permanent bearings. The bridge is 107 m (351 ft) long with a cantilevered abutment span and the two half-spans of 41 m (135 ft).

Steel has its sculptural moments, but generally in the grand manner on bridges of great scale. An exception is the relatively modest grace of the prototype basket-handle bridge, completed in 1953 in bright colours over the Blanco River near Vera Cruz in Mexico. It is a road bridge with a span of 76 m (250 ft). The novelty of the bridge demanded very creditable innovative design. Fehmarn Sound Bridge (on the Hamburg–Copenhagen route) opened in 1963, is a larger example of this style. Its span is 240 m (787 ft) and it carries both road and rail.

UNUSUAL DUTIES

It is generally after the war is over that the bridge-builder has his day, but not always. The needs of an army on the move were met by the famous Bailey bridge. It consists of square panels diamond-braced (a sort of truss concept) that can be pinned together at the corners to form a girder. The girder is assembled on the approach and then pushed out over the space to be bridged. The strength of the bridge can be varied by doubling or tripling the panels, in the depth or in the width of the truss, so the heaviest bridge is a 'triple-triple' with nine panels per bay of the truss. Bailey bridges can be rapidly assembled and launched for spans up to about 75 m (240 ft). Several Bailey suspension bridges with spans up to 122 m (400 ft) were built during the Second World War. Other types of bridge, notably the Callendar-Hamilton, were used for greater spans and heavier loads. Both types have been widely and successfully used in peacetime, the Bailey mainly on constructional sites, and the Callendar-Hamilton mainly in remote and inaccessible places.

An unusual wartime relic is the sinking bridge at Basra, over the Shatt-el-Arab at Margil (Basra), Iraq. There was a shortage of steel and machinery, so a conventional opening span could not be built. A span that could be lowered about 6 m (20 ft) below the surface of the water was, therefore, constructed and successfully used; the opening is 23 m (64·5 ft) clear.

At least one other example of a submersible bridge existed – also built for military purposes. Needham records (see reference on page 232) a rope suspension bridge built in about AD 494 on the Huai River in eastern China with about ten cables, all of which could be lowered beneath the water by turning windlasses. The bridge could not then be burned nor cut, passage across it was denied, and the river was blocked against boats.

The southern approach spans of the Jacques Cartier Bridge – built in 1929 across the St Lawrence at Montreal – were raised in an elaborate sequence of jacking up and heightening the piers, to give headroom for ships when the St Lawrence Seaway was built in the late 1950s. The bridge was kept open to traffic while this was done, as follows:
First stage Temporary access at spans 0 to 2; spans 1 and 2 raised; new abutment built; jacking started at other end, including raising span 14 to its final position, and jacking span 10 to a level position.
Second stage 'Through' truss erected and slid into position for span 10; jacking to final levels up to pier 5.
Third stage Final jacking and heightening of piers.
Spans were raised 0·15 m (6 in) at a time, and supported on concrete blocks; at intervals of 0·61 m (2 ft) the piers were concreted. The roadway was raised 15 m (50 ft) at the Seaway, and a further 9 m (30 ft) of clearance obtained by substituting the new span 10

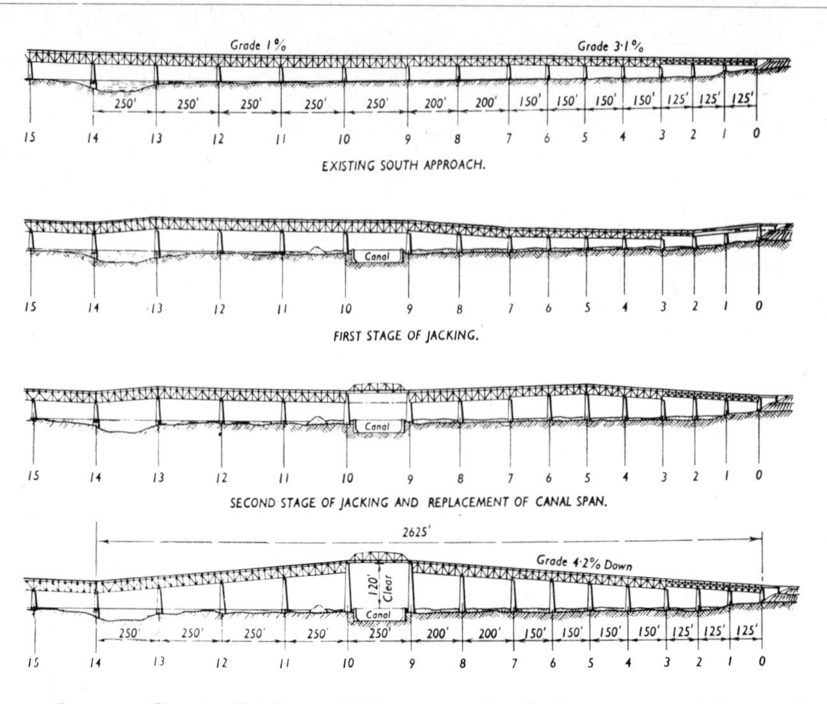

Jacques Cartier Bridge, which crosses the St Lawrence at Montreal, has been modified by raising one end of it. When the St Lawrence Seaway was built, its entrance was sited so that it passed beneath the south end of the bridge, but it was necessary to get headroom for shipping using the Seaway. The bridge was jacked up, and its piers heightened, in a complicated sequence of operations without precedent on such a scale. The diagrams show the bridge 'before and after'. It was raised a maximum of about 25 m (80 ft) and one span was replaced, all without major interruption to the traffic using it.

THE HEAVIEST LOADS

Specifications for the loads that a bridge will carry (called the 'live' or 'imposed' loads) are generally complex. They depend partly on the length of the span. A 'distributed' load (i.e. so much per metre run over the whole bridge) is one criterion, and an indivisible heavy load is another. The latter in railway loads are generally heavier and have more severe dynamic characteristics than highway loads. The bridge of long span designed for the heaviest imposed loading is claimed to be Hell Gate Bridge which crosses the East River in New York City. This bridge is a steel arch of 298 m (977 ft) span, completed in 1916; for sixteen years it was the world's longest arch span. It carries railway traffic, and its live load (distributed) specified in design is 24 000 lb/ft (36 t/m) that is over 10 000 t spread uniformly over the bridge and equivalent to nearly half the dead load (the dead load is the weight of the bridge itself). The load per metre of bridge for Hell Gate is double the rate specified for Sydney Harbour and Quebec bridges, five times that for the Verrazano Narrows Bridge (the heaviest rate for a long-span suspension bridge) and about twenty times the rate for the Forth and Severn bridges.

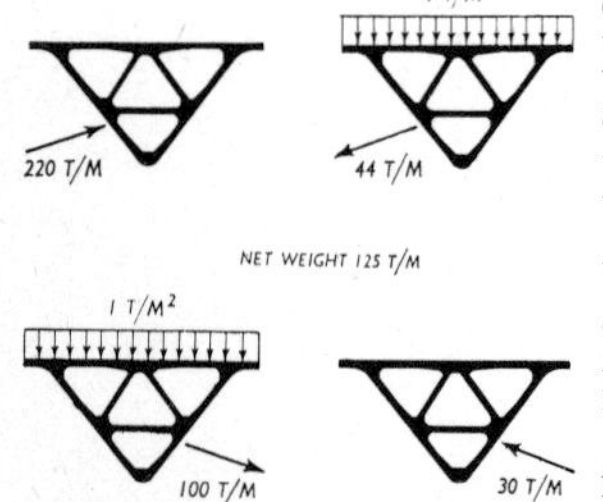

Four design loading conditions for 'Nabla' beams. In every case the dead weight of the beam, viz. 125 t/m, acts as well. The first condition gives a total load of 13 200 t from the seaward side

In round figures the maximum load on Hell Gate Bridge is, therefore, 30 000 t. The Nabla beams of the Haringvliet Barrage in Holland (page 250) carry a heavier live load, according to the assumptions made in design, than

Table 1. The world's longest spans. (Bridges with a main span exceeding 600 m (almost 2000 ft) all of which are suspension bridges, and all but one road bridges; two others are designed for rail traffic on a second deck.)

Bridge	Main span m	ft	Year of completion	Country and location	Other details
Akashi–Kaikyo	1780	5840	Project	Japan, Honshu–Shikoku link at Akashi Strait	Side spans 2 × 890 m (2 × 2920 ft). Double deck for road and rail. Open truss design, truss 37·5 m wide × 18 m deep (123 × 59 ft).
Humber	1410 (centre-lines of towers)	4626	u.c. 1978	England, north-east coast	Side spans of 530 and 280 m (1739 and 919 ft). Clear distance between piers is 1380 m (4527·5 ft)
Verrazano Narrows	1298	4260	1964	USA, New York City	Two side spans each 370 m (1214 ft). Twelve traffic lanes on two decks. Towers 210 m (690 ft) high above mean high water. Bridge is 5 km (3 miles) long
Golden Gate	1280	4200	1937	USA, California, San Francisco	Two side spans, each 343 m (1125 ft). Towers are 227 m (745 ft) high above high water. World's tallest bridge
Mackinac	1158	3800	1957	USA, Michigan, St Ignace–Mackinaw City (crosses between Lakes Michigan and Huron)	Two side spans each of 549 m (1800 ft) give this bridge the longest continuous suspended length in the world – 2256 m (7400 ft), but Akashi–Kaikyo will be longer – 3560 m (11 680 ft). Bridge is over 6·5 km (4 miles) long
Atatürk	1074	3524	1973	Turkey, Istanbul over Bosporus	Follows Severn Bridge in its design principles
George Washington	1067	3500	1931	USA, Hudson River, New York City	Second deck added in 1959–62. An epoch-making bridge of far greater span and capacity than its contemporaries
Formerly: Salazar	1013	3323	1966	Portugal, Lisbon, River Tagus	Two decks. Designed for railway tracks on lower deck. Designed and built by American engineers. Side spans 2 × 483 m (2 × 1586 ft)
Forth (road)	1006	3300	1964	Scotland, near Edinburgh, over Firth of Forth	First 1000 m span suspension bridge in the world not designed and built by American engineers. Towers 156 m (512 ft) high; side spans 2 × 408 m (2 × 1340 ft)
Severn	988	3240	1966	England/Wales, near Bristol, over River Severn	Introduced major advances in design

Table 1. cont.

Bridge	Main span m	ft	Year of completion	Country and location	Other details
Tacoma Narrows	853	2800	1950	USA, Washington	Rebuilt after failure in 1940
Angostura	712	2336	1967	Venezuela, Ciudad Bolívar	Crosses the Orinoco River
Kanmon Strait	712	2336	1973	Japan, links Honshu with Kyushu	The cables consist of parallel wires shop-prefabricated into strands of 91 wires that were then hauled across the catwalk. First time used for such a long span
West Bay (San Francisco–Oakland)	704	2310	1936	USA, California, San Francisco–Yerba Buena Island	Two equal main spans (see tables 4 and 9). Suspended length is 2822 m (9260 ft), longer than Mackinac, but with a central anchor
Bronx–Whitestone	701	2300	1939	USA, New York City	Completed in 23 months; crosses East River
Pierre Laporte	668	2190	1970	Canada, Quebec City	Crosses St Lawrence River
Delaware Memorial	655	2150	1951 1968	USA, Wilmington, Delaware,	Two identical bridges, each 5·6 km (3·5 miles) long
Melville gas pipe-line bridge	610	2000	1951	USA, Louisiana, Atchafalaya River	Side spans of 305 and 198 m (1000 and 650 ft) plus two backstay spans of 198 and 170 m (650 and 558 ft) respectively. Very economical bridge (cost about $2 million). Its narrow pipe deck needs stiffening in the horizontal plane; a transverse beam at each tower carries cables which span on each side, like the main cables turned sideways
Walt Whitman	610	2000	1957	USA, Pennsylvania, Philadelphia	Crosses Delaware River
Tancarville	608	1995	1959	France, Le Havre	Crosses River Seine
Lillebaelt (Little Belt)	600	1968	1970	Denmark	The second bridge crossing the Lillebaelt. Has plated deck like Severn, and inclined suspenders

Note: Ropeways and cableways are excluded from the tables on bridges. For comparison, the longest span of a cableway is 4115 m (13500 ft), the span from the Coachella Valley to Mount San Jacinto of a cableway in California, USA. The longest span for a high-voltage transmission line is 4888 m (16037 ft) across the Sogne Fiord, Norway.

Table 2. Chronological list of the world's longest span.

Bridge	Main span m	ft	Year of completion	Country and location	Other details
Akashi–Kaikyo	1780	5840	Project 1988	Japan	See table 1
Humber	1410	4626	u.c. 1978	England	See table 1
Verrazano Narrows	1298	4260	1964	USA	See table 1
Golden Gate	1280	4200	1937	USA	See table 1
George Washington	1067	3500	1931	USA	See table 1
Ambassador	564	1850	1929	Links USA and Canada (Michigan–Ontario)	Suspension bridge crossing Detroit River (Detroit–Windsor)
Quebec	549	1800	1917	Canada, St Lawrence River	See table 4
Forth (rail)	521	1710	1889	Scotland	See table 4
Brooklyn	486	1596	1883	USA, New York City, over East River	Suspension bridge. Approaches recently modified; the bridge is now able to carry 100 000 vehicles per day, twice the planned capacity of Humber Bridge
Niagara–Clifton	387	1268	1869	USA/Canada, Niagara River	Suspension bridge 3 m (10 ft) wide. Destroyed in a gale in 1889, and replaced by 256 m (840 ft) steel arch in 1899. The arch – Honeymoon Bridge – was destroyed by ice in 1938 and replaced by the present Rainbow Bridge
Cincinnati–Covington	322	1057	1867	USA, Ohio/Kentucky, over Ohio River	A notable advance in suspension bridges. Strengthened in 1898, steel cables being added over the iron ones. In use today
Lewiston–Queenston	318	1043	1851	USA/Canada, Niagara River	Serrell's suspension bridge. Destroyed in 1864. Replaced in 1899 by another suspension bridge that was demolished on completion of the present arch bridge
Wheeling	308	1010	1849	USA, West Virginia, over Ohio River	First long-span wire suspension bridge in USA. Destroyed in a gale in 1854 and rebuilt by Roebling in 1856. In use today

Table 2. cont.

Bridge	Main span m	ft	Year of completion	Country and location	Other details
Fribourg	265	870	1834	Switzerland, over Sarine Valley	Suspension bridge with iron wire cables anchored in rock. Span given was between towers; timber deck was 247 m (810 ft) long. Demolished in 1923
Menai	177	580	1826	Wales, over Menai Strait	Suspension bridge. Cables of wrought-iron links. Extensively reconstructed in 1940. In use today
Chak-sam-chö-ri lamasery	137	450	1420	Tibet, Brahmaputra River	Footbridge with light deck suspended from iron chains. In use in 1878
Trezzo	77	251	*c.* 1380	Italy, Adda River	Fortified arch, the longest masonry span attempted for centuries. Destroyed in war *c.* 1410
Lan Chin	76	*c.* 250	65	China, Yunnan Province, over Lantshang River	Iron chains with timber deck resting on them. Traditionally ascribed date of construction is quoted. Bridge was repaired in 1410. Tyrrell gives span as 100 m (330 ft)
Narni	43	142	14	Italy, River Nera	Four unequal spans; one remained in 1676. Said to be the finest Roman bridge, the arches, typically, being of dry-jointed masonry

The list above is as accurate as major sources permit. The date of Lan Chin Bridge is uncertain, but what is known is that Chinese suspension bridges with iron chains and spans in the ranges 60–120 m (200–400 ft) existed in the Middle Ages. The table continues with bridges that, while not qualifying as the world's longest span, show the development of suspension, timber and masonry bridges, mainly in Western civilisation.

Bridge	Main span m	ft	Year of completion	Country and location	Other details
Union	133	437	1820	England/Scotland, Berwick, over River Tweed	First British suspension bridge to carry vehicular traffic. Span given is between towers; the deck spans 110 m (361 ft). In use today
Colossus (Fairmount)	104	340	1817	USA, Philadelphia, over Shuylkill River	Covered arched timber truss with a rise of 11·6 m (38 ft). Destroyed by fire in 1838. Tested with 22 t wagon drawn by sixteen horses. Succeeded by a wire suspension bridge (109 m – 358 ft span) in use until 1874
Fairmount	124	408	1816	USA, Philadelphia, over Shuylkill River	First wire suspension bridge. Cost of bridge is reputed to have been $125. Had 0·46 m (18 in) passageway and was safe for only eight people. Collapsed under snow and ice load soon after completion, and was replaced by Colossus

Table 2. cont.

Bridge	Main span m	ft	Year of completion	Country and location	Other details
McCall's Ferry	110	360	1815	USA, Pennsylvania, over Susquehanna River, near Lancaster	Arched truss of timber built from the river ice, 'with reckless disregard of the most elementary safety precautions'. Destroyed by ice two years later
Piscataqua River	74	244	1794	USA, New Hampshire (11 km (7 miles) from Portsmouth)	Timber arch with three trusses 5·5 m (18 ft) deep. Rise of arch 8·2 m (27 ft). Whole bridge was 732 m (2400 ft) long, mainly of piled bents and short spans
Schaffhausen	122 or 59	400 or 193	1758	Switzerland, River Rhine	A pier of a former bridge divided the span into two parts, but it would spring off its bearing on the pier under light load. Supported on jacks and repaired in 1783, the only repair in 42 years. Burned by French Army in 1799
Wittengen	119	390	1758	Germany, Baden, over River Limmat	Longest span ever in timber. Burned by the French Army in 1800
Pontypridd	46	150	1750	Wales, near Newbridge, River Taff	Stone arch with rise of 11 m (35 ft)
Iwakuni	45	149	1673	Japan, Isle of Honda	Three equal spans, wooden arches. Five-span bridge. One span was replaced every five years and the whole bridge rebuilt four times in a century. In use today
Port de Martorell	38	124	Roman	Spain, River Noya	One of about a dozen Roman bridges in Spain
Trajan's Bridge	52 (probably)	170	AD 104	Below Iron Gates, River Danube	Destroyed AD 120. Reputedly twenty timber arches each above 100 ft (30 m) on stone piers. About 1250 m (4200 ft) long, 18 m (60 ft) wide and 45 m (150 ft) high
Alcantara	30	98	AD 105	Spain at Alcantara near the Portuguese border, River Tagus	Six semicircular granite arches, the two longest each spanning 30 m (98 ft). In use until 1809. Now restored

Note: Primitive suspension 'bridges' (i.e. a rope but no deck) have been widely used throughout the world, with spans up to 600 or 700 ft (200 or 230 m) according to reliable reports, but like ropeways and cableways, have been excluded from the table. The bridges listed, including the Chinese ones but with the probable exception of the first Fairmount Bridge, are much more civilised and could be crossed without trepidation by the most sedentary person.

Table 3. The world's longest arch spans.

Bridge	Main span m	Main span ft	Year of completion	Country and location	Other details
Arches of steel over 300 m (nearly 1000 ft) in span					
New River Gorge	518·2	1700	u.c. 1976	USA, West Virginia, Fayette County	Rise of arch is 113 m (370 ft); four-lane road bridge. Two-pinned arch; roadway 267 m (876 ft) above river
Bayonne (Kill van Kull)	510·5	1675	1931	USA, Bayonne (New Jersey) to Port Richmond (Staten Island)	N-braced arch with rise of 81 m (266 ft). The thrust of each of the two ribs of the arch at the skewback is about 14000 t
Sydney Harbour	509	1670	1932	Australia, Sydney	First long-span bridge built by British engineers for a generation. Width of bridge 48·8 m (160 ft)
Fremont	383	1255	1971	USA, Portland, Oregon, over Willamette River	An arch of extraordinary grace and slenderness, illustrating the advance in design in steel since the great arches of the early 1930s
Zdakov	380	1244	1967	Czechoslovakia, Lake Orlik (Vlatva River)	Arch is two-hinged, the span between hinges being 330 m (1082 ft)
Port Mann	366	1200	1964	Canada, British Columbia, Fraser River	A tied arch similar to Fremont, flanked by cantilevers of 110 m (360 ft)
Runcorn	330	1082	1961	England, over River Mersey (Runcorn–Widnes)	Replaced a transporter bridge
Birchenough	329	1080	1935	Rhodesia, Sabi River	Very graceful and economical bridge, using an alloy steel
Glen Canyon	313	1028	1959	USA, Arizona, Colorado River	Deck is 213 m (700 ft) above water-level
Lewiston–Queenston	305	1000	1962	USA/Canada (New York/Ontario), over Niagara River	Longest built-in or encastré (as opposed to pinned or hinged) arch span in the world
Hell Gate	298	977·5	1916	USA, New York City, East River	See text

Note: Thatcher Bridge (over the Panama Canal–344 m span) and Laviolette Bridge (over the St Lawrence at Three Rivers – 335 m span) are both shaped like an arch but are cantilever truss bridges.

Table 3. cont.

Bridge	Main span m	ft	Year of completion	Country and location	Other details
Arches of concrete over 250 m (about 820 ft) in span					
Gladesville	305	1000	1964	Australia, Sydney, over Parramatta River	Consists of four ribs of precast elements. Working stress in concrete 15·5 N/mm² (2250 lb/in²). Rise of arch is 40·8 m (134 ft)
Amizade	290	951	1964	Brazil/Paraguay, Foz do Iguacu, crosses Parana River	Called Friendship Bridge. A very graceful arch.
Arrabida	270	885	1963	Portugal, Oporto, over River Douro	Road bridge 26·5 m (87 ft) wide; rise of arch 52 m (172 ft)
Sando	264	866	1943	Sweden, Kramfors, over Angerman River	A slender arch, span one-third longer than any other at that time. Rise of arch 40 m (131 ft). Working stress in concrete 10·7 N/mm² (1550 lb/in²)
Arches of wrought iron over 152 m (500 ft) in span					
Luiz I	172·5	566	1885	Portugal, Oporto, over River Douro	Tied arch with upper and lower roadway. Designed by T Seyrig. Longest span in the world of iron
Garabit	165	541	1885	France, over River Truyère	Very similar to the Pia Maria
Pia Maria	160	525	1877	Portugal, Oporto, over River Douro	Two-hinged crescent-shaped braced ribs 10 m (33 ft) deep at centre, for single-line railway
Grunenthal	156	513	1892	Germany, over Kiel Canal	High-level bridge for road and rail. Crescent-shaped braced arch
Levensau	*c* 170	*c* 560	1894	Germany, over Kiel Canal	High-level bridge for road and rail. Crescent-shaped braced arch
Arches of cast iron over 70 m (about 230 ft) in span					
Old Southwark	73	240	1819	England, London, over River Thames	Longest span ever built in cast iron. Three arch spans —64+73+64 m (210+240+210 ft) – by John Rennie. Rise 1/10 of span. Eight ribs each in thirteen pieces, 1·8 m (6 ft) deep at centre. Bridge was 216 m (720 ft) long and contained nearly 6000 t of metal. Replaced in 1921
Sunderland	72	236	1796	England, over River Wear	Designed by Thomas Paine. Rise of arch 10·4 m (34 ft). Widened 1859. Six ribs with open spandrels each with 105 cast-iron voussoirs, like a masonry arch

Table 3. cont.

Bridge	Main span m	ft	Year of completion	Country and location	Other details
Arches of masonry (stone or brick) over 61 m (200 ft) in span					
Augusta	97	320	–	USA, Utah	A natural arch 9 m (30 ft) wide and 80 m (265 ft) high
Plauen (Elster Valley)	90	295	1903	East Germany, over Syra River	Rise 1/5 of span. Ornamental spandrel, with row of minor arches. The stone of the arch is hard slate, according to Tyrrell
Salcano	85	279	1908	Italy, near Trieste, over River Isonzo	Viaduct with nine approach arches plus river span. Destroyed in First World War and rebuilt
Luxembourg	84	276	Early 20th century	Germany, Petrusse River	Rise 31 m (102 ft). Has spandrel arches and two 21·6 m (71 ft) approach arches; total length 350 m (1148 ft)
Grosvenor	61	200	1833	England, Chester, over River Dee	Longest masonry span in Britain. Rise of arch is 12·8 m (42 ft) and radius 43 m (140 ft). Road bridge still in use

Note: In 1910, there were sixteen metal arched bridges with spans of 152 m (500 ft) or more, the longest being the Clifton arch at Niagara of 256 m (840 ft) (Honeymoon Bridge). At the same date there were nine concrete arches over 61 m (200 ft) in span.
Arches of timber have not been included in this table but some of the longest spans are in table 2.
Long-span timber bridges were typically hybrid arch and truss structures.

Verrazano Narrows Bridge. *Photo: Triborough Bridge and Tunnel Authority*

Brooklyn Bridge

East River Bridges: Brooklyn (foreground) and Manhattan *Photo: New York City Transportation Administration*

Queensboro Bridge. This was the first bridge to use high-tensile steel extensively. The main-span cantilevers are joined at the centre, making the bridge statically indeterminate. *Photo: New York City Transportation Administration*

Williamsburg Bridge. *Photo: New York City Transportation Administration*

Marine Parkway Bridge, New York City. *Photo: Triborough Bridge and Tunnel Authority*

the Hell Gate Bridge, although each beam spans only 56·5 m (185 ft). The worst design load is made up of 125 t/m, the weight of the beam (i.e. about 7000 t on the span) plus a load acting sideways as shown in the sketch of 220 t/m. The latter is a peak figure for the dynamic load caused by waves on the gate, and is equivalent to a total load of 12 500 t on the span. It is erroneous to add arithmetically two loads that act in different directions, so one cannot claim a total load of 20 000 t on this short span; nor can one add the live load on the deck to the total, because it does not occur at the same time as the heavy wave loads. Nevertheless, all these loads are taken account of in design, as the sketches indicate.

THE UGLIEST BRIDGES

Bridges attract immoderate comment because of their appearance and their cost. Some examples of those that are fine architecture have been given. What of the other extreme? It is rather difficult to build an ugly suspension bridge, but Williamsburg Bridge in New York City has achieved that distinction. Among stayed girder bridges, Severins Bridge has a Teutonic stolidity that jars the eye, but critics give it credit, for the purpose of the single tower is to keep the view of Cologne Cathedral clear of any tall structures.

Steel trusses tend to be gawky and hideous to look at, but not invariably so. Sciotoville Bridge (a railway bridge built in 1917 over the Ohio River with one continuous truss over two 236 m (775 ft) spans and an important prototype of the long, continuous truss bridge on US railways), according to Condit, conveys an unparalled sense of mass and power. Steel trusses can, by scale and brute force, achieve a certain éclat if carefully handled. Lacking that care, they become primitive monsters. 'My god, it's a blacksmith's shop' its Consulting Architect is credited with saying when he saw

George Washington Bridge. *Photo: Port Authority of New York and New Jersey*

Bayonne Bridge. *Photo: Port Authority of New York and New Jersey*

Old and new bridges over the River Indus between Rohri and Sukkur, Pakistan. The new bridge – the Ayub arch – was completed in 1962, and spans 246 m (806·75 ft). The old bridge – Lansdowne Bridge, built in 1889 – has been dismantled

Queensboro Bridge, but it has many close rivals. William Morris said that the Forth Bridge was the supremest specimen of all ugliness and proved that there never could be an architecture of iron (*sic*). However, Forth is the most graceful of all the big cantilevers, the curved lines of the tubes, and indeed the tubes themselves, giving the bridge a sweep that the angularity of Quebec Bridge, for example, does not have.

The old Lansdowne Bridge over the River Indus between Sukkur and Rohri in India must surely have been the most unsightly of all the great cantilevers. It was completed almost a year before the Forth Bridge, and during that brief time it was the world's longest cantilever span (250 m (820 ft)). In 1884, *The Engineer* said of it: 'Contemplating the monstrosity of the general design, one would expect that in point of economy and detail construction, a fair degree of excellence had been attained. But neither is this the case. There are many ways of reducing the unsupported lengths of the great uprights and raking struts, and consequently of reducing materials; but as these would involve some calculations of stresses beyond those of the most elementary kind, they were probably not deemed worth the trouble. . . . A derrick, the half of an English roof truss, a Whipple girder, the other half of the roof truss and another derrick are excellent things in themselves, but to string them together up on one line, and thereby making a bridge, is not engineering, nor is it architecture.' The bridge was replaced by a steel arch of undoubted elegance opened in 1962.

The Tower Bridge in London has always evoked partisan comment. The purists detest its sham masonry, and the romantics love the fairy-tale created by that masonry. It has been called, by Waddell, 'the most monumental example of extravagance in bridge construction in the world'.

It seems to be as difficult to build an ugly bridge of prestressed concrete as to build an ugly suspension bridge. The author tried to find an example, but was told by one promoter of the material that suicide is fatal. A more direct critic, however, cited the bridges on the M4 motorway, which he described as 'drooping dinosaurs'.

Starrucca Viaduct, near Susquehanna, Pennsylvania was built in 1848 (see page 40). *Photo: ASCE*

Arvida bridge. The world's only bridge of any size built of aluminium (strictly, light alloy) is the fixed arch of 88 m (290 ft) span, crossing the Saguenay River at Arvida in the province of Quebec, Canada. The arch is dramatic and graceful. Its high first cost is offset by the fact that painting is unnecessary; although the aluminium weathers it does not corrode. Methods of site jointing by riveting were perfected for this bridge. It was completed in 1950 and it weighs nearly 200 tonnes, less than half the weight of a comparable steel arch. During the 1950s, several interesting structures were built of light alloy, but aluminium then became relatively more expensive, and so has been virtually ruled out for structural work, except in a few special cases in which lightness is all important. There exist only a handful of aluminium bridges. *Photo: Alcan*

Above: The ribs of the arch of Fremont Bridge are 1·2 × 1·2 m (4 × 4 ft) in section, and the bridge was devised to keep them slender for aesthetic reasons. By contrast, the combined girder and tie member is 5·5 m (18 ft) deep by 1·7 m (5·5 ft) wide. The mid-span part of the arch was lifted into position in one piece – see chapter 4. Fremont Bridge claims two superlatives – the longest tied-arch span and the greatest lift in bridge construction. *Photo: Oregon State Highway Department*

Below: Rockville Bridge over the Susquehanna River, Pennysylvania, USA, was completed in 1901. It is the world's longest bridge of stone, being 1161 m (3810 ft) in length, with 48 spans. It is exceeded in length by the 6 km (3·75 miles) length of the brick arches of the railway viaduct from London Bridge to Deptford Creek (page 58)

Table 4. The world's longest cantilever spans (over 350 m (about 1150 ft)).

Bridge	Main span m	Main span ft	Year of completion	Country and location	Other details
Quebec	548·6	1800	1918	Canada, St Lawrence River	Rail bridge
Forth (rail)	521	1710	1889	Scotland	Two equal main spans
Nanko	510	1673	1974	Japan, Osaka–Amagasaki	Graceful K-braced (cantilevers) and N-braced (suspended span) truss. See also table 54.
Commodore John J Barry	501	1644	1974	USA, Pennsylvania (Chester/Bridgeport), over Delaware River	Main span has suspended span of 250 m (822 ft). Two 250 m (822 ft) side spans.
Greater New Orleans	480	1575	1958	USA, Louisiana, New Orleans, over Mississippi River	Road bridge. Anchor arms of 260 and 180 m (853 and 591 ft). Length 3626 m (11 896 ft). Second Mississippi bridge at New Orleans
Howrah	457	1500	1943	India, Calcutta, over Hooghly River	Replaced a floating bridge built in 1874 and intended to last 25 years. K-truss with roadway suspended below bottom flange
East Bay (San Francisco–Oakland)	427	1400	1936	USA, California (Oakland–Yerba Buena Island)	Double-deck highway bridge (see tables 1 and 9)
Baton Rouge	375	1235	1968	USA, Louisiana, Mississippi River	Six-lane road bridge
Astoria	376	1232	1966	USA, Oregon, over Columbia River	Claimed as the longest continuous three-span truss in the world, 751 m (2464 ft) long
Tappan Zee	369	1212	1956	USA, New York City, over Hudson River (Nyack–Tarrytown)	World's heaviest cantilever according to Steinman (1956). Bridge is 9·5 km (6 miles) long and is part of the New York Thruway
Longview	366	1200	1930	USA, Washington–Oregon, crosses Columbia River	
Baltimore	366	1200	u.c. 1976	USA, Maryland, crosses Patapsco River	Continuous three-span truss of 805·9 m (2644 ft) will exceed Astoria
Queensboro	360	1182	1908	USA, New York City, East River	See page 34

FROM ART TO MONEY – THE COSTLIEST BRIDGES

The Starrucca Viaduct was built in 1848 and was among the earliest rail links between the Eastern seaboard of the USA and the Midwest. It crosses Starrucca Creek, near Susquehanna, Pennsylvania, and is over 1000 ft (300 m) long, having a series of 50 ft (15 m) semicircular arches supporting two tracks. It is believed to be the first structure in the USA to make use of structural concrete. It was built at a cost of $320 000 and was the world's costliest bridge at that time, but the record did not last long.

The first of the really long railway bridges was the Victoria Bridge over the St Lawrence at Montreal, built by Stephenson in 1860. At that time *The Engineer* pointed out that although other bridges had longer spans

Three generations of London Bridge. Peter de Colechurch's bridge served London from 1176 until 1831, John Rennie's from 1831 until 1971. The facings and balustrades of Rennie's bridge have been re-erected at Lake Havasu, Arizona, USA; the reincarnation shares with Vanbrugh's bridge at Blenheim Palace the distinction of having had a lake put beneath it to make it respectable. The new bridge is architecturally outstanding, and is the finest bridge in London. Some comparative facts

	First Bridge	*Second Bridge*	*Third Bridge*
Cost of building		£425,018 9*s* 2*d**	£5¼ million
Time taken in building	33 years	7 years	5 years
In use for	655 years	140 years	
Weight		130,000 tons	55,000 tons
Spans	21 originally 20 later	5	3
Widest span	21·5 m (70 ft 5 in)	46 m (152 ft)	104 m (340 ft 10 in)
Width of bridge inside balustrades	14 m (46 ft approx.)	20 m (65 ft)	31·6 m (103 ft 10 in)
Number of piers	20	4	2
Navigation clearance above Trinity high water level	5·2 m (17 ft)	8·3 m (27 ft 3 in)	9 m (29 ft 6 in)

* The sum actually paid to the contractor. The published cost was £1,458,311 8*s* 11¾*d*, including fees and land purchases in connection with the bridge and bridge approaches.

'the Victoria Bridge as a whole is incomparably the greatest example of pontine construction extant, as much greater than other large bridges as the Great Eastern is greater than other large ships' (it had 24 × 74 m (242 ft) + 1 × 101 m (330 ft) tubular spans for single-line railway and was 1¾ miles (2·8 km) long with approach spans). Some acid comment followed, however, on the reported cost of the bridge of about £1·5 million: '£75 000 of annual interest is a large tribute to pay for the privilege of crossing the St. Lawrence. . . . The Grand Trunk Company never could and never can afford such a work especially whilst one practically as strong and every way as useful might have been erected for two-thirds of the cost, which even then would have been enormous. As a work of engineering we trust it will be the last in which the outlay is so disproportionate to the

St Nazaire Bridge, the world's longest cable-stayed span, shown just before its completion. It was opened to traffic in the summer of 1975

end.' Tyrrell notes in 1897 its superstructure was replaced by a new, trussed bridge which weighed twice as much, had five times the capacity, but cost only $1·5 million, say about a quarter as much.

The two costs cannot be compared directly because the second one concerns the superstructure only, not the piers. If one assumes that half the cost in the first case was for the superstructure, then the comparison is still startling – ten times the value for money nearly 40 years later. Roadways were cantilevered on to the second bridge and it was extensively modified when the St Lawrence Seaway was built, and it remains in use today.

The record for the costliest bridge is held by the Verrazano Narrows Bridge which cost $178 million – contract cost only, the total cost of the bridge and its approaches being $325 million. This is more than the estimate of US$ (equivalent) 200–250 million for the cost of the Zarate plus Brazo Largo crossings of the Parana in Argentina, and apparently a similar sum for the Niteroi Bridge for which no official figure is available.

The cost of a bridge is seen in better perspective – and in a more relaxed frame of mind! – at the end of its life. Starrucca Viaduct is still in good shape with trains running over it. It must have paid for itself time and time again. Roebling's Niagara Railway Bridge had to be repaired and replaced and although it cost less presents no competition in terms of value for money. The Golden Gate Bridge was desribed as an economic crime when it was proposed – yet how well it has served San Francisco. History seems to be repeating itself with BART (Bay Area Rapid Transit), San Francisco's new public transport scheme. Other expensive bridges have recouped their costs in tolls in quite short periods. Yet another, the steel arch over the Volta River in Ghana is technically excellent, but according to the gossip was built (prior to independence) so that political canvassers could travel easily to the northern part of the country and so influence its politics, but the volume of traffic was negligible. How tricky to make an economic assessment in such a case! Even odder, when Blenheim Palace was built and its estate landscaped by Capability Brown, a great bridge was built and then a lake excavated beneath it, to give the bridge *raison d'être*.

FLOATING BRIDGES

Some rapid feats of construction with floating bridges were achieved in war in ancient times but today floating bridges are considered a *pis aller*. It is of interest, therefore, that a German engineer has suggested that with wide and deep water and bad foundations, when floating bridges may be the only possibility, the drawbacks of sinking under load and movement at the joints between floating elements may be reduced by prestressing against buoyancy – that is tying the bridge down against its own buoyancy – a technique also currently proposed for deep-water structure for the oil industry.

There are half a dozen or so floating bridges of note in existence, including a group in the USA near Seattle. The longest in the world is the second Lake Washington Bridge at Seattle, completed in 1963, which is 2437·8 m (7998 ft) long. Second is the Hood Canal Bridge nearby at Point Gamble which is 2174 m (7131 ft) long.

Delaware Aqueduct. The oldest wire suspension bridge. One of a select group of structures designated by the American Society of Civil Engineers (ASCE) as 'national historic civil engineering landmarks'. It was built in 1848 by Roebling for Delaware and Hudson canal barges to cross above the Delaware River, and was used for that purpose until 1898. It was then converted to a vehicular bridge after a fire had destroyed part of it

SUSPENSION BRIDGES

Links of wrought iron, pinned together to form a chain, were the basis of the suspension bridge as it emerged in the Industrial Revolution and as it was exemplified as a world leader in Sir Samuel Brown's Union and Telford's Menai Suspension Bridges (see table 2). British engineers, however, grew to dislike the suspension bridge, because of its bad behaviour in gales.

Contemporary accounts of the Menai Bridge stress bad troubles from the wind. The bridge was opened at the end of January 1826. On 7 February a gale caused such violent movements that coaches refused to cross. 'The bridge certainly laboured very hard,' Telford said; 24 roadway bars and 6 suspension rods were broken. On 19 February there was a worse gale, 20 suspension rods were broken and 50 bent, and the undulating movements of the bridge were very alarming; a coach eventually crossed safely. The guard of the same coach reported that on 1 March 'Twenty minutes lost at the bridge, it blowing hard and the bridge in great motion which caused the horses to fall all down together and be intangled in each other's harness. Had to set them at liberty one at a time by cutting the harness.'

Wheeling Bridge. The first bridge over the Ohio River and the oldest long-span suspension bridge (see table 2). *Photo: ASCE*

Brighton Chain Pier, effectively a suspension bridge with four 255 ft (78 m) spans completed in 1823, lost a span in a gale in the 1830s apparently more than once. A contemporary artist's impression of that event (see page 49) has a caption that says that the pier was struck by lightning, but the true cause of its collapse was aerodynamic instability, which was endemic in British suspension bridges of that period, and apparently overcome in

Table 5. The world's longest steel plate and box girder spans.

Bridge	Main span m	Main span ft	Year of completion	Country and location	Other details
Cable-stayed girders over 320 m (nearly 1050 ft) in span					
Geislingen	651	2137	Project 1979	West Germany, between Nuremberg and Heilbronn	Construction due to start late in the summer of 1975. The bridge will carry a motorway across a valley. It will be 1130 m (3700 ft) long and will have concrete towers about 240 m (800 ft) high – taller than Golden Gate
Hooghly River	457	1500	u.c.	India, near Calcutta	Multiple stays; 4×91·4 m (4×300 ft) propped side spans; 34·5 m (113 ft) wide
Saint-Nazaire	404	1325	1975	France, Brittany, over River Loire	Three-span main bridge with multiple stays; 14·8 m (48·5 ft) wide, bridge is 3·4 km (2·1 miles) long
Stretto di Rande	400	1312	u.c.	Spain, near Vigo	Three-span bridge, side spans each 147 m (482 ft). Multiple stays
Luling	376	1235	u.c. 1980	USA, Louisiana, over Mississippi River	
Duisburg–Neuenkamp	350	1148	1970	West Germany, over River Rhine	Three stays at each pylon. Side spans are propped. Bridge is 36·3 m (119 ft) wide
West Gate	336	1102	1974	Australia, Melbourne	Collapsed during construction. Has two stays at each pylon and concrete approach spans: 2·6 km (1·6 miles) long; 37·3 m (122 ft) wide
Brazo Largo	330	1083	u.c. 1976	Argentine, over Rio Parana Guazu	Long embankment connects with Zarate Bridge (see table 9)
Zarate	330	1083	u.c. 1976	Argentina, over Rio Parana de las Palmas	
Kohlbrand	325	1066	1974	West Germany, Hamburg	Architecturally a magnificent bridge, with a timeless elegance
Knee	320	1050	1969	West Germany, Düsseldorf, over the River Rhine	'Knee' refers to the shape of the river. Twin pylons at one end of the span with multiple stays
Plate girder bridges (without stays) over 250 m (about 820 ft) in span					
Ponte Presidente Costa e Silva (Niteroi)	300	984	1974	Brazil, Rio de Janeiro, over Guanabara Bay	Three-span high-tensile-steel section of very long bridge (table 9). Side spans of 200 m (656 ft) and suspended section 200 m (656 ft) in centre of main span
Sava	261	856	1956	Yugoslavia, Belgrade, over Sava River	
Zoo	259	850	1966	West Germany, Cologne, over River Rhine	
The Gazelle	250	820	1970	Yugoslavia, Belgrade, over Sava River	The name is appropriate. The depth at the centre of the span is 3·6 m (12 ft) and the span is propped by two inclined struts giving an arched appearance and a modicum of arch action

Note: Spans of this type increased from a record of 76 m (250 ft) in 1939 (Perth Amboy, New Jersey, USA) to 206 m (676 ft) of the Düsseldorf–Neuss Bridge over the River Rhine in 1952. The Cologne–Deutz Bridge over the River Rhine was the trend-setting post-war prototype.

Table 6. The world's longest spans of concrete (except concrete arches which are in table 3).

Bridge	Main span m	Main span ft	Year of completion	Country and location	Other details
Stayed girder bridges over 150 m (about 500 ft) in span					
Brotonne	320	1049·8	u.c. 1976	France, River Seine, Caudebec	Cable-stayed bridge with prestressed concrete deck
Pasco–Kennewick	299	981	u.c.	USA, Washington, Columbia River	
Wadi Kuf	282	925	1972	Libya, Beida	Road bridge with three spans, end spans each 98 m (322 ft)
Tiel	267	876	1975	Holland, over River Waal	Two side spans each of 95 m (312 ft)
Manuel Belgrano	245	804	1973	Argentina, Corrientes	Two side spans, each of 132·5 m (435 ft)
Rafael Urdaneta	235	771	1962	Venezuela, across Lake Maracaibo	Five spans of 235 m. Bridge also has 135 m (443 ft) spans and is 9 km (5·6 miles) long. Originally designed with a main span of 427 m (1400 ft)
Polcevera	208	682	1967	Italy, Genoa	Length of 1100 m (3609 ft). Carries Genoa–Savona motorway
Cantilever bridges over 200 m (about 656 ft) in span					
Koror–Babelthaup	240·8	790	Due to start	Pacific Trust Territory (USA)	The bridge will connect the two islands, Koror and Babelthaup, which lie south-west of Guam and west of Mindanao
Hamana	240	787·4	1973	Japan, Shizuoka	Length 630 m (2067 ft), width 21 m (69 ft). Five spans including a short span at each side
Hikoshima	236	774	1973	Japan, Yamaguchi	Three-span bridge; length 498 m (1634 ft), width 8·5 m (28 ft)
Urado	230	755	1970	Japan, Shikoku, Kochi	Five spans, similar to Hamana, across the neck of Urado Bay, noted for its view of the full moon. Length 601·5 m (1973 ft), width 8·5 m (28 ft)
Uruguay River	209	686	–	Uruguay/Argentina	
Bendorf	208	682	1964	West Germany, north of Coblenz over River Rhine	Main span consists of two cantilevers joined with a shear-pin. Length of bridge is 103 m (3379 ft)

Table 6. cont.

Bridge	Main span m	ft	Year of completion	Country and location	Other details
Prestressed concrete trusses					
Mangfall	108	354	1960	West Germany	Three lattice spans, two of 90 m (295 ft). Carries Munich–Salzburg motorway. First trussed bridge of prestressed concrete.
Saratov	166	545	1965	USSR, over River Volga	Continuous three-span through truss 106+3×166+106 m (348+3×545+348 ft) spans; over-all length 2800 m (9186 ft)
Suspension bridges with steel cables and prestressed concrete decks					
Save River	210	689	1971	Mozambique	Multiple-span bridge with 100+3×210+100 m (328+3×689+328 ft) spans
Tete River	180	591	1972	Mozambique	90+3×180+90 m (295+3×591+295 ft) spans
Rio Colorado	124	407	1974	Costa Rica	Length 204 m (669 ft) in three spans. The cables run on a curve beneath the deck between two inclined piers the tops of which are level with the deck
Mariakerke	100	328	–	Belgium	Self-anchored suspension bridge of prestressed concrete. This type has potential for spans of 300 m (1000 ft)
Stressed-ribbon bridges					
Holderbank–Wildegg	216	710	1964	Switzerland	Bridge carrying a conveyor belt at a cement works (see table 10)
Genf–Lignon	136	446	1971	France, Rhône	Stressed ribbon span just exceeding the hangar at Frankfurt (see table 22)
Freiburg	42	138	1971	West Germany	Three spans, over-all length 136·5 m (448 ft). Footbridge 4·4 m (14 ft) wide
Bircherweid	40	131	1967	Switzerland	Footbridge 2·8 m (9 ft) wide
Peak National Park	34	112	1975	England	Footbridge over railway. First British stressed-ribbon bridge

Table 7. Chronological list of the world's longest simply supported truss span.

Bridge	Main span m	ft	Year of completion	Country and location	Other details
Metropolis	219	720	1917	USA, Illinois, Ohio River	Railway bridge 1067 m (3500 ft) long with seven spans (only one of 720 ft). Parabolic top chord; 33·5 m (110 ft) deep at centre
Municipal	204	668	*c.* 1910	USA, St Louis	Three spans 33·5 m (110 ft) deep making extensive use of nickel steel
Elizabethtown	179	586	1906	USA, over Great Miami River	Truss 24 m (80 ft) deep at centre, 12 m (40 ft) close to supports, parabolic top chord. N-braced with 20 m (65 ft) panels divided by sub-panels
Ohio River	157	515	1876	USA, Cincinnati	Railway bridge, wrought iron, with the first simply supported truss exceeding 500 ft in span. By about 1910, there were at least 25 spans surpassing it on US railways

Union Chain Bridge over the River Tweed at Horncliffe near Berwick, built in 1820, was the first major suspension bridge in Britain. Reconstruction of the deck was completed in 1974 *right* Chains and English tower of Union Chain Bridge. The upper, wire rope cables and their hangers were added in 1902. The original suspension system consisted of 4·6 m- (15 ft)-long forged iron links formed into three pairs of chains, positioned one above the other on each side of the deck. The two sets of chains were placed at 5·5 m (18 ft) centres and supported a deck consisting of two 0·8 m (2 ft 8 in) wide footpaths and a 3·7 m (12 ft 3 in) carriageway. Hangers were wedged into cast-iron saddles which fitted over each joint in the chain, the joints being staggered so that the hangers are at 1·5 m (5 ft) centres along the bridge. The hangers supported a longitudinal flat iron bar which passed through a vertical slot formed in the flattened bottom end of each hanger and was retained in position by cotters and wedges. The timber deck of the roadway was fixed to 0·33 × 0·15 m (13 in × 6 in) timber cross beams 5·8 m (19 ft) long which sat on, and were cleated to the longitudinal flat iron bar. *Photos: Tweed Bridges Trust*

1 2 3 4 5 6

Brighton Chain Pier under duress – view of its partial destruction by lightning on the evening of 15 October 1833, according to the caption on this old print, but the lightning seems to be more romantic than accurate

American ones only by Roebling. His success conditioned American engineering opinion, and the illness was forgotten until 1940. Thus the modern suspension bridge emerged purely as an American invention and technique, brought to maturity by Roeblings, father and son. Their Brooklyn Bridge is the archetype of the modern suspension bridge, which remained basically unchanged from Brooklyn Bridge until the construction of the Severn Bridge in Britain in 1966. Roebling senior considered that a span of 3000 ft (1000 m) was perfectly feasible, at greater cost, but his opinion was not put to the test for half a century until the George Washington Bridge was built.

American suspension bridges went from strength to strength until the Tacoma Narrows Bridge was completed across the Puget Sound in the State of Washington in 1940. From the start, the bridge undulated freely in light winds, and six months after it had been opened the bridge tossed about in a moderate gale until the deck collapsed.

Facing page: Group of pictures showing construction of long-span suspension bridges.
1 Forth Bridge. Erecting catwalks. Cantilevers of the Forth Rail Bridge in the background
2 Forth Road Bridge. Cable spinning
3 Forth Road Bridge. Expansion joint, like a roll-top desk
4 Verrazano Narrows Bridge: positioning cable bands
5 Verrazano Narrows Bridge: cable spinning
6 Verrazano Narrows Bridge: placing cable saddle at top of tower

The stiffening girders of the bridge were deep, narrow and plated girders of I-section, not open trusses, and its shape (like a letter H) and solidity happened to be such that the bridge built up resonant undulations when buffeted by the wind. Since then models of bridges, and later other structures as well, have been tested in wind tunnels when such behaviour is probable. At Forth in Scotland, a slot was left between the two strips of road to prevent differences of pressure above and below the deck, and the stiffening truss was kept very open, so that the wind could blow through it. At Severn, a radical difference appears. The deck structure is a box girder, shaped aerodynamically, rather like the wing of an aeroplane designed for zero lift. The deck, therefore, causes the minimum of disturbance or eddying of the wind, and the aerodynamic force that the wind exerts on the deck is much less – say one-quarter in a gale – than for the Forth.

There is another important feature. If resonant oscillations develop, the structure itself should be able to damp them down. In some suspension bridges this has been attempted by attaching a hydraulic cylinder – a great dash-pot – between deck and cable at the centre of the span. At Severn it

Severn Bridge. *Photo: Freeman, Fox and Partners*

was done by inclining the suspenders, which thus have the appearance of a series of vees. If the bridge starts to oscillate, it moves (for purposes of this simple explanation) longitudinally as well as up and down. When it moves to the left, those suspenders inclined towards the left tighten, and those towards the right slacken, so damping the movement.

There are other improvements, too, in the Severn Bridge. The result is that it has unprecedented slenderness and it was built at a cost substantially less than could otherwise have been expected. Forth was a very economical suspension bridge, yet Severn saved about one-quarter of the steel over Forth and about a quarter of the cost (making a correction so that the comparison is for similar bridges and considering the superstructures only). A rough comparison between Severn and Verrazano Narrows per unit length of bridge and per traffic lane (the Verrazano Narrows Bridge is longer, has a greater central span which would be expected to add to the cost, and twelve traffic lanes against four at Severn) indicates that for equal value for money, Severn should have cost about a quarter as much as the big bridge, but in fact cost less than one-tenth (comparable contract sums \$178 million and £6 million); say twice the value for money, making an arbitrary allowance for the extra length of span – a worthwhile improvement but not in the same class as the Victoria Bridge, Montreal.

A peculiarly American sphere of brilliant engineering was thus captured and improved by British engineers, and the success repeated at Bosporus Bridge and now at Humber Bridge. It will be interesting so see what further developments occur. The extreme limit of span of the suspension bridge is probably about 3000 m (10 000 ft) so there is scope yet for further adventurous innovation in one of the most skilled and daring activities of the civil engineer.

Suspension bridges and arches exert a pull or push sideways at their abutments. The pull of the cables of a suspension bridge is generally resisted by massive anchorages. An alternative for short spans is to build a deck stiff enough to resist the pull of the cables as a strut. Such a bridge is called a self-anchored suspension bridge. The world's longest span of this type is Emmerich Bridge over the Rhine close to the Dutch border. It has a central span of 500 m (1640 ft) and two side spans each of 151·1 m (497 ft) and was completed in 1965.

An engineer inspects the links of the Clifton Suspension Bridge, Bristol. The circular pins in the foreground are folding wedges with which the sag of the cable can be adjusted. *Photo: Howard Humphreys and Sons*

The world's longest span for a chain suspension bridge is the 339·5 m (1114 ft) main span of Florianopolis Bridge at Santa Catherina in Brazil, completed in 1926. The stiffness of the individual links of the chain suspension gives a truss-like character to the suspension, explicitly utilised in this case in an ingenious and economical stiffening truss.

Japanese bridge building. Today's most ambitious group of bridge-building projects comprises three routes each linking the mainland island of Japan, Honshu, with the island of Shikoku. Twenty major bridges are involved, mostly suspension bridges but also other types. The longest span of all – as listed at the head of table 1 – will take thirteen years to build and is on the route Kobe–Naruto; this route also includes the Ohnaruto suspension bridge of 870 m (2854 ft) centre span.

The route from Okayama to Sakaide includes about 14 km (9 miles) of combined road and rail bridges, including two long-span suspension bridges, one of 1100 m (3609 ft) and one of 990 m (3248 ft) centre span.

Alexander 3 Bridge over the Seine in Paris (table 10) is 40 m (131 ft) wide, spans 109 m (358 ft) and links the Champs Elysées with the Esplanade des Invalides.
This sensationally graceful bridge may well be a rival to Waddell's comment (page 36) about the Tower Bridge. An arch thrusts sideways at its abutments; the shallower the arch, the greater the side thrust. Arches are generally built where massive abutments exist or can be made easily. The abutments at this bridge were elaborately contrived by constructing expensive foundation works. The horizontal thrust of the arch is 288 t per metre width of the bridge, i.e. about 11 500 t. Compare this figure with the thrust of 28 000 t of the Bayonne arch (table 3 and page 35) with its relatively elephantine dimensions. Each abutment consists of a single caisson (constructed under compressed air) 44 m (144 ft) in width and about 33·5 m (110 ft) long and 3·4 m (11 ft) deep.
The arch is three-pinned and the arch ribs are of cast steel, supporting a steel deck. The bridge has needed very little maintenance in the 70-odd years of its life, except that some of the decorative work was damaged at the time of the liberation of Paris, and has been restored. The arch and its abutments have behaved satisfactorily. *Photo: Service de la Navigation de la Seine, de la Marne, et de l'Yonne*

Artist's impression of the New River Gorge Bridge, West Virginia, which will be the world's longest arch span by a small margin when it is completed in 1976. The steel of the bridge will not be painted; it will form a patina as it weathers, but will not otherwise corrode. Two cableways, each of 50 t capacity, are erecting the steelwork. The bridge will be 926 m (3036·5 ft) long and 21·6 m (71 ft) wide and will contain about 19000 t of steel. *Photo: West Virginia Department of Highways*

The third route (Onomichi–Imabari) strings together a chain of islands with nearly 10 km (6 miles) of bridges, including six suspension bridges, one of them with a span of 1000 m (3281 ft). The second and third routes will take nine years to build. The start of construction of these bridges was delayed by the unfavourable economic events of 1973 and 1974, but a start is about to be made in 1975 on one of the three routes. Because of this uncertainty, only the longest-span bridge of them all is listed in the tables.

FAILURES AND DISASTERS

The collapse of Tacoma Narrows Bridge is perhaps the best-known bridge failure of the last half century. But there are a host of others to keep it company. Happily, no lives were lost at Tacoma. Other failures have caused severe loss of life, mainly to those building the bridge and only rarely to users of it. There can be few great bridges that have not exacted a toll in lives from accidents during construction. The failures of the first Tay Bridge in Scotland, and of the Quebec Bridge in Canada are well known and fully documented.

More recently, there have been failures caused by brittle fracture which can happen in high-tensile steel, characteristically at a welded joint when the temperature drops. Duplessis Bridge in Canada, some Vierendeel truss bridges in Belgium and, with some qualification, King Street Bridge in Melbourne, Australia, are examples.

The most sensational failures of recent years have been collapses of box-girder bridges: West Gate Bridge, again at Melbourne; Milford Haven in Wales; and at Coblenz in Germany. Extensive enquiries following these failures have led to new procedures for the structural analysis of box girders, which evidently had been 'pared down' a bit too much.

The most common cause of trouble with concrete bridges is the centring or falsework, that is the temporary bridge that is built to support the concrete until it attains its full strength.

Table 8. The world's longest spans of opening bridges of various types.
Bascule bridges are counterbalanced cantilevers that swing upwards to open about a horizontal axis. The Scherzer rolling lift bridge is a cantilever with a counterbalance shaped like a quarter of a wheel; it lifts by rolling backwards along a track.
Swing bridges are pivoted at the centre of the span, and open by turning horizontally. They are called draw spans in the USA.
Vertical-lift bridges have towers at each end, from which ropes are operated to lift the bridge, like the sash of a window.
Transporter bridges are permanently open. A beam at high level spans the waterway, and a trolley or crab, is pulled across it, a car or platform being suspended from the trolley at road level, carrying the vehicles or pedestrians. The transporter bridge is thus a hybrid, neither bridge nor crane.

Bridge	Main span		Year of completion	Country and location	Other details
	m	ft			
Bascule bridges over 90 m (about 300 ft) in span					
South Capitol Street	118	386	1949	USA, Washington, DC, over Anacostia River	Double-leaf, plate-girder bascule flanked by three 48 m (157 ft) spans on either side
Saulte Ste Marie	102	336	1941	USA/Canada, over ship canal	Double-leaf railway bridge
Black River	101·5	333	1940	USA, Ohio, Lorain	Double-leaf highway bridge
Tennessee River	94·5	310	1917	USA, Chattanooga, Tennessee	
Jose Leon de Carranza	90	295	1969	Spain, Cadiz	Carries Seville–Cadiz motorway. Double-leaf bascule 96·8 m (318 ft) between trunnions
Record span for a single-leaf bascule					
St Charles Air-Line	79	260	1919	USA, Chicago, 16th Street	A railway bridge
Most famous bascule bridge					
Tower Bridge	79	260	1894	England, London	
Swing bridges over 152 m (500 ft) in span ('span' means the total length of the moving structure)					
al Firdan	168	552	–	Egypt, over Suez Canal	Twin swing spans. The distance given is pivot to pivot in this case
Fort Madison	160	525	1927	USA, Iowa, Mississippi River	Railway bridge
Willamette River	159	521	1908	USA, Oregon, Portland	Railway bridge 537 m (1762 ft) long with five spans
East Omaha	158	519	1903	USA, over Missouri River	Railway bridge

Table 8. cont.

Bridge	Main span m	ft	Year of completion	Country and location	Other details
George P Coleman Memorial	152	500	1952	USA, Virginia, Yorktown, over York River	Two equal swing spans. Bridge is 1143 m (3750 ft) long
Vertical-lift bridges					
Arthur Kill	170	558	1959	USA, New York City, Staten Island–New Jersey	Railway bridge. N-braced truss, can be lifted 41 m (135 ft). Towers are 65·5 m (215 ft) high. Raising takes two minutes
Cape Cod Canal	166	544	1935	USA, Massachusetts	Railway bridge, can be lifted 41 m (135 ft)
Delair	165	542	1960	USA, New Jersey	
Marine Parkway	164	540	1937	USA, New York City	The most beautiful yet, according to Steinman. Bridge is 1226 m (4022 ft) long. See page 35
Elbe	106	348	1973	West Germany, Hamburg	Europe's longest lifting span; for road and rail. Three-span trussed bridge 300 m (984 ft) long
First vertical-lift bridge					
South Halstead Street	40	130	1894	USA, Chicago	Built by J A L Waddell
Scherzer rolling lift bridges					
Chicago River	78 (between piers) 84 (length of span)	255 (between piers) 275 (length of span)	c. 1910	USA, Chicago, Illinois	Double-leaf bridge at entrance of Grand Central Station
Chicago Ship Canal	38	123·5	c. 1912	USA, Chicago, Illinois	Four single-leaf bridges side by side, each 9 m (29 ft) wide, form the widest and heaviest bridge of this type
Longest transporter span					
Volgograd	874	2867	1955	USSR, Volgograd	Temporary bridge, 1403 m (4604 ft) long used in construction of hydro-electric power station. Basically a suspension bridge

Note: Barton swing aqueduct is the only swing bridge that carries a canal over a canal (Brindley's Bridgewater Canal over the Manchester Ship Canal). It was built by Sir Leader Williams in 1893. It is swung full of water, the weight of the swing span being 1450 t and of the water it contains 800 t. It spans a clear passage of 27 m (90 ft) and is 71·5 m (234·5 ft) long.

Oléron Bridge, linking the island of that name with the French mainland in Brittany, carried a stage further the dramatically successful bridge-building techniques established by Freyssinet in his post-war Marne bridges when he combined precasting with prestressing. Oléron Bridge is a prestressed concrete box girder, structurally continuous over groups of spans, with a length of 2900 m (9514 ft) mainly in 80 m (262 ft) spans. It was erected rapidly – 10 m of bridge a day – from precast elements each 3·3 m (11 ft) long, using a travelling erection girder and epoxy resin to glue the joints. *Photo: Campenon-Bernard*

Table 9. The longest bridges (as opposed to bridge spans) in the world. (Approach embankments are excluded from the length quoted, and the bridge is in one length unless noted otherwise in the last column.)

Bridge	Length of bridge km	miles	Year of completion	Country and position	Other details
Bridges over 7 km (nearly 4·5 miles) long all of which are in North or South America					
Lake Pontchartrain No. 2	38·422	23·87	1969	USA, Louisiana, near New Orleans, Metairie–Lewisburg	Precast prestressed concrete. 17·1 m (56 ft) spans
Lake Pontchartrain No. 1	38·352	23·83	1956	USA, Louisiana, near New Orleans, Metairie–Lewisburg	Precast prestressed concrete. Part of Greater New Orleans Expressway. 25·6 m (84 ft) spans
Lake Pontchartrain trestle	34·6 (original length) 9·2 (shortened length)	21·5 5·7	1883 1890–6	USA, Louisiana, near New Orleans, South Point–North Shore	Railway trestle of timber. Crossing 6 miles of lake and 16 miles of swamp, on bents 4·6 m (15 ft) apart. Later filled in. Included two draw spans each of 76 m (250 ft)
Chesapeake Bay bridge/tunnel	28·5	17·7	1964	USA, Virginia	Total length includes tunnels and artificial islands
Sunshine Skyway	24·5	15·2	1954	USA, Florida, Tampa Bay	Length includes six causeways and five steel and concrete bridges. 263 m (864 ft) cantilever span
Lucin Cut-Off (Great Salt Lake)	19·3	12	1904	USA, Utah, Lakeside	Replaced in 1960 by an embankment. Originally 27 miles (43·4 km) shore to shore including 15 miles (24 km) of embankment
Ponte Presidente Costa e Silva (Niteroi)	13·9	8·7	1974	Brazil	See table 5. Length over water 8·9 km (5·5 miles)
Chesapeake Bay	12·9	8	1952 1972	USA, Virginia	Two identical bridges nearly 4 miles (6·5 km) over water. 488 m (1600 ft) span suspension
San Mateo–Hayward	11·67 11·27	7·25 7	1929 1967	USA, California (San Francisco Bay)	The later bridge replaced the earlier one
Seven Mile	11·27	7	1912	USA, Florida	Knight's Key-Pacet Key. Built as a railway bridge, converted to a road bridge in 1938.
Zarate–Brazo Largo	10 (railway approaches) 6·5 (road approaches)	6·25 4	u.c. 1976	Argentina, near Buenos Aires	Two bridges crossing the two arms of the Parana River; total length including embankments 30 km (18·6 miles) (see table 5)

Table 9. cont.

Bridge	Length of bridge km	miles	Year of completion	Country and position	Other details
Tappan Zee	9·7	6	1956	USA, New York over Hudson River	296 spans. Carries New York Thruway (see table 4)
Lake Pontchartrain	8·7	5·4	1965	USA, Louisiana	Twin bridges, Pointe aux Herbes–Howze Beach
San Francisco–Oakland	8·4	5·2	1936	USA	Total length including tunnel about 160 m (525 ft) crossing Yerba Buena Island (see tables 1 and 4)
Rafael Urdaneta (Maracaibo)	8·3	5·2	1962	Venezuela	See table 6
Albermarle Sound	8·2	5·1	1910	USA, North Carolina, Skinner's Point/Mackay's Ferry	Railway trestle of timber. Bents 3·8 m (12·5 ft) apart with 5 × 15 m (50 ft) girder spans, one 43 m (140 ft) rolling lift span and a 29 m (94 ft) swing span
Huey P Long	7·1	4·4	1935	USA, Louisiana, over Mississippi River	Railway bridge with road section 3 km (1·86 miles) long. 241 m (790 ft) cantilever main span. Heroic continuous-truss bridge of the 1930s. Longest railway bridge in the world (trestles excepted)
The longest bridges regionally					
CHINA					
Nanking	6·8	4·2	1968–9	Over Yangtze River	9 × 160 m (525 ft) main spans. Diamond-truss bridge, double-track rail with road above. Road section about 4·5 km (2·8 miles) long
Great Bridge	1·7	1·06	1957		Double-track rail; road above
Hwang Ho	102 spans.		1905	Yellow River	Peking–Hankow Railway
INDIA					
Upper Sone	3·1	1·9	1900	Sone River	93 spans of 30 m (100 ft). Railway bridge between Calcutta and Delhi. Lower Sone Bridge has 28 × 46 m (150 ft) spans
Godavari	2·7	1·7	1900	Godavari River	Railway bridge between Calcutta and Madras
AFRICA					
Lower Zambezi	3·67	2·28	1934	Mozambique, Mutarava	33 spans of 80 m (262 ft). Total of 46 spans plus 56 trestle spans. Longest railway bridge in the world when completed

Table 9. cont.

Bridge	Length of bridge km	miles	Year of completion	Country and position	Other details
EUROPE					
Oeland	6·06	3·77	1972	Sweden	155 spans 34–130 m (112–426 ft) of prestressed concrete
Eastern Scheldt	5·02	3·12	1965	Holland	52 × 95 m (312 ft) spans and bascule with 2 × 20 m (66 ft) leaves
BRITAIN					
Railway viaduct	6	3·75	1836	England, London/ Bridge to Deptford Creek	878 brick arches carrying London and Greenwich Railway. World's longest brick structure. Some arches have been replaced. In use today
Tay (rail)	4·2	2·62	1878	Dundee	Collapsed 1879
	3·3	2·04	1887		Replacement bridge
The longest urban elevated roads					
West Side Highway	8·8	5·5	–	USA, New York City	The Battery to 72nd Street
Brooklyn–Queens Expressway	6·4	4	–	USA, New York City	Williamsburg Bridge to Queens Boulevard in Queens
Pulaski Skyway	6·7	4·17	1932	USA, New Jersey	Approaches west end of Holland Tunnel, New York City. River cantilever has 168 m (550 ft) span
Western Avenue Extension	4·1	2·55	1970	England, London	Prestressed concrete, precast in segments up to 137 t in weight. Longest elevated road in Europe. Greatest spans 62 m (203 ft)
Bolzano	2·58	1·6	u.c.	Italy, Ravenna–Modena motorway, crossing Bologna	Prestressed concrete with 75 spans of 34·5 m (113 ft) and 22·2 m (73 ft) wide

When does an elevated road become a bridge? Western Avenue is a feat of bridge engineering far superior to Lake Pontchartrain No. 2. Elevated roads are listed separately in this table because they cannot be classified unambiguously as bridges in all cases, yet they can be considerable works of bridge engineering.

Summary of Table 9
The world's longest bridge is a trestle structure of precast prestressed concrete at Lake Pontchartrain, 38·422 km (23·87 miles) long. The world's longest timber bridge was also a trestle at Lake Pontchartrain, originally 34·6 km (21·5 miles) long. The world's longest steel bridge is the Chesapeake Bay Bridge 12·9 km (8 miles) long. The two long South American bridges have steel river spans but concrete approach spans.

The longest railway bridge in the world (trestles excepted) is Huey P Long Bridge, 7·1 km (4·4 miles) long, since the 10 km (6·25 miles) length of railway bridging at Zarate–Brazo Largo is effectively two bridges. The longest steel-trussed bridge in the world is the San Mateo–Hayward Bridge. The world's longest structure of brick is the railway viaduct from London Bridge to Deptford Creek. The longest of stone is the Rockville Bridge in Pennsylvania, shown in the illustration on page 38.

Table 10. The slenderest bridges; the shallowest arches; structural continuity.

Type	Bridge	Span m	Span ft	Depth m	Depth ft	Ratio span: depth	Other details
The slenderest bridges							
Stressed ribbon	Holderbank–Wildegg	216·4	710	0·26	0·85	832	See page 6
	Genf–Lignon	136	446	0·4	1·3	340	See table 6
	Peak National Park	34	111·5	0·16	0·5	215	
Suspension	Severn	988	3240	3	10	324	See table 1
	Humber	1410	4626	4·5	14·76	314	
Stayed-girder: concrete	Pasco–Kennewick	299	981	2·1	7	140	See table 6
steel	Duisburg–Neuenkamp	350	1148	*c.* 3·8	*c.* 12·5	*c.* 92	See table 5
Concrete footbridges: arched	Housatonic River	30·5	100	0·23	0·75	133	Reinforced concrete arch, 1910, at Stockbridge, Mass., USA. Rise of arch 3 m (10 ft); depth at springing 0·8 m (2·5 ft)
	Langelinie	20	66	0·2	0·67	100	See page 16
prestressed beam	St James's Park	21	70	0·36	1·17	76 (structural) 60 (over all)	Depth includes 7·6 cm (3 in) of finishes
Prestressed concrete	Esbly	74	242·75	0·86	2·83	85·7	See page 13
	Medway	152	500	2·2	7·2	69	
	Bendorf	208	683	4·4	14·4	47	See table 6
Steel girder (unstayed)	Niteroi	300	984	7·4	24·3	40·5	See table 5
	The Gazelle	250	820	3·6	12	69	See table 5
	Cologne–Deutz	185	606	3·0	10	61	

Table 10. cont.

Material	Bridge	Span m	Span ft	Rise m	Rise ft	Ratio span: rise	Other details
The shallowest arches							
Cast steel	Alexander 3	109	357·6	6·28	20·6	17·36	See page 52
Concrete	Revin	122	400	9·75	32	12·2	Over River Meuse, Belgium
Brick	Maidenhead	39	128	7·47	24·5	5·2	Longest span in brick
Cast iron	St Louis	64	210	5·79	19	11	Paris, 1862, longest cast-iron span in France
For comparison: N-braced steel	Bayonne	510·5	1675	81	266	6·3	See table 3
Concrete	Gladesville	305	1000	40·8	134	7·5	See table 3

Bridge	Length of continuous structure m	ft	Number of spans	Other details
Structural continuity				
Astoria	751	2464	3	Longest three-span continuous truss, 188+375+188 m (616+1232+616 ft) (table 4)
Naples Viaduct	1280	4200	15 to 20, each of 91 m (300 ft) or less	Twin steel box girders, completed in 1975. Tangenziale section of the Genoa–Calabria motorway
Western Avenue	1280	4200	Spans of 28–35 m (90–115 ft)	Prestressed concrete – Section 1 of the scheme. Prestressed sequentially to achieve complete continuity, including an overhead roundabout which is free to expand laterally
Western Avenue	1158	3800	About 20, mainly of 62 m (204 ft)	Section 5, curving over railway. Completed 1971 (see table 9)
Tinsley Viaduct	1033 915	3388 3052	20 upper deck 18 lower deck	Double-deck steel box girder, Sheffield England, completed 1967
Saratov	710	2329	5	Prestressed concrete truss spans (see table 6)

dams

THE WORLD'S HIGHEST DAMS

Details of dams more than 200 m (656 ft) in height. The height is measured from the lowest foundation to the top, and may be rather greater than the visible height from lowest ground level. The length given is measured along the top of the dam (i.e. round the curve for an arch). Reservoir capacities are gross volume up to normal top water level; 1 cubic kilometre (km^3) is 1000 million cubic metres.

Figures in bold type give the height of the dam and the year the dam was completed; u.c. means under construction. Further data on most of the dams in this list are given in tables 11–16.

Nourek Dam 317 m (1040 ft) u.c. – 1977 on the Vakhsh River in Tadjikistan, USSR, close to the Afghanistan border. An earth-fill embankment with a clay core. The reservoir is intended for hydro-electric power, irrigation and water-supply and will store 10·5 km^3 (about 2·5 $miles^3$) of water. The length of the dam is 730 m (2395 ft). It has been under construction since 1962. An outstanding feature of this and several other recent high dams in Russia is that they have been designed to resist severe earthquake shocks. According to reports in 1974, construction was lagging badly. The dam had reached half its full height but only one-third of the total volume was in place. The reservoir was in use, with one-seventh of its ultimate volume of water, and some power was being generated. The completion date given above is the most recent estimate of the year in which the dam will reach its full height, taking account of the delays. The whole scheme, which includes water for irrigating more than 1 million ha (2·5 million acres) of land in the Amu-Darya catchment will be completed in 1979. A higher dam is to be started in the next Five Year Plan, at a site 70 km (44 miles) upstream of Nourek on the same river. This dam – Rogun Dam – will have a height variously reported as 325 m (1066 ft) and 323 m (1060 ft), and will be built in ten years, applying the lessons and experience learned at Nourek.

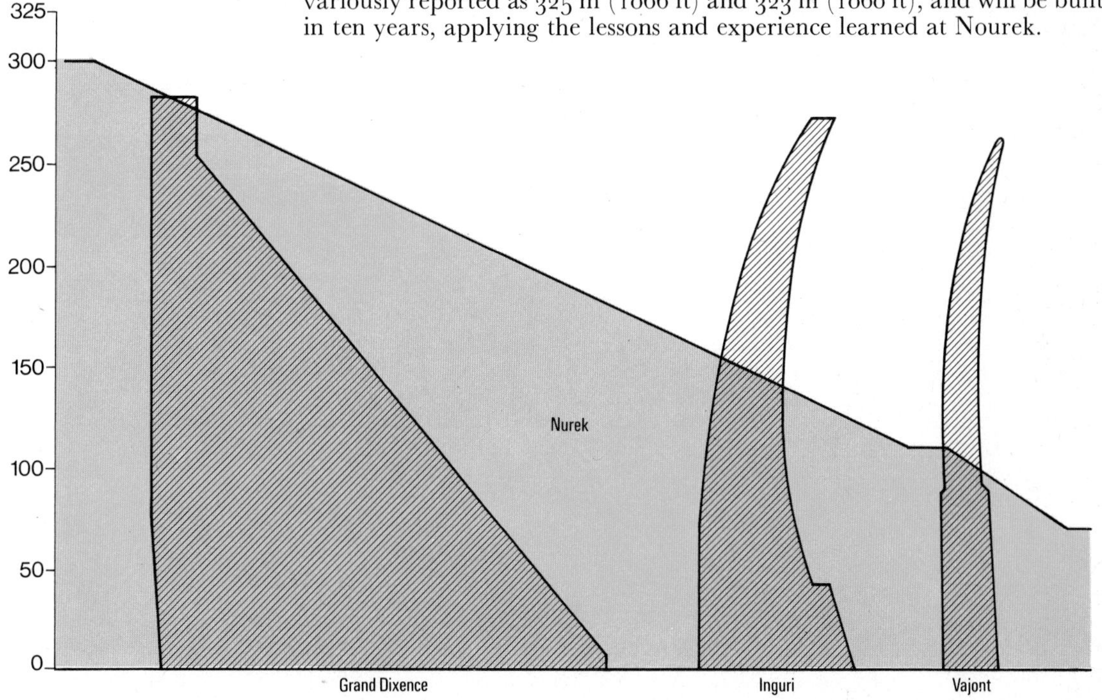

Grande Dixence Dam, 285 m (935 ft) 1962 impounds a tributary of the Rhône in Valais, Switzerland, and is at present the world's highest dam. It is a gravity dam, built of concrete. It was constructed in three stages, and a special study was made of the joints between the stages, so that load would be transferred across them satisfactorily when the next stage was built. The dam is high in the Alps, and impounds a seasonal hydro-electric reservoir (0·4 km³ (0·1 miles³)). Seasonal hydro-electric reservoirs in the Alps are small in comparison with those formed by some other large dams, but they are of critical value, nevertheless, to the power systems that they serve. The length of the dam is 695 m (2280 ft).

Inguri Dam, 271 m (889 ft) u.c. on the Inguri River in the north Caucasus, Georgia, USSR, is an arch dam that will form a reservoir with a storage capacity of about 1 km³ (0·25 mile³) for hydro-electric power generation and flood prevention. Length 758 m (2487 ft); volume of concrete 3·8 million m³ (almost 5 million yd³). The slow progress made in building this dam was heavily criticised in *Pravda* in July 1974.

Vajont Dam, 262 m (858 ft) 1961 on the Vajont River, a tributary of the Piave in north-eastern Italy (Veneto) is a doubly curved or cupola arch. Its reservoir was engulfed by a gigantic land-slip in 1963 (see page 79). The dam remains, unharmed but useless. The steepness of the valley is indicated by the length of the dam – 190 m (623 ft) – which although measured along the curve of the arch is less than the height of the dam. This shape, together with the slenderness of the dam, account for the fact that it contains only 0·35 million m³ (0·46 million yd³) of concrete.

Mica Dam, 244 m (800 ft) 1973 on the Columbia River in British Columbia, Canada, is an embankment of sand and gravel with a core of glacial till. Its volume is 32 million m³ (42 million yd³) and the length of its crest is 792 m (approximately 0·5 miles). Filling of the reservoir commenced on 1 April 1973 and is expected to be completed before commissioning the first generating sets at Mica, which is scheduled for late 1976. The reservoir will store about 24·5 km³ (6 miles³), of which 60 per cent will be usable for production of power at site and downstream in Canada and the United States.

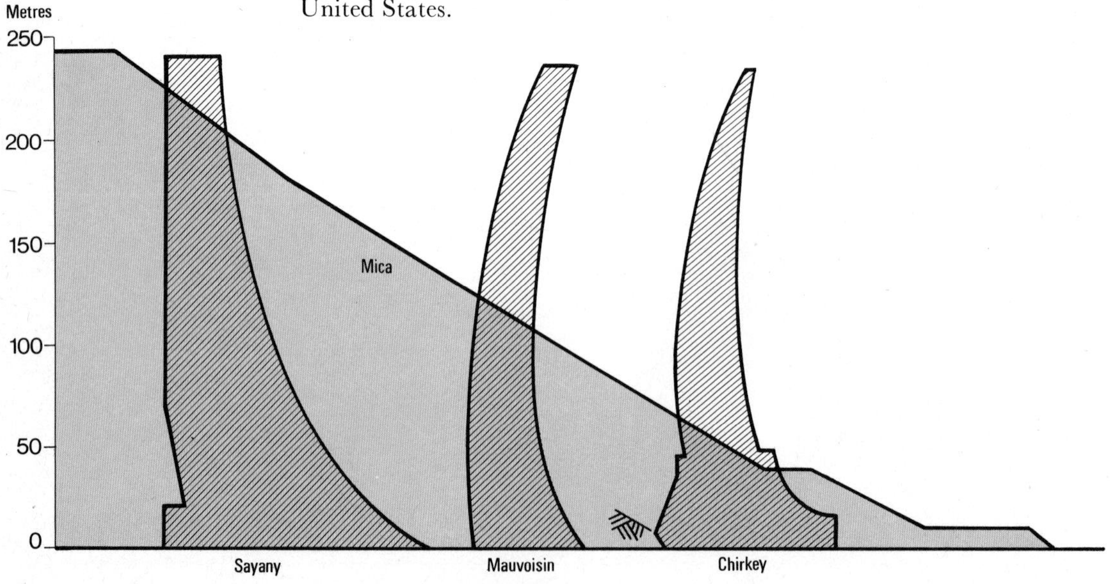

Sayany (or Sayano-shushens-kaya) Dam, 242 m (794 ft) u.c. on the Yenisei River, USSR, is an arch-gravity dam of unprecedented size, having double the volume of its US rival, Auburn Dam, now also under construction. It will have a length of 1068 m (3504 ft) and will form a reservoir of 31 km³ (7·5 miles³) for power and navigation. The full reservoir will impose a load of about 18 million t on the dam. Its construction was also under attack from *Pravda* in 1974 for slow progress.

Mauvoisin Dam, 237 m (778 ft) 1957 on the Dranse de Bagnes River, a south-bank tributary of the Rhône, just to the west of the Dixence, in Valais, Switzerland, also impounds a seasonal reservoir for hydro-electric power. Tunnels, aqueducts and pumping stations diverting the rainfall and snow melt from a great area of the Swiss Alps feed the two reservoirs. Mauvoisin Dam is rather smaller than Grande Dixence – their lengths are 520 m (1706 ft) and 695 m (2280 ft) respectively – but the volume of concrete it contains is very much less – 2·03 million m³ (2·66 million yd³) or about a third that of the Grande Dixence – because it is an arch dam, which is much more economical in material than a gravity dam.

Chivor Dam, 237 m (778 ft) u.c. is under construction on the Bata River about 90 km (56 miles) north-east of Bogotá in Colombia. The dam is rock-fill embankment with a clay core, and is part of a hydro-electric scheme. Also called Esmeralda Dam.

Chirkey Dam, 236 m (774 ft) u.c. on the Sulak River, north Caucasus, USSR, is a doubly curved arch with a length of 338 m (1109 ft) and a volume of concrete of about 1·25 million m³ (1·6 million yd³). It will have a reservoir of about 2·8 km³ (0·7 mile³). It is sited in an earthquake-prone region.

Oroville Dam, 235 m (770 ft) 1968 impounds a multiple-purpose reservoir on the Feather River in California, USA. The dam is an earth embankment. Its length is 2109 m (6920 ft) and the capacity of its reservoir 4·4 km³ (about 1 mile³).

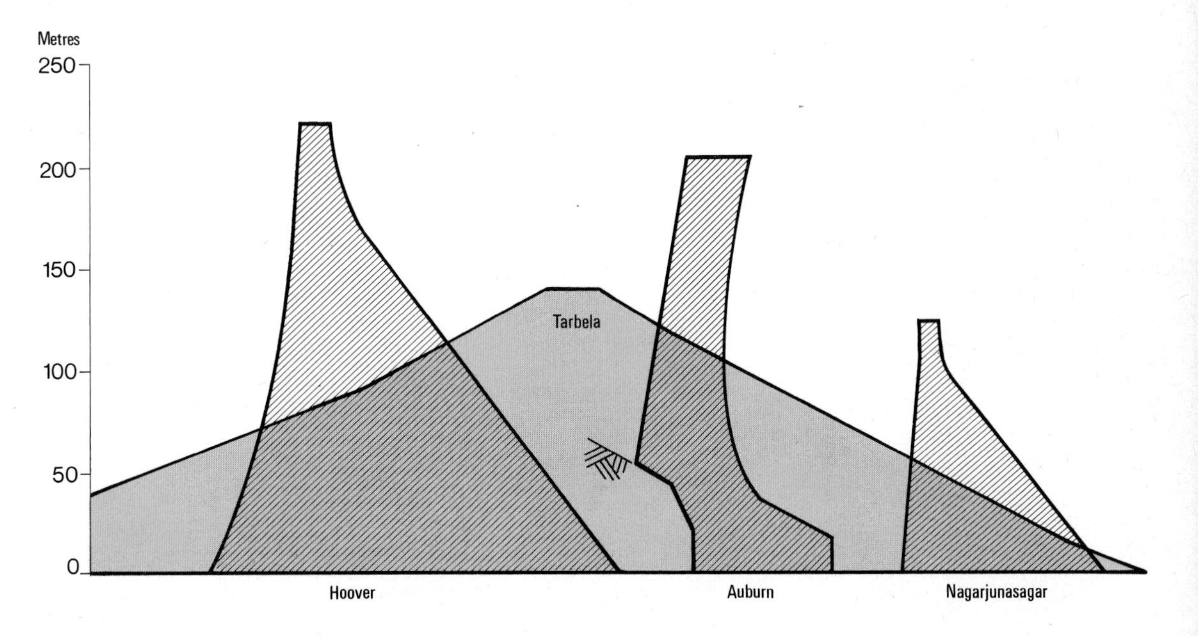

Chicoasen Dam, 232 m (761 ft) u.c. – 1980 a rock-fill embankment that will impound the Rio Grijalva in southern Mexico, for the generation of hydro-electric power. It will contain 15 million m³ (20 million yd³) of fill and will form a reservoir of 2·5 km³ (0·6 mile³) in volume.

Contra Dam, 230 m (755 ft) 1965 is another Swiss arch dam for hydro-electric power. It impounds the Verzasco River in the Province of Ticino, at a site where the steepness of the valley gave a length of 380 m (1247 ft) to the dam, and a volume of concrete of 0·66 million m³ (0·86 million yd³). The dam has a thickness of 7 m (23 ft) at the top and 25 m (82 ft) at the base, this remarkable slenderness making it one of the most daring, if not the most daring, arch dams in the world according to its designers (compare with Crystal Dam).

Bhakra Dam, 226 m (742 ft) 1963 on the Sutlej River near Chandigarh in India is a gravity dam. Its length is 518 m (1699 ft) and the capacity of its reservoir is nearly 10 km³ (over 2 miles³). This reservoir is the key to regulation of the Punjab rivers now utilised by India (for irrigation and power) under the Indus Waters Treaty. The value each year of the output from Bhakra Dam (that is the electrical energy sold and the crops sold from the land irrigated) is about equal to the capital cost of building the scheme.

Hoover Dam, 221 m (726 ft) 1936 (formerly Boulder Dam) on the Colorado River, USA, where it forms the border between Nevada and Arizona, was for 22 years the world's highest dam, and was a generation ahead of other large concrete dams. It is an arch-gravity dam, which as the name implies, is a hybrid structural form relying partly on the weight of the dam (gravity) and partly on arch action between the rock abutments (like the arch of a bridge turned on its side) to resist the great load of the reservoir. The height of the dam above the lowest ground level is 176 m (577 ft), it is 379 m (1243 ft) long, and it contains 3·36 million m³ (4·40 million yd³) of concrete, a volume then unprecedented and unequalled until 1940 when Grand Coulee Dam was completed. Comparing again, its

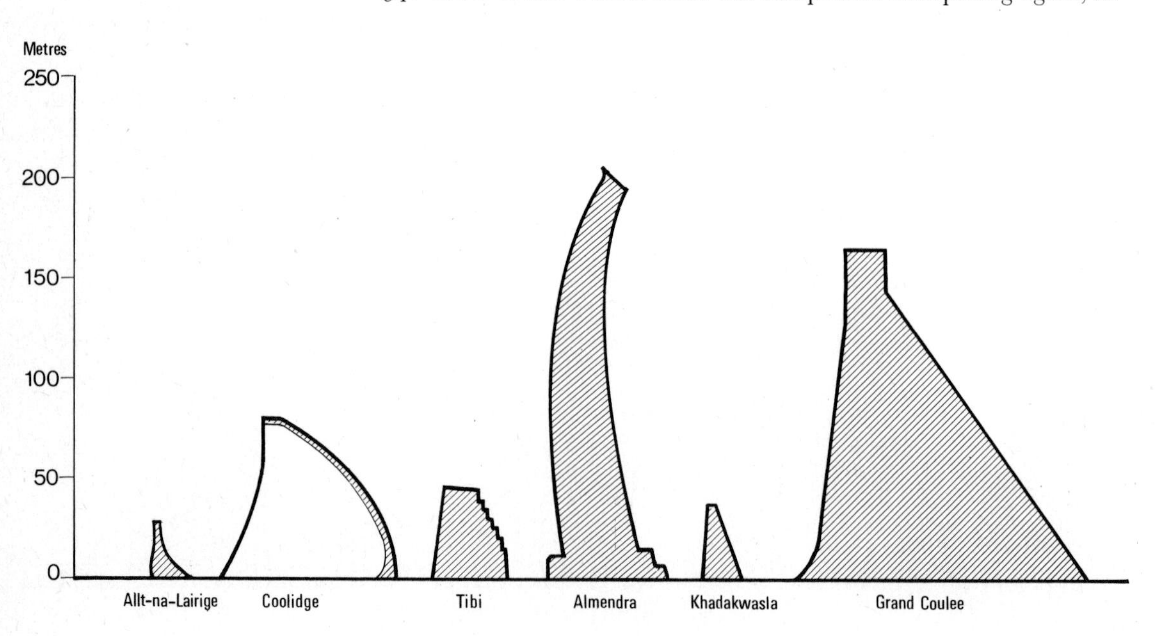

Hoover Dam and Lake Mead from the Arizona side. The four towers are intakes to the power stations which are grouped downstream of the dam. The spillways on each bank lead into tunnels which discharge beyond the limit of the picture. The arch-gravity form of the dam is shown very clearly. *Photo: US Bureau of Reclamation*

length is less than the length of Mauvoisin Dam, but its volume is two-thirds as much again. Hoover Dam impounds a reservoir storing 38 km³ (9 miles³) again unequalled for decades in size, which is multiple-purpose – that is it serves for navigation, flood control, recreation and irrigation; but it permits massive generation of hydro-electric power and that is its primary function.

Mratinje Dam, 220 m (722 ft) u.c. Yugoslavia, a concrete arch dam that will impound the Piva River near Fŏca (Črna Gora). The length of the dam will be 268 m (879 ft).

Dworshak Dam, 219 m (718 ft) u.c. now nearing completion in Idaho, USA and Toktogoul Dam below prove that the gravity dam has not yet been superseded among the world's highest dams. Dworshak's design includes several features that have permitted rapid and economical construction.

Glen Canyon, 216 m (710 ft) 1964 is an arch dam on the Colorado River in Arizona, USA. It impounds a hydro-electric reservoir of 33 km³ (8 miles³) storage capacity. The length of the dam is 475 m (1558 ft).

Toktogoul Dam, 215 m (705 ft) u.c. on the Naryn River, Kirzighia, USSR. A gravity dam for hydro-electric power and irrigation; 412 m (1352 ft) long and having a volume of concrete of 2·7 million m³ (3·5

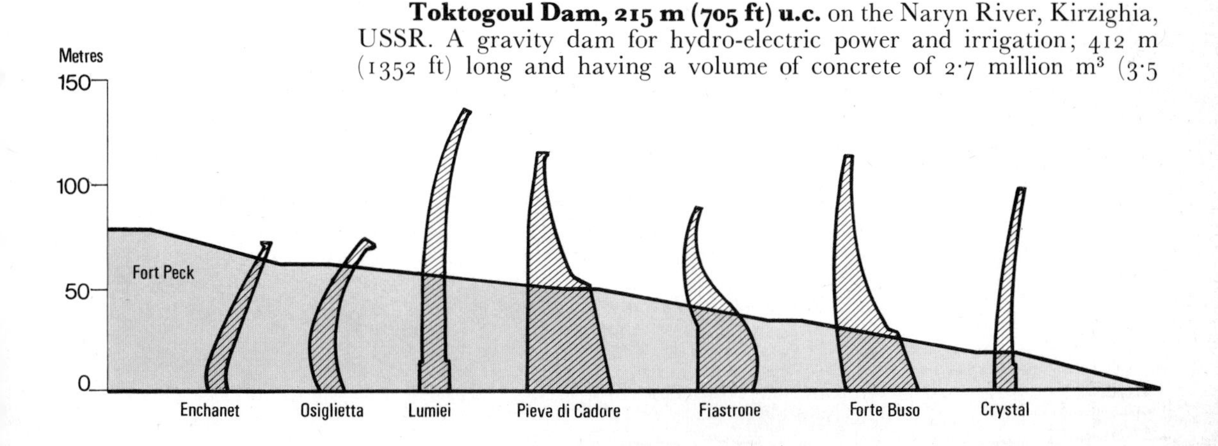

Daniel Johnson Dam (Manicouagan 5) in Quebec shown shortly before its completion

Akosombo Dam, Ghana. The power station is on the left, and the spillway on the right with a secondary dam beyond

million yd³). Its reservoir will store nearly 20 km³ (5 miles³) of water, and will permit the irrigation of 0·5 million ha (over 1 million acres) of land in the Syr-Darya catchment.

Daniel Johnson Dam, 214 m (703 ft) 1968 on the Manicouagan River, a north-shore tributary of the St Lawrence in the Province of Quebec, Canada. The dam is a multiple-arch structure and is the world's largest of that type. Its reservoir gives enormous storage but also creates a head for the generation of power. Length of dam: 1306 m (4285 ft); volume of concrete 2·26 million m³ (2·96 million yd³).

Auburn Dam, 209 m (685 ft) u.c. is a doubly-curved arch to be completed in the mid 1970s in northern California, USA. Its length is 1265 m (4150 ft). It will be longer than Sayany Dam, but will contain only about half the volume of concrete of the Russian dam. It will be 60 m (196 ft) thick at the base and 12 m (40 ft) thick at the top (compare with Contra Dam).

Luzzone Dam, 208 m (682 ft) 1963 on the Brenno di Luzzone in Ticino Province, Switzerland, is an arch dam impounding a hydro-electric reservoir of 87 million m³ (114 million yd³) capacity. Length of dam: 530 m (1739 ft); volume of concrete: 1·35 million m³ (1·77 million yd³).

Keban Dam, 207 m (679 ft) 1974 a rock-fill embankment in Turkey, regulates the headwaters of the Firat (Euphrates) River, and permits the generation of large quantities of electricity that is transmitted mainly to Ankara and Istanbul. Reservoir: 31 km³ (7·5 miles³); dam: 1097 m (3599 ft) long containing 15 million m³ (19·6 million yd³) of fill.

Reza Shah Pahlavi Dam, 203 m (666 ft) 1963 on the Dez River in Iran and the last entry below are part of a scheme including fourteen dams to restore the agricultural productivity of Khuzestan Province.

Almendra Dam, 202 m (662 ft) 1970 Spain. An arch dam for hydro-electric power, regulating the Tormes, a tributary of the Douro.

Reza Shah Kabir Dam, 200 m (656 ft) u.c. Concrete arch, 381 m (1250 ft) long. It impounds the Karun River.

Table 11. The world's largest reservoirs (over 40 km^3 (10 $miles^3$) in volume).

A rounded figure is given for the total volume of the reservoir. It includes 'dead storage' below the lowest outlet in the dam, and the space reserved as a silt store, but not the volume corresponding to the freeboard between normal top water level and the parapet of the dam.

Conversions: 1 $mile^3$ = 4·168 km^3 1 km^3 = 1 000 000 000 m^3
= 0·2399 $miles^3$
= 810 683·5 acre feet
= 1 307 903 000 $yards^3$

u.c. means 'under construction' 1 km = 0·62 miles

Dam	Volume of reservoir km³	miles³	Year of completion	Year reservoir first filled	Details – length, breadth, surface area	River and Country
Bratsk	169·2	40·59	1964	–		Angara River, Irkutsk, USSR
Aswan (Saad el Aali)	164	39·36	1970	c. 1976	Extends 500 km (300 miles)	Nile, Egypt
Kariba	160	38·38	1959	1963	Area 5180 km² (1991 miles²)	Zambezi, bordering Rhodesia and Zambia
Akosombo	148	35·5	1965	1969	Length: 400 km (250 miles). Area: 8482 km² (3275 miles²). Shore line 7250 km (4500 miles). Seasonal drawdown: 2·5–3 m (8–10 ft) equivalent to 1685 km² (650 miles²)	Volta, Ghana
Daniel Johnson	142	34·07	1968			Manicouagan, Quebec, Canada
Krasnoyarsk	73·3	17·58	–	–		Yenisei, Siberia, USSR
WAC Bennett (Portage Mountain)	70	16·79	1967	First power 1968	T-shaped reservoir Surface area 1658 km² (640 miles²)	Peace River, British Columbia, Canada
Zeya	68·4	16·41	u.c.			Eya, East Siberia, USSR
Wadi Tharthar	67	16	1956		A sump for river floods	Tigris, Iraq
Ust-Ilim	59·3	14·23	u.c.			Angara, Irkutsk, USSR
V I Lenin	57	13·68	1953	Soon after completion	Largest reservoir in Volga–Kama cascade	Volga (Kuibyshev), USSR
Bukhtarma	53	12·71	1960	–		Irtish, Altay, USSR
Cabora Bassa	52	12·5	1975	Sluices closed 1974	Will extend over 2700 km² (1040 miles²). Average depth 23 m (75 ft)	Zambezi, Mozambique
Irkutsk	46	11·04	1956	–		Angara, Irkutsk, USSR

The world's largest artificial lake measured by surface area is Lake Volta formed by Akosombo Dam. Its area is 4 per cent of the area of Ghana: 80 000 people were resettled from this area.

Natural lakes regulated by dams. The world's largest lake, Lake Victoria, is regulated by Owen Falls Dam on the Victoria Nile in Uganda. Variation of the level of the lake by 3 m (10 ft) gives storage of 202 km^3 (48·5 $miles^3$), more than any of the reservoirs listed in this table. A portion of this storage is reserved for Owen Falls but most of it is compensation water for Egypt. Other large lakes regulated by dams are Lake Ontario and Lake Geneva.

Lumiei Dam in the Carnic Alps, Italy. A cupola arch in a valley of extraordinary steepness

There are 25 dams in this list, of which 14 are arch dams, 4 are gravity, 1 is a multiple-arch dam (all of concrete), and 6 are embankments of earth or rock.

The cross-sectional sketches indicate the economy of material (concrete) gained by the arch, especially the slender arches. For dams in the range of heights below 200 m, embankments are by far the most numerous in recent construction, since it often proves cheaper to move great quantities of rock or earth than to fix shutters and then mix and place a much smaller quantity of concrete. The arch dam is the apogee of daring and brilliance, matched throughout the whole range of structures – indeed of all the artefacts of man – only by the long-span suspension bridge. The embankment dam, by comparison, is a triumph of logistics over intellect, but its success has extended the range of economic construction to wide valleys on the world's largest rivers, where the dam is nothing less than a minor mountain.

ARCH DAMS

The extreme of daring is the cupola or overhung arch, exemplified at its greatest – in spite of its bitter tragedy – by Vajont Dam. Another example in Italy is Lumiei Dam, 136 m (446 ft) high and the world's highest arch at the time it was completed. The first major cupola was Osiglietta (76 m (249 ft) high), completed in 1939. There are several comparable dams in France, notably Enchanet and Couesque. The thinnest arch dams are the experimental dams built in the 1950s at Tolla (Corsica) and Gage (France), though they were anticipated by a remarkably similar experimental dam – Stevenson's Creek Dam – built in the USA in 1926. Some notable arch dams were built in Australia in the early years of this century.

The arch of greatest radius (161 m (528 ft)) – a structural feat as daring as extreme slenderness – is Pieve di Cadore Dam, which forms a reservoir on the Piave River in the same system that Vajont served. This claim to distinction is about to be overtaken by the great arch that the Russians are building at Sayany, which appears to be a record-breaker of awe-inspiring proportions, and by its American rival, also under construction, Auburn Dam. The radius of the arch at Auburn is 427 m (1400 ft) in its central section and 1219 m (4000 ft) for the outer segments.

Cheeseman Dam is a gravity arch of masonry, completed in 1904, for the water supply of Denver, Colorado. It is one of the ASCE's national historic engineering landmarks (table 16)

Tarbela Dam *Photo: TAMS*

BUTTRESS DAMS

The prototype of these structures is the multiple-arch dam, in which a series of buttresses perform the same function structurally as the native rock in the case of the single arch. The Ambursen dam, an inclined slab resting on buttresses, was developed from the timber crib dam. Round-headed, diamond-headed and massive buttress dams and hollow dams, are all attempts at using gravity (i.e. the weight of the dam) with greater economy in material than the plain gravity dam. Of these, the highest one Farahnaz Pahlavi, Iran, has a height of 107 m (315 ft). But it will soon be overtaken by Zeya Dam, USSR (165 m (541 ft) high), at present under construction.

DAMS BENEATH GLACIERS

The ultimate in collecting Alpine run-off for hydro-electric power is achieved when water is collected as it melts beneath a glacier. A tunnel is driven to the underside of the glacier, and the ice excavated so that a little dam can be built. The first of these was at Tré-la-tête, a glacier not far from Mont Blanc, which feeds La Girotte Reservoir. The dam consists of a steel-faced crest over which the ice passes, and a transverse channel in which the newly melted water collects and flows into the tunnel.

RIVER BARRAGES

River barrages consist mainly of a series of sluice-gates that can be raised to let a flood pass, or lowered to raise the level of the river. There is no clear-cut distinction between dams and barrages. The world's greatest barrages are listed in chapter 3. See also chapter 3 for spillways for large dams, and chapter 2 for cut-offs, curtains and blankets.

SEA DAMS

The greatest sea dams are in Holland and are described in chapter 4. When a dam is built to enclose an area of the sea, the main difficulty is to close the ever-narrowing gap through which the tide flows twice a day. Thus the 'cubature' – the volume of flow through the gap on each tide – is the best measure of the difficulty of the job, rather than height, volume, etc. (see table 15).

Table 12. The world's largest dams; embankments of earth or rock (over 40 million m³ (52 million yards³) in volume).

Dam	Volume of dam Millions of m³	Volume of dam Millions of yd³	Year of completion	Height of dam m	Height of dam ft	River and Country
Tarbela comprising:	142·2	186	1975	143	470	Indus, Pakistan
main dam	121·6	159				
side valley dam	15·3	20				
saddle dam	1·9	2·5				
cofferdam	3·4	4·5				
Mangla Reservoir comprising:	106·6	139·4	1967			Jhelum, Pakistan
Mangla Dam	64·6	84·5		116	380	
Jari Dam	32·4	42·4		71	234	
Sukian Dam	9·6	12·5		15	50	
Fort Peck	96	125·6	1940	76	250	Missouri, Montana, USA
Eastern Scheldt	70*	92	u.c.	46	151	Rhine Delta, Holland
Oahe	70	92	1963	75	246	Missouri, South Dakota, USA
Gardiner (South Saskatchewan)	66	86	1968	68	223	South Saskatchewan, Canada
Oroville	60	78	1968	236	774	Feather, California, USA
San Luis	59	77	1967	116	382	California, USA
Nourek	58	76	u.c.	317	1040	Vakhsh, Tadjikistan, USSR
Garrison	51	67	1956	64	210	Missouri, North Dakota, USA
Vado Hondo	50	65	Project	127	418	Zenta, Salta Province, Argentina
Dickey	47	62	Project	105	345	St John, Maine, USA
Cochiti	47	62	u.c.	77	253	Rio Grande, Santa Fé, New Mexico, USA
Kiev	44	58	1964	19	62	Dnieper, Ukraine, USSR
Gorky†	44	58	1955	28	92	Volga, USSR
WAC Bennett (Portage Mountain)	44	57	1968	183	600	Peace, British Columbia, Canada
Aswan (Saad el Aali)	43	56	1970	111	364	Nile, Egypt

* Includes 7 million m³ (9 million yd³) of artificial dunes contiguous with the dam. It is now uncertain that the dam will be completed in this form.
† Partly concrete.

Missouri River Dams
1 Fort Peck Dam, Montana
2 Garrison Dam and Lake Sakakawea, North Dakota
3 Oahe Dam, South Dakota
The tall structures at the power stations are surge tanks. *Photos: US Army Corps of Engineers*

EXTENT OF DAM-BUILDING

The number of large dams in the world before 1800	450
The number built 1800–49	117
1850–99	527
1900–49	3583
1950–62	2778
1963–71	5377
Total number completed at end of 1971	12832
Under construction at that date	1282
Projects at that date	1342
Grand total	15456

It is estimated that at present one-tenth of the stream flow of the wörld's rivers is regulated by dams. By the year 2000, two-thirds could be regulated.

The ten countries with the greatest numbers of dams are:

USA	5058	Great Britain	471
Japan	1856	Italy	396
India	1114	Brazil	373
Spain	596	France	355
Canada	475	Australia	304

What is a large dam? It is defined by ICOLD (International Commission on Large Dams) as a dam over 15 m (50 ft) high, or between 10 and 15 m (30 and 50 ft) high if:

- Its crest is more than 500 m (1640 ft).
- Its reservoir has a capacity greater than 100 000 m³ (220 million Imp gal).
- It can discharge more than 2000 m³ (70 000 ft³) per second.
- It has especially difficult foundations.

This definition forms the basis of classification in ICOLD's *World Register of Dams* which has been referred to extensively in preparing this section of the book.

Including small dams, the total number of dams is estimated to be 30 000 to 50 000 in the USA plus 25 000 in the rest of the world.

Roselend Dam, high in the French Alps, shown when construction was nearly complete. *Photo: Electricité de France*

NOTABLE DAMS

Roselend Dam in the French Alps has a structural form of unusual boldness. A deep gorge is closed by an arch with a sloping crest that rises, at the centre, to a higher level than the abutments; buttresses bear directly on the arch, and extend across the shallow wings of the valley. A gravity dam would have needed 2·3 million m³ (3 million yd³) of concrete, but the dam, as built, contains 0·84 million m³ (1·1 million yd³). The dam is 150 m (492 ft) high, and was completed in 1962; it provides seasonal storage 1550 m (5000 ft) above sea-level for hydro-electric power.

Nagarjunasagar Dam (table 13) is a gravity dam of masonry with a height of 125 m (409 ft) flanked at each end with earthen embankments. Its total length is nearly 5 km (3 miles) and its reservoir is about 11·5 km³ (2·77 miles³) in volume. Design of the dam was evolved so that it could be built with hand labour, with minimal use of mechanical plant.

Baipaza Dam, on the same river as Nourek Dam in the USSR, was formed on 28 March 1968, by blasting or 'directed explosion' as the Russian engineer responsible for this remarkable feat termed it. Thirteen chambers were excavated in the very steep right bank of the valley. They were charged with 1860 t of explosive, and about 1·2 million m³ (say 1·5 million yd³) of rock (as measured *in situ*) were moved by virtually a single explosion to form the body of the dam, which is 60 m (197 ft) in height. A much larger feat of this kind is planned (see chapter 2).

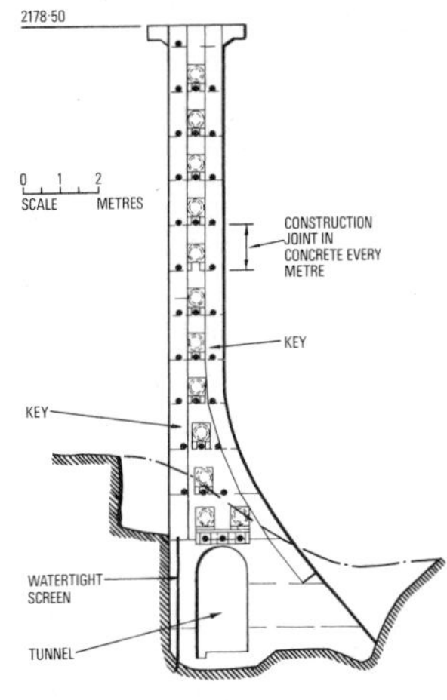

Tourtemagne Dam. The dots are horizontal prestressing cables and the dotted circles with squares round them are flat jacks, that were expanded in each joint to maintain the prestressing force.
A similar dam is under construction at Nambe Falls, New Mexico, USA

Djatiluhur Dam in Indonesia is circular. The power station lies within the dam and a tail-race tunnel leads from beneath it. A separate embankment impounds the reservoir. This layout was adopted because of very difficult ground conditions.

Allt na Lairige Dam, in the Highlands of Scotland, is a prestressed dam. The effect of gravity can be increased by anchoring cables deep in the rock beneath the dam, running the cables vertically to a second anchorage at the top of the dam, and then tensioning them. This technique was applied to repair a dam in the 1930s, but it was first used in the design of a new dam at Allt na Lairige, completed in 1956. Economy of material in the cross-section results. The prestress at Allt na Lairige is 56 t/ft run of the dam.

Tourtemagne Dam, Valais, Switzerland, is the first, and the only, prestressed arch dam. The dam is a thin arch, and the purpose of the prestress is to counteract secondary thermal stresses which the severe climate, and emptying and filling the reservoir at all seasons with glacial water, impose.

Crystal Dam, under construction, on the Gunnison River in Colorado, USA, is claimed as the thinnest arch dam. When completed it will be 98 m (320 ft) high, 3 m (10 ft) thick at the crest and 9 m (29 ft) thick at the base.

Blackbrook Dam, 8 km (5 miles) west of Loughborough, is the only dam in Britain on record as damaged by an earthquake. The earthquake occurred on 11 February 1957, and its epicentre was only 4 miles (over 6 km) from the dam. The parapet was damaged and there was some cracking. Leakage increased from about 2000 to nearly 1 million l (440 to about 220 000 Imp gal) per day, but after a period, reverted to normal.

Above: Mica Dam on the Columbia River in British Columbia, Canada, is the world's highest embankment or 'fill' dam as it is called. In this recent picture, the reservoir is quite high, but it will not fill until 1976. Water is seen discharging from the outlets of the two diversion tunnels that bypassed the river during construction of the dam. On the opposite bank, and not seen in the picture, are the six intakes to the generating sets in the underground power station that is beneath the bank to the left of the dam in this picture

Below: Grand Coulee Dam. The third power station – referred to in chapter 3 – is to the left. When completed, it will bring the installed hydro-electric capacity of Grand Coulee up to about 10 GW, as large as that of any power station yet under construction in the world. In addition water is pumped from the reservoir to irrigate great tracts of land. When the dam was completed in 1942, irrigation was the major use foreseen

HISTORICAL FACTS

The beaver anticipated man; fossilised relics of beavers' dams exist. Palaeolithic man built stone fish dams. The first true dam on record is the Sadd-el-Kafara (table 16). It was (and is – the central section was washed away, probably soon after its construction, but the rest remains) a massive structure, consisting of two walls of masonry with gravel fill between them. Its length was 106 m (348 ft), and its width across the base 84 m (276 ft). However, this dam is an oddity. It was probably built to store water for use at an alabaster quarry nearby. The basin irrigation then practised in Egypt did not require dams, and with one exception, there are no others recorded in Egypt until the modern era.

Irrigation requiring dams probably started in the Tigris and Euphrates valleys. Dams made of earth, timber and reed were probably typical; they had to be maintained every season. These practices spread to India and through the East. More permanent structures also appeared, for example a rock-fill dam, about 1300 BC near Homs, Syria and Sennacherib's two masonry dams on the Khosr River (just above its confluence with the Tigris at Mosul) dating from 694 BC; these dams still exist as ruins. Marib Dam in the Yemen, was built in 750 BC, and twice heightened, the second time to 14 m (47 ft) in 325 BC.

The Romans were competent and resourceful dam-builders, but have left few dams of great size, the largest of the three Subiaco Dams (table 16) being an exception. The art was practised in the Middle East and Arab worlds throughout the Middle Ages.

Japan has a long history of dam-building, and a record of about 30 embankment dams over 15 m (50 ft) in height built between about AD 500 and 1600. During the Eddo period (1603–1867) over 500 dams in that category were built that still exist. The oldest major dam in Japan is Kaerumataike Dam which dates from AD 162. Like all others, it is an earthen embankment; it is 17 m (56 ft) high and has a length of 260 m (853 ft).

Sri Lanka has a similar history, but dating from about the 5th century BC. Irrigation schemes near the ancient city or Anuradhapura were based on earthen embankments of similar height to the Japanese ones, and like them, forming reservoirs of modest size. However, one dam with a length of 18 km (11 miles) is reported, and a maximum height of 21 m (69 ft). One example is Panda Wewa Dam, dating from about 450 BC, and in working order until the early 19th century. It was an embankment 7 m (22 ft) high and about 3 km (1·75 miles) long, and formed a reservoir with a capacity of about 11 million m^3 (15 million yd^3).

Several tens of thousands of irrigation dams were built in India in medieval times. The origins of this tradition are believed to be very early. A reservoir extending over 650 km^2 (250 $miles^2$) was impounded near Bhopal in the 11th century. The Moti-Talav earth dam near Mandya in Mysore, 24 m (79 ft) high, dates from about the 10th century and is still in use.

The first historically proven arch dam is considered to be a flood-control and water-supply structure built on the Turkish-Syrian border during the reign of the Byzantine emperor, Justinian I (AD 527–65) by Chryses of Alexandria, '. . . In the form of a crescent, in order that its arch, which was turned against the stream of the water, might be better able to resist its violence.'

Tibi Dam, near the village of that name, on the Monegre River in the province of Alicante, Spain, is the prototype of the modern age of dam building. It was completed in 1594 and was the highest dam in the world for nearly three centuries. It is in use for irrigation – its original purpose. It is 20·5 m (67 ft) thick at the crest and 33·7 m (111 ft) at the base, its volume being 36400 m³ (47600 yd³). It is faced with dressed masonry and has a core of rubble masonry set in lime mortar. On the left of the dam can be seen a spillway that was added in 1738 and today is closed with stoplogs. In 1943 a new outlet tunnel was driven through the rock on the right bank abutment

The earliest multiple-arch dam on record is at Meer Alum near Hyderabad in India; it was built in about 1802 and is still there.

Netzahualcoyotl Dam (table 14) was named after the engineer who built it. Its purpose was to prevent floodwater from reaching the ancient city of Tenochtitlan. The site is now an urban zone within Mexico City. The dam was built of wood, rock and mud crowned with a strong wall of masonry. In parts it was very deep, and was more than 4 fathoms (presumably 7 m (24 ft)) in width and had sluice-gates. It prevented 'Filthy water from meeting with the fresh water on which the city was founded'; 20000 people worked to build it. It was destroyed by Cortez in 1521.

Modern dam-building started in Spain, with France and Britain soon active too. Tibi Dam can perhaps be taken as the starting-point, and the prominence of failures in the Spanish dams listed in table 16 is not representative. Furens Dam was the first to be designed from a structural analysis, and Zola Dam (built by, and named after, Emile Zola's father) demonstrated the advantages of the arch dam. Zola Dam was completed in 1854 with a height of 42·5 m (139 ft). New standards of watertightness and safety were introduced with Vyrnwy Dam, a masonry dam 44 m (144 ft) high and 357 m (1172 ft) long, completed in North Wales in 1892, which at that time formed the largest reservoir in Europe (59 million m³ (13000 million Imp gal)).

Hoover Dam was a generation ahead of its time; it was matched in daring by the great bulk of Grand Coulee Dam. The biggest dams up to, say, about 1960, were generally of concrete. Economy could be achieved at some sites by building slender arch dams, and in these daring and beautiful structural forms, dam-building has probably already passed its peak. Since then the rock-fill dam has rapidly become popular for most of the largest schemes.

Puentes Dam, completed in 1884, was built 30 m (about 100 ft) downstream of the site of an earlier dam that failed catastrophically in 1802, causing over 600 deaths. The position of the old dam can be seen

The rapid increase throughout the world in the number of dams built, and their increasing size and competence, during the last quarter of a century, has been the most striking phenomenon in the history of dams. A mature stage has now been reached. The world's largest rivers can be controlled and there are probably not many sites where dams higher than those now in service could ever be required. Design of the dam is complex and now demands a group of specialised designers; one gifted individual is no longer enough. If the predictions about the regulation of stream flow on page 72 actually occur, then the great age of dam-building may well have passed, or be on the wane, by the end of this century, and could prove to be as well defined as the epoch of the Gothic cathedrals.

FAILURES AND DISASTERS

A heavier responsibility is carried by the designers and builders of large dams than for any other form of construction with the possible exception of nuclear power stations. Failure is potentially able to cause a catastrophe comparable with a major natural disaster. Failure or malfunction of some kind occurs once every 1500 to 1800 dam-years (the life of a dam is reckoned to be from 50 to 250 years) which means about ten incidents a year with 15000 dams in the world. ICOLD records more than 500 incidents, of which 200 were failures. Of these, 290, of which 90 were failures, occurred on large dams during the period 1900–65. Incidents are monitored and analysed; valuable lessons can sometimes be learned from them.

Puentes Dam. An ambitious structure with a height of 50 m (164 ft) built of masonry and completed in 1791. When the river-bed was excavated, a canyon filled with alluvium and extending to great depth was discovered. It was believed that the masonry could be founded on timber piles and cribs, the piles driven into the alluvium, across the 17 m (56 ft) width of the canyon, at the central, deepest part of the dam. However, when, in 1802, the reservoir was filled to a height of 47 m (154 ft) for the first time, it blew out the plug of alluvium and the reservoir emptied in an hour.

Part of St Francis Dam, California, after its failure in 1928. *Photo: Engineering News Record*

Bouzey Dam. This dam, near Belfort in France, failed in 1895. The failure was caused by uplift – that is the hydrostatic pressure of the reservoir acting upwards on the dam – a phenomenon not previously known, but since taken into account in the design of dams.

Dolgarrog Dam. One of only two British dam disasters that have entailed loss of life occurred at Dolgarrog Dam in North Wales in 1925. As a result of this disaster, new legislation – the Reservoirs (Safety Provisions) Act – was introduced. It was the earliest, and to date the best example of safety legislation in the world for dams and reservoirs.

St Francis Dam. A gravity dam, 56 m (184 ft) in height and completed in 1926 for the water-supply of Los Angeles in California, USA. The reservoir was filled to spillway level for the first time on 5 March 1928, and on 12 March the dam failed. The failure was complete and catastrophic, and blocks of concrete weighing 10 000 t or so were carried several thousand feet downstream. A Commission decided that the dam failed because of weaknesses inherent in the foundation. It had no cut-off wall beneath the upstream face, no grouting, and nothing to relieve uplift pressure. The dam was sited on a fault, between conglomerate and schist. The former rock became soft and plastic when wet; the latter was laminated and would slip like a pack of cards when load was applied parallel to the laminations.

Panshet and Khadakwasla Dams. India's oldest masonry dam, and the first large-scale dam of that kind in the world, is Khadakwasla Dam in the Mutha River catchment, upstream of the city of Poona. It was completed in 1870, after ten years of construction under the direction of Colonel Fife of the Royal Engineers. Shri NGK Murti, in an ICOLD paper, drew attention to the slenderness of the dam by comparison with more recent masonry dams, and described the dam as 'a saga of daring and dash by Col. Fife'. The dam gave excellent service until 1961. In July of that year, the Panshet Dam failed and released about 200 million m^3 (260 million yd³) of water at a time when the reservoir at Khadakwasla was already full with the spillway-gates open. The dam was overtopped for its full length and the water rose 9 ft (3 m) over the top of the dam. Shri Murti recorded: 'Observers have reported that on that fateful day, they felt the dam was vibrating. The overflow scoured most of the earth at its downstream toe. Yet this wonderfully lean structure withstood a water-level 14 ft (4 m) above the design level and the dynamic forces unleashed by the overflow for nearly four hours, and then failed, not at the place of maximum height, but at a place where there was a step in the foundations and when the overflow was about 3 ft (90 cm) above the top of the dam.'

The dam has since been rebuilt. The failure of Panshet Dam made 80 000 people homeless. Deaths were estimated from 30 to 100, and could have been enormously greater if Khadakwasla Dam had not stood for four hours.

Malpasset Dam. The first and only failure of an arch dam occurred in December 1959, when Malpasset Dam was swept away in an instant, and destructive floods caused heavy loss of life in the town of Fréjus (about 80 km (50 miles) from Nice) in the south of France. The cause of the failure was complex and a rare combination of unusual circumstances was met. Uplift pressure was again the destructive agent, but this time acting at a depth against a band of rock that became impervious under load. A wedge-shaped piece of ground beneath the abutment of the dam (formed by a fault and this band of rock) lifted, so destroying the stability of the dam.

Vajont Dam. The most destructive of all reservoir failures, having a ferocity that is beyond words to describe, occurred in October 1963, when the side of a mountain, Mount Toc, slid into the reservoir of Vajont Dam, travelling at a speed of about 25 m/s (about 90 km/h or 60 miles/h) over a distance of 400 m (1300 ft). The volume of material in the slide was about 250 million m³ (330 million yd³), which is of the order of ten times larger than any other landslide in historical times. Water was displaced violently over the top of the dam – about 40 million m³ (50 million yd³) of it – and obliterated the village of Longarone and three other villages. Some 2000 lives were lost. The dam withstood the tremendous load imposed on it –

Pieve di Cadore Dam

Table 13. The world's largest concrete dams (over 4 million m³ (5·25 million yd³) in volume).

Dam	Volume of concrete Millions of m³	Millions of yd³	Year of completion	Height of dam m	ft	Type of dam	Country	River and location
Sayany	9·12	11·92	u.c.	242	794	Arch gravity	USSR	Yenisei River
Grand Coulee	8·09 8·53 (after completion of third power station)	10·59 11·16	1942 1974	168	550	Gravity	USA	Columbia River, Washington
Zeya	8	10·46	u.c.	115	377	Gravity buttress	USSR	Eya River, East Siberia
Itaipu	7·6 (main dam only)	10	u.c.	175	580	Gravity	Brazil/ Paraguay	Parana River, Iguazu Falls
Folson	6·89	9	1955	104	340	Gravity	USA	American River, California
Shasta	6·66	8·71	1945	183	602	Gravity	USA	Sacramento River, California
Grande Dixence	5·96	7·79	1962	284	932	Gravity	Switzerland	Dixence River, Valais
Nagarjunasagar	5·6*	7·3*	1974	124	406	Gravity*	India	Krishna River, Macherla, Andhra Pradesh
Kentucky (Gilbertsville)	5·3	6·9	1944	63	206	Gravity	USA	Tennessee River, Paducah, Kentucky
Dworshak	4·97	6·5	u.c.	219	717	Gravity	USA	Orofino River, Idaho
Bratsk	4·8†	6·3†	1964	125	410	Gravity	USSR	Angara River
Auburn	4·6	6	u.c.	209	685	Arch	USA	American River, California
Ust-Ilim	4·4†	5·8†	u.c.	105	344	Gravity	USSR	Angara River
Krasnoyarsk	4·4	5·7	1972	124	407	Gravity	USSR	Yenisei River
Bhakra	4·1	5·4	1963	226	740	Gravity	India	Sutlej River, Punjab

* Masonry, not concrete; the dam also contains 2·35 million m³ (3·1 million yd³) of earth embankment.
† Concrete volume only; Ust Ilim Dam also contains 8·7 and Bratsk 12·2 million m³ (11·4 and 16 million yd³) of embankment respectively.

Table 14. The world's longest dams (more than 15 km (10 miles) in length). (Embankments along sea-coasts or river-banks are not included.)

Dam	Length km	Length miles	Year of completion	Height of dam m	Height of dam ft	Country	River and location
Kiev	54·1	33·6	1964	19	62	USSR	Dnieper, Ukraine
Wadi Tharthar	52	32·2	1956	6·5	21	Iraq	Tigris
Dneprodzerzhinsk	36·3	22·6	1964	35	115	USSR	Dnieper, Ukraine
Zuider Zee	32·5	20·2	1932	7·5	25	Holland	Sea dam, in two parts separated by Wieringen Polder
Finnish Gulf	25·8	16	Project	–	–	USSR	For flood protection of Leningrad; plans approved
Hirakud	25·4	15·8	1956	62	202	India	Mahanadi, Orissa
Roseires	16·9	10·5	1966	66	217	Sudan	Blue Nile
Kanev	16·1	10	u.c.	25	82	USSR	Dnieper, Ukraine
Riga	15·2	9·4	u.c.	33	108	USSR	Dougava, Latvia
Ancient dams of great length							
Lake Hungtze	100	62	16/17th century	Moderate		China	Kiangsu; dike to impound lake
Padawiya	18	11	12th century	18	59	Sri Lanka	60 km (38 miles) north-east Anuradhapura; embankment dam
Veeranam	16	10	1037	Moderate		India	Cuddalore, Madras, embankment dam
Netzahualcoyotl	16	10	1449	6	20	Mexico	Lake Texcoco (see text)

estimated to be 4 million t – by a wave of 260 m (850 ft) above reservoir-level, or roughly twice the height of the dam. But the dam's prodigious structural strength – a tribute as much to elaborate strengthening of the rock abutments as to the sturdiness of the dam itself – made little if any difference to the catastrophe.

Filling the reservoir – its depth was of the order of 200 m (650 ft) – introduced hydrostatic pressures deep into the rocks of the mountainside, acting against gravity. A slip surface, following the bedding planes of the rock, was identified, down which the whole mass had slid undisturbed.

May Dam. Not all failures are violently destructive. The reservoir of May Dam near Konya in Turkey has never filled. It is situated on alluvium overlying karstic limestone, and about 30 sink-holes have formed in the bottom of the reservoir. Virtually all the water that flows into the reservoir escapes through these sink-holes and flows away underground.

Table 15. Major sea dams.

Dam	Tidal cubature at mean tide. Millions of cubic: m	yd	Average tidal range m	ft	Maximum depth, mean sea-level to sea-bed m	ft	Length of dam km	miles	Year of completion or closure	Country	Purpose
Eastern Scheldt	1100	1439	2·8	9·2	35	115	9	5·6	u.c. 1978*	Holland	Primary dam of Delta scheme
Brouwershavense Gat	300	392	2·3	7·5	30	98	5	3·1	1971	Holland	Primary dam of Delta scheme
Haringvliet	260	340	2·6	8·5	12	39	5	3·1	1970	Holland	Primary dam of Delta scheme
Veerse Gat	70	91	2·9	9·5	17	56	3	1·9	1961	Holland	Primary dam of Delta scheme
Zuider Zee	575 reduced to 20 just before closure	752 reduced to 26 just before closure	1·0	3·3	7	23	32	20	1932	Holland	Enclosure of Ijsselmeer
Rance	184	240	8·5†	27·9†	18	59	0·7	0·4	1966	France	Tidal power
Plover Cove	28	37	2·4	7·8	13	44	2	1·2	1967	Hong Kong	Water-supply
Scapa Flow (four eastern entrances)	not available		4‡	13‡	17	56	2·3 (total)	1·4 (total)	1944	Scotland (Orkney)	Naval defence
Eider	50	65	3	10	10	33	5	3·1	1973	West Germany	Sea defence

* As originally planned. † 13·5 m (44·3 ft) at equinoctial spring tide. ‡ Maximum.

Note: The earliest major tidal closure on record was in AD 1270. A gap 360 m (1200 ft) wide was closed, despite powerful tidal currents, to seal off an area of the Lower Maas River, 50 km (30 miles) long, near Dordrecht in Holland. With a second dam at the other end, a polder 50000 ha (124000 acres) in area was created. The polder was lost in one night, 150 years later, 65 villages with 10000 inhabitants being submerged.

Table 16. Chronological list of the world's highest dam.

Dam	Height – lowest foundation to crest m	ft	Year of completion	River, province or country	Details
Nourek	310	1017	u.c.	USSR	See text
Grande Dixence	284	932	1962	Switzerland	See text
Vajont	262	858	1961	Italy	See text
Mauvoisin	237	778	1958	Switzerland	See text
Hoover	221	726	1936	USA	See text
Chambon	136	446	1934	River Romanche, France	Gravity dam. One of the early major Alpine hydro-electric dams
Owyhee	127	417	1932	USA, Oregon, Adrian	Arch dam for irrigation and flood control
Diablo	119	390	1929	USA, Washington, Skagit River	Arch dam for power
Schräh	111	364	1924	Switzerland, Schwyz, Aa River	Gravity dam for power
Arrowrock	107	351	1916	Boise River, Idaho, USA	Concrete arch dam for irrigation
Buffalo Bill	99	325	1910	Wyoming, USA	Also called Shoshone Dam. Concrete arch dam for irrigation
New Croton	91	299	1905	USA, New York, Croton River	Gravity dam for the water-supply of New York City
Cheeseman	72	236	1904	USA, Colorado. South Platte River	See caption, page 68
Puentes II	71	233	1884	Rio Guadalentin, Spain	For irrigation
Furens	50	164	1866	France	See historical note
Guadarrama	93	305	1788–99 (construction)	Spain	Intended for navigation. The dam collapsed when it reached a height of 57 m (187 ft) and was abandoned
Puentes I	50	164	1791	Rio Guadalentin, Spain	For irrigation. Failed catastrophically in 1802
Tibi	46	151	1594	Spain, Alicante, Rio Monegre	For irrigation. In use today
Mudduck Masur	33	108	15th century	India, Madras	Earth dam, apparently the highest among medieval Indian dams
Kebar	25	85	c.1300	Iran, 170 km (106 miles) south-west of Tehran	In existence but completely silted. The oldest existing arch dam
Almonacid	c. 21	c. 70	13th century	Spain, Rio Augasvivas	Later heightened to nearly 30 m (100 ft). Today completely silted and used as a bridge
Daimonike	32	105	1128	Japan, near Nara	Earthen embankment

Table 16. cont.

Kala (Balula)	24	79	5th century	Sri Lanka	Earth dam 6 km (3·7 miles) long 40 km (25 miles) south of Anuradhapura
Subiaco	39 (probably)	130	*c.* AD 50	Italy, River Aniene	One of the three dams creating ornamental lakes beside Nero's Villa. Local builders stole the stone and the dam collapsed in 1305
Cornalbo	24	79	*c.* AD 2nd century	Spain, Badajoz River	For the water-supply of Merida. Earth dam 200 m (656 ft) long
Gukow River	30	98	*c.* 240 BC	China, Shansi Province	Stone crib dam 300 m (984 ft) long
Numrood	(not known)		*c.* 1000 BC	Iraq	Built at head of Euphrates Delta to feed the Nahrawan Canal system on the left bank, and the Dujail and Ishaki Canals on the right. Swept away by a flood in AD 629
Sadd el Kafara	11	37	*c.* 2750–2500 BC	Egypt, Wadi el Garawi	Near Helwan, 32 km (20 miles) south of Cairo

Note: Kebar and Almonacid are included for interest, and because of the uncertain accuracy of the data.

Arrowrock Dam on the Boise River, Idaho, was the highest dam in the world when it was completed in 1916. *Photo: US Bureau of Reclamation*

Furnas Dam. An unusual hazard during construction was first experienced at Furnas Dam on the Rio Grande in Brazil. An explosion wrecked the tunnel intake, and an uncontrolled flood through the tunnel resulted. The cause was rotting vegetation in the reservoir. Methane dissolves in water only under pressure, and in this reservoir, which was deep and which had not been cleared of trees before flooding, the water at depth became saturated with methane. Leakage passed through the gates into the tunnel, with a drop in pressure to atmospheric (the tunnel was under construction and was dry) so releasing methane which gradually built up to an explosive concentration.

Mohne and Eder Dams, in the Ruhr Valley in Germany, were destroyed with catastrophic floods, by the Royal Air Force, in 1943. Both were gravity dams of cyclopean masonry, about 40 and 48 m (131 and 157 ft) high respectively and were attacked when the reservoirs were full. A breach about 60 m (200 ft) wide and 17 m (56 ft) deep was formed in each dam by the famous 'bouncing bomb'; some 1300 people were drowned and 55 of the cream of the RAF's air crews were lost.

Scorpe Dam, an earth embankment 69 m (226 ft) high was also attacked. It sustained eleven direct hits in two attacks and craters 12 m (39 ft) deep were formed in the dam from 6 ton bombs. The reservoir had been drawn down and the dam did not fail, but it gave serious trouble for the next fifteen years or so, until extensive repairs had been completed.

South Fork Dam, an earth-fill dam 23m (76 ft) high, near Johnstown, Pennsylvania, USA, failed in 1889, causing the world's worst dam disaster in terms of loss of life, the number of deaths, 2209, slightly exceeding those of the Vajont disaster. Unlike other failures mentioned here, however, the failure was not significant in an engineering sense. The dam had been built without engineering supervision, and failed when the reservoir overfilled and flowed over the top of it.

towers, masts and tall buildings

Great heights in a tower or building signified, until recent times, religious expression or imperial monumentality. There were a few exceptions like the follies of English gentlemen and towers for making lead shot. Then came the tall chimney, the skyscraper, and towers and masts for telecommunications and television (table 17). The last may be cantilevers (that is free-standing – a cantilever may be equally on its side as a bridge or upright as a tower) or stayed (that is pinned at the base, and held up by stay or guy-ropes).

THE WORLD'S TALLEST BUILDINGS (over 250 m (820 ft) in height).

Sears Tower, Chicago, Illinois, USA **442 m (1974)** (1450 ft – 109 storeys); height to top of TV antennae is 548·6 m (1800 ft). Headquarters of Sears Roebuck Company, department stores. Its structure does not consist of columns and beams, but is a bundled tube in which nine square tubes are nested together and rise to different heights. See page 93 for a fuller explanation.

World Trade Center, New York City, USA **411·5 m (1973)** (1350 ft – 110 storeys). It consists of twin towers. The walls of each tower are closely braced Vierendeel trusses that resist wind loads as well as support the weight of the tower with the help of a few columns round the services core of the tower. This is an example of the hull and core structure also referred to on page 93.

Empire State Building, New York City, USA **381 m (1931)** (1250 ft – 102 storeys); height to top of TV tower is 448·6 m (1472 ft). Doyen of tall buildings; its original height is given. The riveted steel structural frame of the building is typical of the period, and exemplifies a type of structural design that remained basically unchanged for many years. The steel frame was erected very speedily, over 50 000 t being erected in six months in 1930.

The Chicago skyline with the Sears Tower prominent in the centre. Its steel frame is clad with black aluminium and bronze-tinted glare-reducing glass. The John Hancock, Standard Oil, and First National Bank buildings can also be seen. *Photo ASCE*

The John Hancock Center, Chicago

Standard Oil Building, Chicago, Illinois, USA **346 m (1973)** (1136 ft – 89 storeys).

John Hancock Center, Chicago, Illinois, USA **343·5 m (1968)** (1127 ft – 100 storeys). An early example of a hull and core structure. The diagonal bracing transmits load to the enormous corner columns. The building is said to enjoy more continuous 24-hour use than any other, and its designers consider that it may be the answer to practical mid-city living.

Chrysler Building, New York City, USA **319 m (1930)** (1046 ft – 77 storeys). Built by Walter P Chrysler, but never a property of the Chrysler Corporation.

60 Wall Tower, 70 Pine Street, New York City, USA **290 m** (950 ft – 67 storeys).

First Canadian Place, Toronto, Canada **289 m (u.c. 1976)** (949 ft – 72 storeys). First framed-tube structure in Canada. Height given is above ground; the main structure goes 14 m (47 ft) below ground level. Topped out in April 1975. The steel structure was erected in fifteen months (about 40 000 t of steel), erection reaching a rate of three storeys a week.

Place Ville Marie, Toronto, Canada **72 storeys.**

First National City Corporation (Citycorp Center) New York City, USA **279 m (u.c. 1976)** (914 ft – 59 storeys). A tower carried on massive columns about seven storeys high, leaving an open space at ground level.

40 Wall Tower, New York City, USA **274 m** (900 ft – 71 storeys).

Water Tower Place, Chicago, Illinois, USA **262 m (1975)** (859 ft – 76 storeys). The world's tallest concrete building. Has high-strength concrete (62 N/mm^2 – 9000 lb/in^2) at its base. The tower consists of shear walls and closely spaced perimeter columns forming a tubular structure. Second highest concrete building is One Shell Plaza (page 90). An all-concrete building 232 m (760 ft) high is under construction in Australia.

United California Bank Building, Los Angeles, USA **262 m (1974)** (858 ft – 62 storeys).

Transamerica Pyramid, San Francisco, USA **260 m** (853 ft).

First National Bank Building, Chicago, Illinois, USA **259 m (1969)** (850 ft – 60 storeys).

US Steel Corporation Building, Pittsburgh, Pennsylvania, USA **256 m (1971)** (841 ft – 64 storeys). This building is triangular in plan. It has big steel box columns outside the building. They are of weathering steel (the steel is not painted and its surface rusts into a protective patina) and are filled with water. The water can be circulated by pumps and it contains 625 t of anti-freeze. Its purpose is to cool the columns if the building should catch fire.

The Chrysler Building, New York City, complete with gargoyles and romantic spire

The tallest building in Britain will be the National Westminster Bank's headquarters tower in the City of London when that structure reaches its full height **183 m (u.c.)** (600 ft – 49 storeys). The estimated date of completion for the tower is 1978, but that date is, of course, later than the

Table 17. The world's tallest structures (masts or towers over 500 m (1640 ft) in height).

Name or description	Height m	Height ft	Year of completion	Country and location	Other details
Warszawa radio mast	645·38	2117	1974	Poland, Plock (north-east region)	Mast of tubular, galvanised steel, with fifteen guy-ropes
Stayed television mast	629	2063	1963	USA, North Dakota, between Fargo and Blanchard	Operated by KTHI–TV. Steel mast erected in 30 days. Designed to sway 4·2 m (13·9 ft) in a gust of 192 km/h (120 mph)
CN Tower	553	1815	1975	Canada, Toronto	World's tallest free-standing tower. The main concrete structure is 446 m (1464 ft) high. The tower serves for communications and 'observations'
Ostankino TV tower	537	1762	1967	USSR, Moscow	Also a free-standing concrete tower, the main structure of which is 385 m (1263 ft) high. The tower was heightened by 4 m (13 ft) in 1973. There is a meteorological observatory high in the tower
Stayed television mast	533	1749	1963	USA, Knoxville, Tennessee	Steel mast operated by WBIR–TV
Stayed television mast	533	1749	1962	USA, Columbus, Georgia	Steel mast operated by WTVM–TV and WRBL–TV
Stayed television mast	510	1673	1960	USA, Cape Girardeau, Missouri	Steel mast operated by KFVS–TV

N V Nikitin, the designer of the Moscow television tower, reported to a congress in 1968 that the design figure for the static deflection of the tower was 5·8 m (19 ft) corresponding to a wind experienced for 1·2 hours in one year, and the maximum amplitude of vibration 1·4 m (4·58 ft). He expected that the maximum deflections of the tower would be less. No one had felt any unpleasant vibrations. A vast programme of observations of the behaviour of the tower was in progress.

topping out. According to unofficial reports, the tower could hold one world record – that of the highest cost per m² of floorspace. A figure that has been quoted is £750/m² (£70/ft²) and a very high standard of prestige plus complex construction are given as reasons for the high figure. No official cost estimate is available. A corresponding figure quoted for the Sears Tower is about £160/m²; Sears Tower was built speculatively at the lowest possible cost, however, and there are a few years of inflation between the two.

TALL CONCRETE CHIMNEYS

The International Nickel Company's chimney at Sudbury, Ontario, Canada **381 m (1970)** (1251 ft), is the world's tallest chimney. Like the CN Tower it was slip-formed. Its diameter at the top is 15·8 m (52 ft) and at the base 35·4 m (116 ft).

Burgo Paper Mill, Mantua, Italy. The suspended roof and the walls are sealed together, but are separate structurally, and the roof is free to move

This constructional view clearly shows that the Burgo roof is structurally, a self-anchored suspension bridge. The cables are chains, the links being pinned at each point of attachment of a suspender

The Mormon Tabernacle at Salt Lake City, Utah, USA, completed in 1868, a heroic timber structure spanning 40 m (132 ft) and consisting of arched lattice trusses forming a dome of overall size 46 × 76 m (150 × 250 ft)

Mitchell Power Station, Cresap, West Virginia USA **368 m (1969)** (1206 ft), is second in height to the Sudbury chimney. Generally, the purpose of these chimneys is to dissipate flue gases with a high content of sulphur, and to be tall enough to pierce a low-lying blanket of temperature inversion in the atmosphere, should one occur. This one serves 1600 MW of generating capacity, is 11·3 m (37 ft) in diameter at the top, and 29 m (95 ft) at the base. It was slip-formed.

Magna, near Great Salt Lake, Utah, USA **366 m (1975)** (1200 ft), at the copper-smelter of the Kennecott Copper Corporation. Slip-formed and tapered, 38 m (124 ft) in diameter at the base. Example of the tall chimneys that are 'popping up all over the place' in the USA – at about 80 a year, average height about 200 m (600 ft) – to reduce air pollution.

Europe's tallest chimney, at Puentes, Spain, is **350 m** (1148 ft) high.

Drax Power Station, Yorkshire, England **259 m (1969)** (850 ft), is of cylindrical (i.e. untapered) shape, 26 m (85·75 ft) in diameter. It has the greatest capacity of any chimney. Three flues inside the shell can discharge the gases from burning up to about 36000 t of coal (i.e. 4000 MW of electrical capacity) in a day. The outer shell was built with hanging platforms, and the three inner flues with slip forms.

NOTABLE STEEL STRUCTURES

Gateway Arch, St Louis, Missouri, USA **192 m (1966)** (630 ft). A catenary in shape, equal in extreme width to its height. It is a stressed skin structure, consisting in cross-section of a double-walled equilateral triangle, with concrete between the two skins in the lower half and cellular stiffening in the upper. The outer wall is of stainless steel 6 mm (0·25 in) thick and the inner of mild steel 10 mm (0·37 in) thick. Lifts pass through the hollow centre to and from an observation gallery at the crown.

Woolworth Building, Broadway, New York City, USA **232 m (1913)** (761 ft – 55 storeys). The tallest building in the world when it was completed; has Gothic architectural detailing and heavy, portal-braced steel frame.

Maine-Montparnasse Tower, Paris, France **210 m (1974)** (689 ft – 58 storeys). The tallest in Europe after the Eiffel Tower. Has a concrete core surrounded by a steel framework. In plan, a truncated ellipse 62 × 39 m (203 × 128 ft). Built over Métro line.

The airship hangar at Akron, Ohio has a span of 99 m (324 ft). See table 23

Table 18. Chronological list of the tallest structure in the world.

Name	Height m	ft	Year of completion	Country and location
Warszawa radio mast	645·38	2117	1974	Plock, Poland
KTHI–TV mast	629	2063	1963	Fargo, North Dakota, USA
WVTM–TV and WRBL–TV mast	533	1749	1962	Columbus, Georgia, USA
KFVS–TV mast	511	1676	1960	Cape Girardeau, Missouri, USA
WGAN–TV mast	493	1619	1959	Portland, Maine, USA
KSWS–TV mast	491	1610	1956 (fell in gale 1960)	Roswell, New Mexico, USA
KW–TV mast	479	1572	1954	Oklahoma City, USA
Empire State Building	381	1250	1931	New York City, USA
Chrysler Building	319	1046	1930	New York City, USA
Eiffel Tower	300·5	986	1889	Paris, France
Washington Memorial	169	555	1884	Washington, DC, USA
Cologne Cathedral	156	513	1880	Cologne, West Germany
Rouen Cathedral	148	485	1876	Rouen, France
St Nicholas Church	145	475	1874	Hamburg, West Germany
Saint-Pierre-de-Beauvais	153	502	1568	Beauvais, France (fell soon after)
Notre Dame	142	465	1439	Strasbourg, France
St Paul's	149	489	1315	London, England (destroyed by lightning 1561)
Central spire, Lincoln Cathedral	160	525	1307	Lincoln, England (fell in storm in 1548)
Great Pyramid of Cheops (Khufu)	147	481	*c.* 2580 BC	El Gizeh, Egypt
Snefru North Stone Pyramid	104	342	*c.* 2600 BC	Dahshur, Egypt

Note: Estimated height of the Pharos Lighthouse at Alexandria, Egypt, erected 300 BC and fell AD 1326, is 113 m (370 ft) according to one source, 183 m (600 ft) according to another.

VAB (vertical assembly building) for Saturn 5 rockets can accommodate four rockets. The building is 160 m (524 ft) high. Air conditioning prevents clouds and rain inside the building

Ingalls Building, Cincinnati. A photograph taken in 1906 of the first tall building with a reinforced concrete frame

Right: Three generations of architectural style. The Woolworth Building (1913) in the centre, with its gingerbread Gothic; on the left, between-the-wars and on the right the post-war styles

TALL CONCRETE BUILDINGS AND TOWERS

Emley Moor TV Tower, near Huddersfield, Yorkshire, England, is the tallest free-standing tower in Britain **329 m (1971)** (1080 ft). It replaced a guyed steel tower that collapsed in 1969 when overloaded with ice that built up during freezing fog. The tallest stayed mast in Britain is the 387·7 m (1272 ft) cylindrical steel tower at Belmont, Lincolnshire.

One Shell Plaza, Houston, Texas, USA is the world's tallest structure of light-weight concrete **218 m (1969)** (715 ft – 60 storeys). The concrete is of high strength (6500 psi – 45 N/mm^2 – at 28 days) and has three-quarters the weight (1850 kg/m^3 – 116 lb/ft^3) of ordinary concrete.

Marina City Towers, Chicago, USA, were the world's tallest concrete structures at the time of their completion **182 m (1963)** (597 ft – 65 storeys).

Ingalls (Transit) Building, Cincinnati, USA was the first skyscraper of reinforced concrete **16 storeys – 1903.** It is now one of the ASCE's national historic civil engineering landmarks.

The world's largest roof of aluminium. The roof is a space frame covering an exhibition hall at Anhembi Park, São Paolo, Brazil. It is square, with a side length of 260 m (853 ft) and a span of 60 m (197 ft) between columns. *Photo: Alcan*

Table 19. The world's tallest structures by material of construction.

Material of main structure	Height m	ft	Name	Year of completion	Country and location	Other details
Structural steel	645·38	2117	Stayed radio mast	1974	Poland, Plock	
Concrete	446	1464	CN Tower	1975	Canada, Toronto	Height of concrete structure only is given
Light-weight concrete	218	715	One Shell Plaza	1968	USA, Houston, Texas	See text
Iron	300·5	986	Eiffel Tower	1889	France, Paris	Original height. Now heightened by antennae to 321 m (1053 ft)
Masonry (stone)	169	555	Washington Memorial	1884	USA, Washington, DC	Paradoxically, this obelisk can claim to be the tallest masonry structure yet to have the highest iron frame of its day. Iron columns support the lift-shaft and stair-well but the main structure is of masonry
Masonry	156	513	Cologne Cathedral	1880	West Germany, Cologne	Included for the purist – nothing but masonry
Brick	102	334	Sienna Town Hall – Torre del Mangia	1348	Italy, Sienna	Extreme height to top of lightning-conductor is quoted. Top 10 m (30 ft) or so is a stone belfry: Built by Minuccio and Francesco di Rinaldo
Timber	160	526	Radio tower	1932	West Germany, near Munich	Braced, tapering tower square in plan, built of pitch pine. In service

Note: The spire of Lincoln Cathedral (table 18) was partly masonry, partly timber.

THE STRUCTURAL FRAME

Evolution of the skyscraper depended on the use of a structural skeleton – that is a frame of beams and columns that carried all the loads, and reduced walls to a 'cladding' by taking over their load-bearing function. Traditional timber framing while both precursor and stimulator of the structural frame was not a fully comparable explicitly-load-bearing concept.

The world's first building with a true structural frame is the boat store at Sheerness in England (Kent) completed in 1860. This building has a four-storey three-bay frame, 64 m (210 ft) long, 41 m (135 ft) wide and 16 m (52 ft) high, with cast-iron columns and secondary beams (spanning 4 m (13 ft)) and riveted wrought-iron primary beams (spanning 9 m (30 ft)). Bolted connections between the beams and columns are rigid and give the building stability. The Chocolat Menier Factory at Noisiel-sur-

The Federal Reserve Bank of Minneapolis, Minnesota, USA is a unique building that cannot be pigeonholed in any category. It consists of a gigantic portal frame, spanning 84 m (275 ft) from which the building hangs like an inn sign. The suspension is a catenary framed in steel, within the rectangular shape of the hanging building. The building is 18 m (60 ft) deep and has ten office floors in the suspended structure and service floors within the depth of the cross-member of the portal frame. The purpose of the long span – effectively a bridge ten storeys deep, which is unique among the world's structures – is to avoid putting columns into 'secure' areas of vaults and lorry ramps below. The design permits six more storeys to be added above the building as it is illustrated here. They would be supported by an arch, the reverse in shape of the catenary. The horizontal thrust outwards from the arch would balance the horizontal thrust inwards from the catenary. *Photo: ASCE*

Marne in France, completed in 1872, was also claimed as the first framed building. Both buildings are still standing, and both were isolated developments. Also, a warehouse at Saint-Ouen Dock, Paris, erected in 1864–5, had all loads carried by an iron frame and was of fireproof construction.

Modern practice started in earnest in the United States. It was pioneered in Chicago and spread to New York, where the earliest skyscrapers were hybrids structurally – of masonry, iron and steel – depending partly on load-bearing walls. A complementary development of equal significance was the invention of the lift in 1857. The first fully framed all-steel structure – 'the ultimate step in the creation of the modern skyscraper' – was the Rand-McNally Building erected in Chicago in 1889.

This type of construction reached its zenith with the Empire State Building, and continued without alteration until the last decade or so. The modern vogue is for 'hull and core' – a box-like core in the centre of the building accommodates lifts and services and is very strong structurally. The floors span between the core and the outer structure – the hull. The Sears Tower's bundled tubes are a most subtle structure, but along the same line of thought. In structural terms, the building consists of nine tubes – or 'hulls' to be consistent with the terminology just used – each 23 m^2 (75 ft^2) and joined together in three rows to form one large square – the bundle. The common walls that these tubes share increase the rigidity of the bundle, and so permit a structure that is economical in steel. According to height and other requirements, several similar structural concepts had been applied. Such economy is essential for the success of very high buildings. It has been gained by the use of high-strength steels (up to 100 per cent stronger), welded joints, and better structural design (including the ideas just described) and detailing as illustrated by the figures in table

A 185 m (600 ft) high circular office tower block, the main building of the Australia Square development in Sydney is Australia's highest building, and was the tallest in the world to be constructed in light-weight concrete until the completion of One Shell Plaza

Table 20. Comparative weights of steel

Building	Date	No. of storeys	Ratio height:width	Gross area millions of ft^2	Weight of steel in lb/ft^2
Empire State	1931	102	9·3	2·75	42·2
Chrysler	1930	77	8·5	1·1	37·0
World Trade	1973	110	6·9	9·0	37·0
Sears	1974	109	6·4	4·4	33·0
John Hancock	1968	100	7·9	2·8	29·7

20. The higher the building, the higher one would expect the weight of steel per square foot to be, but the World Trade Center, although a third as high again, has the same amount as the Chrysler Building, and the John Hancock Center has about 70 per cent of the Empire State Building.

Buildings of 150 storeys may be expected within the next ten years, buildings of 100 storeys will become more common, and office buildings now in the range of 30–50 storeys will be built 40–70 storeys high. This is the opinion of US designers, based on the new competence just described. One view common in Britain is that the 'social limit' of height has already been exceeded.

BRICK STRUCTURES

Load-bearing brickwork is used for buildings up to about twenty storeys in height in modern construction. The tallest modern brick buildings are four twenty-storey structures of load-bearing reinforced brickwork at Denver, Colorado, called 'Park Lane Towers'. The walls are 0·25 m (10 in) thick and the towers are designed to resist earthquake shocks. Their height of 63 m (206 ft) is appreciably less than that of the Sienna Tower listed in table 19. A twenty-storey building of reinforced clay blocks at Toulouse is the tallest in Europe. The tallest of non-reinforced brickwork is an eighteen-storey block at Schwamendingen, Switzerland, and the runner-up one of seventeen storeys at Portsdown Hill, Portsmouth.

The tallest building of load-bearing concrete blocks is claimed to be a hotel at Walt Disney World, Florida. Its height is 58 m (192 ft); the joints between the blocks are of epoxy adhesive, not mortar.

Dollan Baths, East Kilbride, Scotland. The arched barrel-vault shells span 99 m (324 ft) between abutments at ground level, that is just too short a span to qualify for inclusion in table 22. The building contains an Olympic swimming pool, and gives a good idea of the scale of the record structures listed in the tables. Note that the span is along the length of the building. *Photo: Cement and Concrete Association*

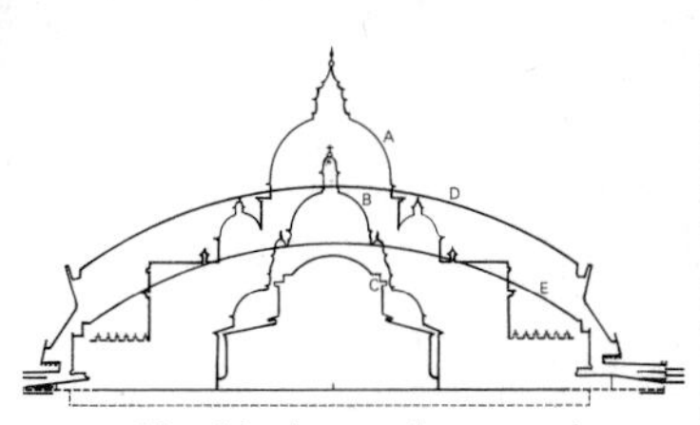

Notable domes, drawn to the same scale

A St Peter's, Rome
B St Paul's, London
C S. Sophia, Istanbul
D Superdome, New Orleans
E Astrodome, Houston

The Palasport, Milan, will be completed in June 1976. It has a circular, saddle-shaped roof with a clear span of 128 m (420 ft) – the world's longest span for a cable net. This form of construction is clearly shown in the two pictures. The massive box girder, 6 by 2·5 m (20 by 8 ft) in cross-section, circular in plan, and following the curve of the saddle in elevation, has the cables anchored to it, and resists their inward pull. It is freely supported, able to twist and flex, yet also held down, for in some circumstances an upward force can occur. The cables are spaced 2 m (6·6 ft) apart

The Assembly Hall at Champaign, Illinois, of the University of Illinois has the world's longest span of light-weight concrete (the concrete has a density of 1685 kg/m^3 – 105 lb/ft^3 – and a minimum 28-day strength of 27·6 N/mm^2 – 4000 lb/in^2). The roof of the hall is a thin-shell folded-plate dome, 121·9 m (400 ft) in diameter and rising 19 m (62 ft) above the ring beam on which it is supported. The concrete of the dome is only 0·09 m ($3\frac{1}{2}$ in) thick over most of the surface (excepting the ribs at top and bottom of the corrugations and the central compression ring) and the maximum depth of the corrugations is 2·3 m (7·5 ft). The dome was cast in place; when the concrete had matured, the ring beam was post-tensioned by wrapping 2503 turns of stressed wire around it. A compressive load of about 5000 t was thus induced in the ring beam, so reducing its diameter by about 0·05 m (2 in) and causing the dome to rise by about the same amount at its centre, so freeing itself from the centering which could then be dismantled. The hall has many novel features, including the 'gridiron' hung from the roof that permits rapid changes in the internal layout, to suit different types of events. The internal view shows the hall arranged for a concert

The Louisiana Superdome (table 21) remains the world's building of greatest free span that is completely and permanently roofed over. The left-hand picture shows the great dome being constructed, and the right-hand one how the building dwarfs its surroundings. *Photo: Sverdrup and Parcel and Associates Inc.*

Table 21. The world's roofs of longest span (over 152 m (500 ft)).

Building	Clear span m	ft	Year of completion	Country and location	Other details
Texas Stadium	240·3 × 190·5 (elliptical)	788·5 × 625	1971	USA, Irving, Texas	Six steel arches forming a dome carried by twelve concrete column-like frames. The centre 1 ha (2·5 acres) of roof 65 m (213 ft) above the playing-field has been left open. Structurally, it is not clear whether the major axis of the ellipse is the span
Louisiana Superdome	207·3 (diameter)	680	1975	USA, New Orleans	Dome, 83 m (273 ft) high at centre. Steel structure is a complex array of trusses acting as a space frame. Encloses 57 000 m³ (2 × 10⁶ ft³) of space and covers 3·2 ha (8 acres)
Centre Nationale des Industries et Technologies (CNIT)	206 (in three directions)	675·8	1958	France, Paris, La Défense	See caption for description, page 110
King County Stadium	201·5 (diameter)	661	u.c.	USA, Seattle, Washington	Dome, consisting of a thin concrete shell stiffened by concrete ribs
Astrodome	195·7 (diameter)	642	1964	USA, Houston, Texas	Steel dome similar to Superdome, 63 m (208 ft) high at centre and covering 3 ha (7·5 acres)
Narita No. 1 Hangar	190 × 90	623 × 295	1972	Japan, Tokyo International Airport	Space grid, with tied portal truss spanning 190 m. Roof was lifted into position (see table 54)
Burgo Paper Mill	163 (length)	535	1962	Italy, Mantua	Structurally, this roof is a self-anchored suspension bridge and is not a cable roof. The two side spans each extend over 43 m (141 ft)
The world's buildings of longest span in the 1850s					
Train shed, New Street Station	64	211	1854	England, Birmingham	Iron roof
Train shed, Grand Central Station	51	166	1856	USA, Chicago	Vaulted timber roof with wrought-iron ties. Destroyed in Great Fire fifteen years later

Table 22. Roofs spanning more than 100 m (328 ft) (those in table 21 excepted).

Name	Span m	Span ft	Year of completion	Country and location	Other details
Parcels-sorting building	146·8	481·6	1966	West Germany, Munich	Two-pinned, ribbed concrete arch
Poliedro de Caracas	143 (diameter)	469	1975	Venezuela, Caracas	Geodesic dome with rise of 38 m (125 ft) built of tubular aluminium members (depth of truss 1·5 m (5 ft)) and fabricated in USA. The world's greatest span of aluminium
Olympic Games Sports Palace	132	433	1968	Mexico, Mexico City	Steel-framed dome, infilled with saddle-shaped triangulated frames of aluminium
Palazzo del Mostre	130 (in three directions)	426·5		Italy, Turin	Similar in form to the CNIT building, except that the shells cantilever out to give a rhombic plan. Area covered 14625 m² (157000 ft²)
Kloten	128 × 125	420 × 410	1975	Switzerland, Zürich	Triple-layer space-grid roof of Jumbo Jet hangar. Has four points of support and weighs 8000 t
University of Urbana	122	400	1962	USA, Illinois	Assembly Hall. Folded-plate thin-shell concrete dome with post-tensioned ring beam
Coliseum	114	375	1964	USA, Georgia, Athens	See caption
Galeries des Machines	111	365	1889	France, Paris	Three-pinned steel arch roof for Paris Exhibition
Dome of Discovery	111	365	1951	England, London	Aluminium dome, built for the Festival of Britain and subsequently demolished. World's longest span of aluminium until completion of Poliedro de Caracas
Convocation Center	110	361	1968	USA, Ohio University	Light steel-framed dome
Special Events Center	107	351	–	USA, University of Utah	Timber dome. World's longest existing span of timber
Car repair shop (railway)	106	348	1958	USA, Louisiana, Baton Rouge	First major dome by Fuller's geodesic method. Clear height 37 m (120 ft). Used only 567 t of steel
Movable shed	106	348	1970s	Japan	Rail-mounted portal-framed structure 45 m (148 ft) long and 48·3 m (158 ft) high above rail, straddling building dock for 1-million-ton ships, and providing shelter for welding, etc.

Table 22. cont.

Name	Span m	Span ft	Year of completion	Country and location	Other details
Marignane Hangar	102	334	1962	France, Marignane	Concrete barrel-vault arched shells, cast on the ground and jacked up to their final position. Hangar is 60 m (197 ft) in width
Brabazon Assembly Hall	101	330	1949	England, Filton, near Bristol	Two-pinned cambered braced portal frames of steel with a tie at haunch level. Three main spans; area covered over 3 ha (8 acres)
Cable roofs spanning more than 100 m (328 ft).					
Lufthansa Jumbo Jet hangar	130	426·5	1970s	West Germany, Frankfurt	This roof is a stressed ribbon, structurally similar to bridge of that type
Palasport di Milano	128 (diameter)	420	u.c.	Italy, Milan	Saddle-shaped net with massive peripheral box girder of steel
Oakland-Alameda Coliseum	128 (diameter)	420	1966	USA, California	A concave shell with radial cables and concrete roof cladding. Light and elegant in appearance with open-braced perimeter wall
Madison Square Garden Stadium	123	404	1967	USA, New York City	Structurally similar to Oakland-Alameda Coliseum, but carries heavy loads
Oklahoma City Arena	122 × 98 (elliptical)	400 × 322	1964	USA, Oklahoma	Saddle-shaped net with concrete cladding
Arizona State Fair Arena	116 (diameter)	381	1966	USA, Arizona	Similar structure to Oklahoma City Arena
Salt Lake County Auditorium	110 (diameter)	361	1961	USA, Utah	Concave, 'bicycle-wheel' roof
New York State Pavilion	107 × 80 (elliptical)	351 × 262	1965	USA, New York City	Bicycle wheel roof for New York World Fair
Scandinavia	105 (diameter)	344	1971	Sweden, Gothenburg	Sports stadium. Saddle-shaped net with roof cladding of galvanised steel sheet

Table 23. The world's largest buildings.

Name	Cubic content of building (or floor area) Millions of m³	Millions of ft³	Year of completion	Country and location	Other details
Boeing Assembly Hall	5·8	205·5	1968	USA, Washington, Everett	Main assembly plant for Boeing 747 (Jumbo Jet)
Vertical Assembly Building	3·67	129·5	1965	USA, Florida, Merrit Island	For Saturn 5 rockets. Has doors 140 m (460 ft) high
Goodyear Airship Hangar	1·56 (3·38 ha)	55 (364000 ft²)	1930s	USA, Ohio, Akron	World's largest hangar – 358 m (1175 ft) long; 99 m (325 ft) wide; 61 m (200 ft) high
Merchandising Mart	1·5	53	1929	USA, Chicago	Eighteen-storey building, the world's largest at the time of its completion

Greatest floor areas.

Name	ha	millions of ft²	Year of completion	Country and location	
South State Mall	86·4	9·3	u.c. 1976	USA, New York, Albany	Government centre
World Trade Center	81·2 (rentable office)	8·74	1973	USA, New York City	Largest office building in the world – similar twin towers
Autolite-Ford Parts Redistribution Center	28·8 (ground area)	3·1	1971	USA, Brownston, Michigan	Largest ground area covered by any building
The Pentagon	60·3	6·5	1943	USA, Virginia	Largest ground area of any office building
Sears Tower	41·8 (gross area)	4·5	1974	USA, Chicago	16500 people use the building daily

Note: 1 hectare (ha) = 10000 m² = 107636 ft² = 2·47 acres.

THE REINFORCED CONCRETE STRUCTURE

Portland cement was patented by Joseph Aspdin in 1824. It is the basic material on which modern concrete depends. It was first used in civil engineering construction by Brunel for the Thames Tunnel. For many years Lambot, a French market-gardener, was credited with the invention of reinforced concrete, with Coignet, a French engineer, second in the field. Lambot made flower-pots and boats of reinforced concrete, some of which still exist. However, the inventor of reinforced concrete is now considered to be W B Wilkinson of Newcastle upon Tyne. He built a house of reinforced concrete in 1865, in which concrete beams were reinforced with colliery rope (i.e. wire rope) which was positioned in the tensile zone of

The two main structures for the Montreal Olympics are the Stadium and the Velodrome. They are constructed of big precast concrete elements, post-tensioned together. The Stadium is elliptical in shape extending in plan over 480 and 280 m (1575 and 919 ft) respectively, on its major and minor axes. Its clear span is apparently greater than any of the buildings listed in table 21, but it has not been possible to confirm the span before going to press. Structurally it is a dome, with 34 radial ribs and an upper ring 60 m (197 ft) above the ground that contains services for light, sound and ventilation. Beside it stands a tower, 168·4 m (552 ft) high, and inclined over the dome a horizontal distance of 60 m (197 ft). The tower is to carry the retractable covering of the elliptical central space; this space is left open in fine weather, but can be covered at other times. The membrane (weight about 200 t; surface area 18 000 m² – 194 000 ft²) that covers it is stored at the top of the tower, and is lowered down the tie ropes and unfolded at the same time. It is sealed round the rim by a compressed-air-operated mechanism, and can be positioned in 15 minutes. The tower will contain two restaurants and sixteen storeys devoted to various sports, etc., with swimming pools below ground. The Velodrome is an arched structure, and has a clear span of about 170 m (558 ft). It consists of a group of arch ribs, arranged in pairs, each pair forming a lozenge-shape, and springing from common abutments. There are three abutments at one end, and one at the other, and three of the 'lozenges' form the basic structure of the roof, the ribs being interconnected by secondary precast concrete elements, mainly Y-shaped. The abutments are anchored into rock with ties, the load on the ties at the single abutment being 23 500 short tons (21 300 t). The arch ribs were erected on centering consisting of steel scaffolding. They were post-tensioned. It was then necessary to jack the roof clear of the centering, raising it 0·1 m. This was done with 226 flat 1000-short-ton jacks, the weight lifted being 41 000 short tons (37 190 t) and the thrust generated 80 000 short tons (72 500 t)

the beam, and splayed out at each end to anchor the wire. That is, the basis of conventional reinforced concrete theory was followed. His house was demolished in 1954, the structure being in good condition at that time. Another example of an early structure of reinforced concrete is Sway Tower, also known as 'Peterson's Folly', in Hampshire. Peterson settled in Hampshire in 1868, and built the tower soon afterwards. The concrete was cast in place using a climbing-shutter, but some precast pieces were used for stairs and cornices.

The dates of these buildings may be compared with those of early concrete bridges (see page 14), noting that the earliest bridges were arches, giving Wilkinson a clear lead in his use of reinforced concrete beams.

LIVING IN HIGH BUILDINGS

Almost the entire population of Madrid (3 million people) live in high-rise flats; most of them prefer this type of housing. About half the population of Singapore (2 million) live in high-rise buildings, and half the population of Hong Kong (4 million) live above the tenth floor. Densities of population reach 5000 persons per ha (2000 per acre) on Hong Kong's housing estates, and at one building 24 000 per ha (9800 per acre).

clear spans within buildings

Large areas inside a building uninterrupted by columns are as much a challenge as long spans in bridge-building. They can be achieved with roofs that are effectively a series of bridges side by side, or by space structures. If a large opening is needed in one side of such a building – a door in an aircraft hangar for example – a further challenge is introduced.

A space structure must be geometrically complete in three dimensions if it is to sustain loads. The arch and the beam are two-dimensional. The dome is a three-dimensional arch and the slab is a three-dimensional beam. Until modern times, the only type of space structure, with one or two exceptions, was the dome. Arch dams, specifically the cupola dams described earlier, are space structures. Space frames of steel can be like a truss in three dimensions (space grids or lattice grids).

The commonest application for concrete is to devise a space structure that is a thin shell, with the concrete always in compression. The opposite of this is the cable net, which is devised so that the cables are always in tension. Broadly, the main types of space structures consist of triangulated grids, domes, barrel-vaults, folded plates and saddle-shaped roofs.

The saddle-shaped roof became popular because it can be visually striking and it offers constructional advantages. Geometrically it is described as anticlastic (that is concave in one direction and convex in the direction

Maintenance hangar at Frankfurt Airport. Sagging cables carry a thin precast concrete roof. Ten anchor weights of 1000 t each are carried on the framework at each side of the hangar to react against the tension in the cables. Straight cables stiffen the roof. *Photo: Dyckerhoff und Widmann K.G.*

A view at roof level of the Brabazon Assembly Hall, Britain's building of longest span (table 22). The Dome of Discovery exceeded it, but like the Brabazon is a thing of the past, and the aircraft under construction below the roof is Concorde

at right angles to it) or as a hyperbolic paraboloid, a shape generated by rotating an inclined straight line round the circumference of a circle. The shape can thus be formed from straight pieces (or cables stretched tight) and therein lies the constructional advantage. Structurally it resolves into a shell in compression (or, if the reader will permit a slight over-simplification, alternatively cables in tension) plus edge beams in tension or compression. The same shape is found in the waisted cooling-tower, water-towers, and even waste-paper baskets. It is said to be the only shape recognised from structural analysis and so applied; all the others were chosen intuitively and analysed later. The most ambitious example is the cable roof of the Palasport, Milan.

Cable structures – which are classified separately from suspended structures such as suspension bridges – are of three kinds, namely cable beam, cable net, and freely hanging. They are considered to be applicable with advantage to moderate spans as well as the long spans listed here.

STEEL SPACE-FRAME GRIDS

The world's largest two-way two-layer rectangular steel space-frame grid forms the roof of the Osaka Expo 1968 Building in Japan. The roof measures 292 × 108 m (958 × 354 ft) and is 30 m (98 ft) high. It weighs 4800 t and was jacked up into position from ground level.

The roof structure covering the hangar erected in 1970 at London Airport for the Boeing 747 constitutes the largest diagonal lattice grid in the world. The roof contains about 1500 t of steel. The hangar is 34 m (116 ft) high and has an overall length of 171 m (561 ft) reducing to 140 m (459 ft), and a width of 84 m (276 ft). The clear door opening is 138 m

The world's largest building of timber is a storage building for asbestos fibre ore at Deception Bay in the far north of Quebec Province, Canada. It is 93 m (305 ft) wide in one clear span and 232 m (760 ft) long, covering an area of about 2·1 ha (5·3 acres). Glued laminated three-pinned arches are spaced 6 m (20 ft) apart; they are of I-section 2 m (78 in) deep. The building is 44 m (144 ft) high at the crown of the arch. It was completed in 1971. It is exceeded in span by the braced timber dome of 107 m (350 ft) span of the University of Utah's special events centre. *Photo: Gower Yeung and Associates Ltd New Westminster, BC*

The Coliseum of the University of Georgia is a modern interpretation of the structural form of the crossing of a Gothic cathedral, but the architectural expression of that form is entirely original and exceptionally graceful. A pair of parabolic arch ribs carry four barrel-vault roofs. The arch ribs span 156 m (510·5 ft) between abutments, but are propped by four fan-shaped columns which reduce the clear span to the figure given in table 22. The roof consists of precast concrete triangles, held in the roof by beams cast in place. The roof covers 1·1 ha (2·75 acres)

(453 ft) wide by 23 m (75 ft) high. The roof is supported by only eight columns and is itself capable of supporting almost 700 t of equipment, of which 300 t can be applied at any number of positions.

The roof consists of four main parts: a horizontal, low-level diagonal grid; a spine girder; a high-level diagonal grid; and a fascia girder.

A second, similar roof was erected beside the original one in 1973.

HISTORICAL FACTS

The earliest example of a dome is reputed to belong to the 14th century BC, and to have been discovered by the archaeologist H Schliemann when he excavated at Mycenae. The arch, the vault and the dome were known in Sumerian Mesopotamia according to Woolley. The invention of the arch is, however, generally credited to the Etruscans in the 5th century BC.

St Sophia at Istanbul, the smallest dome in the compound sketch on page 95, was built in its present form in the 6th century AD, and so is roughly contemporary with An Chi Bridge, described and illustrated in chapter 4. The dome has a clear span of about 30 m (almost 100 ft), but the clear space within the building is almost twice as long as this because the main dome is flanked by semi-domes. The interior is opened out around the main columns that carry the dome.

The hammerbeam roof of Westminster Hall at the Palace of Westminster in London is perhaps the world's most famous timber roof, and the finest example of this particular kind of construction. It was completed in 1402 – at that time timbers of the size needed were readily available – and the hammerbeams or trusses consist of three sets of timbers, side by side, that were held together by oak pins in the original construction. The roof has a clear span of 20·5 m (67 ft)

The Great Arch of Ctesiphon, about 15 miles (24 km) from Baghdad, is a catenary-shaped arch spanning about 27 m (90 ft) with a rise of about 34 m (112 ft), built of brick in about AD 550. Corrugated, thin-shelled structures of similar form, built in the modern era, have taken their generic name from this prototype.

The Leaning Tower of Pisa, completed in 1350, is 55 m (180 ft) high and 19 m (62 ft) in diameter. Its inclination from the vertical is 1 in 10, and has worsened to that figure in recent years.

The first iron dome, modelled on contemporary timber practice, was 39 m (127 ft) in diameter and was constructed of wrought iron at the Corn Market in Paris in 1811.

The first concrete shell roof was built at Jena in Germany in 1924. The optical firm, Zeiss, needed a 'sky' for their newly invented planetarium and a thin concrete shell was devised to meet their needs at a reasonable cost. As it turned out, an important new structural technique had been invented.

A dome 270 m (886 ft) in diameter consisting of about 24 steel ribs with a central ring member was designed completely just before the Second World War, to cover Munich Railway Station. When war broke out, in 1939, the scheme was abandoned.

chapter 2

BELOW GROUND

Excepting the perils of water alone – of impounding it, of keeping it out of construction beneath water, and of resisting storms and floods – the hazards of going underground are the greatest the civil engineer has to face. Tunnelling has always been a dangerous and unpredictable business. In recent years much greater mastery over the vagaries of the ground encountered in tunnelling has been gained. Hazard has not been removed, but it has been greatly reduced. Costs have tended to ease in proportion, and it seems that a great tunnelling renaissance is on the verge of happening.

Greater competence and sureness in handling other sub-surface problems have also been very evident in recent years. Foundations to resist enormous loads are built with speed and confidence at sites that would have deterred engineers previously. In the centres of cities, much less disruption is caused by even major foundation works. In these sub-surface works, water is again a major cause of difficulty and hazard.

Mining has been carried out since prehistoric times. Tunnelling probably started with cave-dwelling. Flints – the essential material for Stone Age man – were won from shafts excavated in chalk. Timber piles are as old as history. Some notable tunnels from ancient times are mentioned in the tables, but otherwise this chapter starts roughly with the Industrial Revolution.

tunnels and underground workings

A tunnel is an underground passage, big enough for a man to move along or work in. That is a rough definition, serving to distinguish tunnels, more or less, from pipes and shafts. Mining is outside the scope of this book, but tunnels and shafts have to be built to gain access to, and to work minerals underground; such works are mentioned here.

Tunnels are classified as:
- in soft ground;
- in rock;

but in practice there is a transition, not a sudden jump, from one class to the other. Table 24 gives Moh's scale of hardness. The lower chalk, through which the Channel Tunnel would have been driven, has a hardness of 1·8–2·2, which is about the upper limit of 'soft ground'.

Table 24. Examples from Moh's scale of hardness
The scale is for minerals, but can be applied, with certain precautions, to rocks and soils. A substance in a given category is capable of scratching one in the next lower category. Mild steel is in category 6, glass 5, a finger-nail about 2·5.

Hardness	Mineral	Range
10	diamond	range of precious stones (10–8)
9	sapphire	
8	topaz	
7·5	zircon	range of percussive drilling (7·5–2·5)
7	quartzite	
6	traprock	
5	schist	range of rotary drilling (5–1)
4	granite	
3·5	dolomite	
3	limestone	
2·5	galena	
2	potash	
1·5	gypsum	
1	talc	

A further classification is by method of construction, thus:

- cut and cover: a trench is dug, the tunnel built at the bottom of it, and then the excavated material replaced over the roof of the tunnel;
- immersed tube, also called a 'trench tunnel': this is a subaqueous version of cut and cover. The tunnel is built as prefabricated elements, which are floated one by one and sunk in a trench excavated in the river-bed. The elements are joined together *in situ*, and their temporary ends, now within the tunnel, removed. The trench is filled in over them;
- the driven tunnel: a tunnel cut from the solid soil or rock.

Note on Delaware Aqueduct (table 25)

Three long aqueducts – Delaware, Catskill, and Croton – supply New York City with water.

The Delaware Aqueduct, a deep-pressure tunnel, consists of:

	km	**miles**
Rondout–West Branch Tunnel	72·4	45
West Branch Reservoir By-Pass Tunnel	3·2	2
West Branch–Kensico Tunnel	35·4	22
Kensico Reservoir By-Pass Tunnel	3·2	2
Kensico–Hill View Tunnel	22·5	14
Total	136·7	85
City Tunnel Number 2; completed in 1936, connects to the Hill View end of the Delaware Aqueduct; including Hill View By-Pass, its length is:	32·2	20
Giving a continuous deep rock tunnel of length	168·9	105

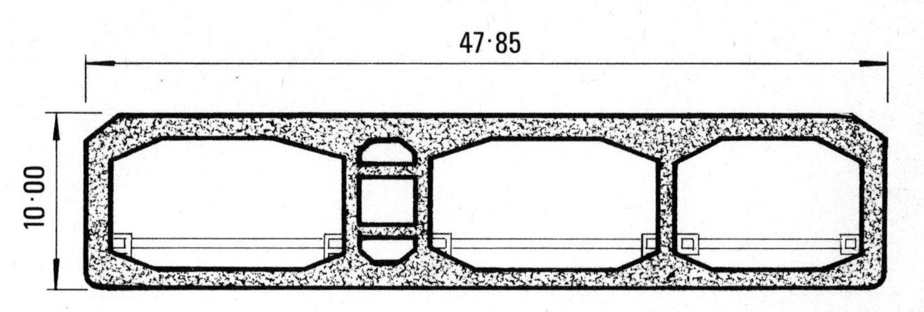

Cross-section of the John F Kennedy Tunnel, Antwerp, which carries two three-lane roads (international motorway standard) and a double-track railway and has three subsidiary apertures. The largest element weighed about 50 000 t; it was floated into position with a freeboard of only 0·1 m (4 in) and when sunk into position, the hydrostatic force on the free end (about 10 000 t) was utilised to push the element against the tunnel already in position. The traffic capacity of this tunnel is enormous, and elaborate interchanges are necessary on both banks to serve it with road and rail traffic. The Elbe Tunnel at Hamburg, West Germany, completed in 1974, was built with elements (three two-lane apertures) almost as big as the large one at Antwerp, and another Dutch tunnel for the Oude Maas is very similar in size (four two-lane roads). It will be completed in 1977; the big elements are being towed 12 km (7·5 miles). In New York City an unusual tunnel is being built with square elements, with two railway lines above and two below, like a Battenberg cake. There is also under construction in Tokyo a tunnel with two three-lane road apertures, roughly equal in size to the three large road tunnels just mentioned

At each reservoir, water can be discharged or can continue flowing down the tunnel. The valves that control the water are at the surface, and at each junction of two tunnels in the list above, there is a shaft taking water up to the surface, and beside it another shaft taking water down to the next tunnel. This feature of the construction leads the purist to question the strict veracity of the term 'continuous deep tunnel'. If these tunnels are considered as separate tunnels, then the Orange–Fish wins pride of place as the world's longest tunnel.

The Delaware Aqueduct has, in addition, the East Delaware, West Delaware, and Neversink Tunnels at its upstream end, separated from the main tunnel by Rondout Reservoir.

The main 85-mile length of tunnel was driven mainly in sandstone from 31 shafts at an average rate of 9·6 m (31·6 ft) per day.

City Tunnel Number 1 is 29 km (18 miles) long and was completed in 1917.

Main drainage of Mexico City (table 25)

A main drainage system for Mexico City involves great lengths of deep tunnels, 28–45 m (90–150 ft) below street level in the city. It consist of three principal tunnels:

	Length		Diameter		Flow		Area served	
	km	miles	m	ft	m^3/s	ft^3/s	ha	acres
Central interceptor	25	15·5	4 and 5	13 and 16·4	90	3178	11 000	27 000
Eastern interceptor	27	16·8	4 and 5	13 and 16·4	110	3884	20 000	49 000
Central outfall	50	31	6·5	21·3	200	7063	—	—

The Central Outfall has a maximum depth of 220 m (720 ft) below the surface. It discharges into the Rio del Salto. By July 1974, 67·4 km (42 miles) of tunnel had been excavated, including practically all of the Central Outfall. The total cost of this scheme is equivalent to about US $320 million and it is one of the greatest Mexican civil engineering ventures. Its benefits far outweigh its costs.

Subaqueous tunnels (table 28)

Tunnels driven beneath the beds of rivers or estuaries (and more recently beneath the sea) do not figure prominently among the world's longest tunnels. However, if a table could be devised according to difficulty, they would be very prominent in it, the most arduous feats of tunnelling,

Casting the concrete elements of the J F Kennedy Tunnel in a dry-dock

with a few exceptions, being in this category. The Mersey Rail Tunnel was one of the few early subaqueous tunnels that could be called 'straightforward' and the term is relative even then. Its road counterpart (see table 27) was the greatest feat of its day (1934) in this class. The Severn Tunnel was completed only after herculean efforts, involving the pumping of enormous quantities of water. (However, within the last few years a small tunnel was driven under the Severn in a businesslike, economical manner, without attracting any attention.) Of those not tabulated, the New York City tunnels, especially the East River tunnels, were hazardous and difficult to a degree. Notable early tunnels include the first Blackwall Tunnel under the Thames (1891–7) and the St Clair Tunnel between Sarnia, Canada (Ontario) and Port Huron, USA, a shield-driven compressed-air tunnel completed in 1891.

Right: A train running through the BART immersed-tube tunnel in San Francisco

Far right: Launching the steel shell of one of the 57 elements of the BART tunnel

Table 25. The world's longest tunnels (over 10 km (6·2 miles) in length except tunnels for hydro-electric schemes).

Name	Length between portals km	miles	Equivalent diameter m	ft	Year of completion	Country and location	Other details
Delaware Aqueduct	168·9	105	3 to 5·9	10 19·5	1944 (limited use)	USA, New York	Water-supply tunnel for New York City (see text)
Orange–Fish	82·42	51·2	5·3	17·5	1974	South Africa	Irrigation tunnel connecting Orange and Fish rivers
West Delaware	70·8	44	5 approx.	16	1960	USA, New York	Water-supply tunnel for New York City
Channel	57·3	35·6	7·1 each (two main tunnels) 4·5 (service tunnel)	23·3 14·8	Cancelled in 1975 after starting first contracts	England–France	Submarine railway tunnel from Sangatte near Calais to Westenhanger near Dover. Length beneath sea 37 km (23 miles)
Seikan	53·85	33·47	9·6 high 8·0 wide (also small service tunnel)	31·5 26·2	u.c. 1972–79 site investigation from 1946	Japan, Honshu to Hokkaido beneath Tsugaru Channel	Rail tunnel with one double-track tube for very-high-speed line. World's longest submarine tunnel, 23·3 km (14·48 miles) beneath the sea. Maximum depth of water 140 m (459 ft)
Central outfall	50	31	–	–	1975	Mexico, Mexico City	Longest of three main drainage tunnels (see text)
Thames–Lee	30·3	18·8	2·5	8·1	1960	England, London (Hampton–Walthamstow)	Aqueduct from Thames to Lee Valley reservoirs. London's second 'New River'. Built with wedged unbolted precast concrete lining
Shandaken	29·1	18·1	3·6	11·8	1923	USA, New York	Part of New York City's Catskill Aqueduct
San Jacinto	29	18	4·9	16	1938	USA, California	Longest tunnel of the Colorado Aqueduct (see chapter 3). Total length of tunnels 148 km (92 miles)
Kielder	28·2	17·5	2·9	9·5	u.c. 1980	England	First stage of Tyne to Tees Aqueduct

The Astrodome (table 21) – 'An admirable work of structural science if not of architecture' – Condit. A possible limit to the span of a structure of this type is given as 425 m (1400 ft)
Right: CNIT building, Paris. Three double-skinned, corrugated, fan-shaped half-arched shells of reinforced concrete. The half-arches meet at the centre of a dome-like shape, and the structure is supported at three points that form an equilateral triangle. The side-length of this triangle is the span quoted in table 21. CNIT claims the shell roof of longest span, and the greatest surface area of any building supported at a single point (i.e. 0·7 ha; the building covers 2·1 ha – 5·2 acres). Each half-arch, if it had been completed along its axis, would have spanned 238 m (780 ft). The half-arches are linked by ties below ground level and the two ties at each support each carry a load of 3750 t. *Photo: CNIT*

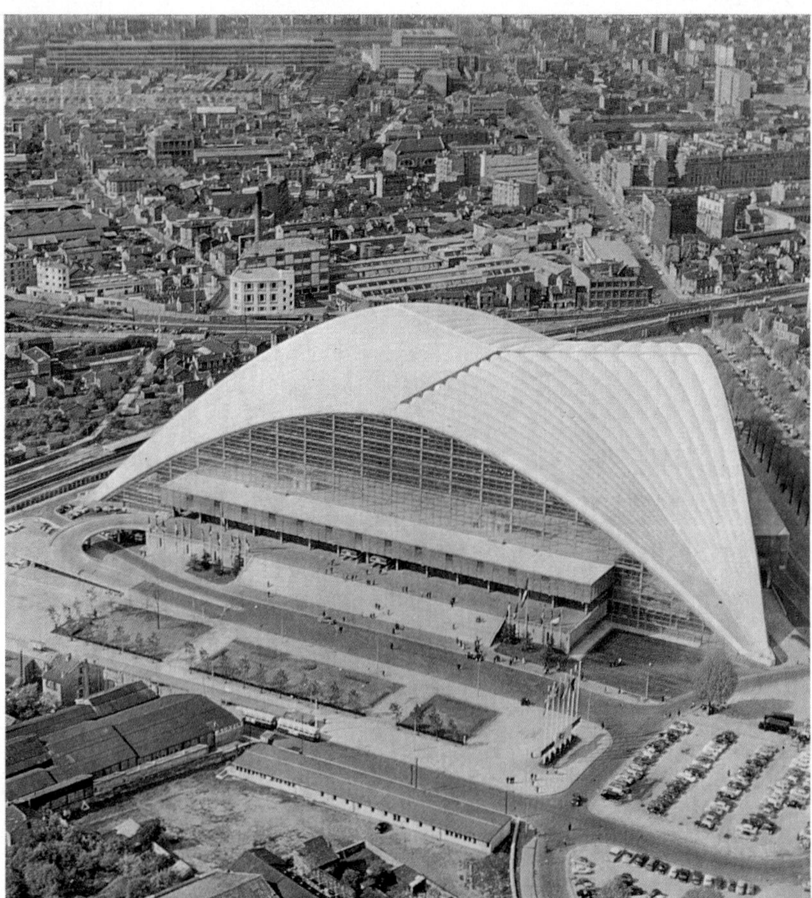

Table 25. cont.

Name	Length between portals km	miles	Equivalent diameter m	ft	Year of completion	Country and location	Other details
Northern Line	27·8	17·3	3·7 each (running tunnels) 6·8 (station tunnels)	12 22·2	1939	England, London (East Finchley (via Bank to Mordern)	Longest continuous tunnel of any underground railway network. Constructed mainly in London Clay with cast-iron bolted lining
Third water tunnel	22	13·7	7·3 (maximum)	24	u.c. 1970–77	USA, New York City	Deep tunnel (213 m (700 ft) maximum) to bring water to centre of city
Florence Lake	20·9	13	–	–	1925	USA	
Alva B Adams	21	13	2·97	9·75	1946	USA, Colorado	Colorado–Big Thompson scheme's 'Continental Divide' Tunnel
Simplon 1 Simplon 2	19·8 19·82	12·3 12·31	5·5	18	1906 1922	Switzerland–Italy	Both are single-line railway tunnels under the Alps and claim the greatest depth below the surface – 2135 m (7005 ft) – of any tunnel. Simplon 1 took eight years to build and was for many years the greatest feat of its kind. Simplon 2 is 20 m (66 ft) longer than Simplon 1
Ely Ouse	20	12	2·5	8·3	1969	England, Cambridgeshire	Water tunnel, part of diversion to reservoirs in Essex
Shin–Kanmon	18·7	11·6	8 to 10	26 to 33	1975	Japan, crosses Kanmon Strait	Tunnel for very-high-speed railway including 880 m (2887 ft) submarine crossing
Great Apennine	18·5	11·5	–	–	1934	Italy (Florence–Bologna Line)	Double-track railway tunnel
Clear Creek	17·3	10·7	5·3	17·5	1962	USA, California	Longest tunnel of Bureau of Reclamation's Central Valley Project (see chapter 3)

Table 25. cont.

Name	Length between portals km	miles	Equivalent diameter m	ft	Year of completion	Country and location	Other details
St Gotthard (road)	16·3	10·1	–	–	u.c. 1980	Switzerland	About half driven in 1974; 1977 is estimated time for completion of driving
Bowland Forest	16	10	2·6	8·5	1955	England, Lancashire	Part of Haweswater Aqueduct from Lake District to Manchester
Rokko	16	10	–	–	–	Japan	Railway tunnel
Henderson	15·8	9·85	–	–	1975	USA, Colorado, Cabin Creek	Railway tunnel in Rocky Mountains
St Gotthard	14·9	9·26	for double-line		1882	Switzerland	Early Alpine railway tunnel. Cost 800 lives between 1872 and 1880
Lötschberg	14·5	9·03	–	–	1913	Switzerland, Bernese Alps	Double-track railway tunnel in rock
Blanco	13·8	8·6	2·6	8·6	1969	USA, Colorado and New Mexico	San Juan–Chama scheme of Bureau of Reclamation
Hokkuriku	13·8	8·6	–	–	1962	Japan	
Appalachia	12·9	8	5·5 to 6·7	18 22	1942	USA, Tennessee	
Fréjus (Mont Cenis road)	12·7	7·9	4·5 high 10 wide	14·7 32·8	u.c. 1978	France–Italy	
Mont Cenis	12·79 (original length) 13·68 (present length)	7·95 8·5	8	26	1871 1881	France–Italy	First Alpine tunnel. Started in 1857. Driven through rock. Hand-drilled for first three years at 0·23 m (9 in) per day. Compressed air drills were then introduced
Cascade	12·6	7·8	–	–	1929	USA, Washington (Spokane–Seattle line)	Longest US railway tunnel until completion of Henderson Tunnel

Table 25. cont.

Mont Blanc	11·6	7·2	9	29·5	1965	France–Italy (Chamonix–Valle d'Aosta)	World's longest road tunnel, has two-lane road
Flathead	11·3	7	5·5 wide 7·2 high	18 23·5	1970	USA, Montana	Railway tunnel for diversion at Libby Dam. East portal is closed by a door after a train has passed, and the fumes blown out of the tunnel with a fan
Kubiki	11·3	7	–	–	1970	Japan	
Tecolote	10·3	6·4	2·13	7	1956	USA, California, near Goleta	Irrigation tunnel of the Cachuma scheme
Arlberg	10·3	6·4	–	–	1884	Austria, west of Innsbruck	Double-track railway tunnel
Moffat	10	6·2	–	–	1928	USA, Colorado (Denver–Salt Lake City)	Railway tunnel, 2822 m (9257 ft) above sea-level
Talow	10	6·2	2·8	9·2	1967	Iran	Aqueduct from Latiyan Reservoir to Tehran
Kanmon	10	6·2	–	–	1958	Japan, Honshu to Kyushu	Claimed as (see table 29) world's longest subaqueous road tunnel. Construction started in 1939. Two rail tunnels, each about 4 km (2·5 miles) long, under Kanmon Strait were completed in 1942 and 1944 respectively

Table 26. The world's longest hydro-electric tunnels (over 16 km (10 miles) in length).

Name	Length		Equivalent diameter		Year of com-pletion	Country and location	Other details
	km	miles	m	ft			
Arpa–Sevan	48	29·8	–	–	u.c.	USSR Armenia	For augmenting the water of Lake Sevan
Rendalen	28·5	17·7	2·4	7·9	u.c.	Norway	Partly lined; flow 60 m^3/s. Driven from four adits and forebay into which the tunnel discharges

Table 26. cont.

Name	Length km	Length miles	Equivalent diameter m	Equivalent diameter ft	Year of completion	Country and location	Other details
Lochaber	24·1	15	4·6	15·2	1930	Scotland	Connects Treig Dam with Lochaber power station, running beneath the north slopes of Ben Nevis
Valtellina	22·6	14	4·94	16·2	1959	Italy	Premadio–Valgrisona
	16	9·9	4·9	16	1951		Forni–Braulio
Vinstra	22·4	14	–	–	1951	Norway	Main supply tunnel; scheme has four tunnels
Eucumbene–Tumut	22·3	13·9	6·4	21	1959	Australia	Longest tunnels of Snowy scheme – (see chapter 4)
Murrumbidgee–Eucumbene	16·7	10·4	3·35	11	1961		
Norddalen–Iptovtn	c. 20	c. 12·5		–	1975	Norway, near Narvik	Longest of three major tunnels of the Skjomen scheme, 330 MW
Juktan	20	12·4	8·8–10	29–33	u.c. 1977	Sweden	Pumped storage scheme; head-race tunnel 5 km (3 miles) to underground power station, then tail-race tunnel 15 km (9 miles)
Lar–Kalan	20	12·4	3	9·8	u.c. 1979	Iran	Diversion from Lar Valley to Fahranaz Pahlavi Reservoir for power and the water-supply of Tehran
Moawhango–Tongariro	19·2	11·9	–	–	1974	New Zealand	Tunnels of the Tongariro scheme
Whakapapa–Tawhitsikuri	16·6	10·3	–	–			
Chute-des-passes	18·8	11·7	9·36 and 10·4	31 34	1959	Canada, Quebec	Includes head- and tail-race tunnels, the power station being underground at about the mid-point
Belledonne	18	11·2	5·3	17·4	project 1978	France, Haute-Savioe	Arc to Isère diversion through Cheylois power station. Tunnel to be driven without intermediate adits and under a maximum overburden of more than 2000 m (6560 ft). Maximum temperature of 43 °C (110 °F) predicted

Table 26. cont.

Mont Cenis–Villarodin	18	11·2	5·5 to 4·2	18 13·8	project 1978	France, Haute-Savioe	Pressure tunnel connecting reservoir and power station near Modane on River Arc. These two entries are part of a scheme that will supersede the existing Isère–Arc scheme
Almendra	18	11·2	–	–	1970	Spain	Pumped storage/regulating scheme linking Duero and Tormes rivers
Castelbello	16·8	10·4	2	6·6	1947	Italy	
Kemano	16·2	10·1	7·62	25	1953	Canada, British Columbia	Two similar tunnels
Oospat	16·2	10·1	3	9·8	–	Bulgaria, Rhodope Mountains	Longest tunnel of Bulgaria's hydro-electric developments

Table 27. Hydro-electric tunnels of largest cross-section (over 150 m² (1600 ft²); others for comparison. Cross-sectional shapes vary from rectangular to pear shaped, circular, etc.).

Name	Cross-section area		Equivalent diameter or cross-sectional size		Length		Year of completion	Country and location	Other details
	m²	ft²	m	ft	m	ft			
Stornoforrs	390	4198	16 × 26·5	52 × 87	3950	12 959	1958	Sweden	Unlined tail-race tunnel. Flow 850 m³/s (30 000 ft³/s)
Pirttikoski	350	3767	16 × 22	52 × 72	2560	8399	1960	Finland	Tail-race tunnel. Flow 500 m³/s (17 650 ft³/s)
Harrsele	260	2799	15 × 18·4	49 × 60	3400	11 155	1957	Sweden	Flow 500 m³/s (17 650 ft³/s)
Harsprånget	190	2045	13 × 15	43 × 49	2900	9514	1952	Sweden	Tail-race tunnel. Flow 400 m³/s (14 125 ft³/s)
Kilforsen	{ 176 208	1894 2239	15 16·3	49 53	2680 3680	8793 12 073	1953	Sweden	Head- and tail-race tunnels. Flow 375 m³/s (13 240 ft³/s)

Table 27. cont.

Name	Cross-section area		Equivalent diameter or cross-sectional size		Length		Year of completion	Country and location	Other details
	m^2	ft^2	m	ft	m	ft			
Aswan	176	1894	15	49	282	925	1964	Egypt	Six similar tunnels. 11 000 m^3/s (388 500 ft^3/s) is total design flow
Notable tunnels of large cross-section									
Krangede	116	1249	11 × 13·4	36 × 44	1450	4757	1936	Sweden	Prototype of 'Swedish' layout with long, unlined, tail-race tunnel of large size
Rove	250	2691	22 × 15·4	72·2 × 50·5	7240	4·5 miles	1927	France, Marseille–Rhône Canal	Largest cross-section in the world when built. Canal tunnel with 3 m (9–10 ft) depth of water. Two 8 m (26·3 ft) wide barges can pass
Yerba Buena Island	–	–	23 × 17·7	76 × 58	164·5	540	–	USA, San Francisco	See San Francisco–Oakland Bay Bridge
Mersey	140·5	1512	13·4	44	3·43 km 4·62 (including branch tunnels)	2·13 miles 2·87	1934	England, Liverpool–Birkenhead	Road tunnel (longest in UK). Carries 11 m (36 ft) road for four lanes of traffic, the longest road tunnel of its day. Second tunnel opened in 1971. Largest cross-section for a tunnel lined with cast iron
John F Fitzgerald Expressway	–	–	61 wide	200	–	–	1958	USA, Boston	
Glen Canyon Dam	–	–	34·4 wide	113	–	–	1959	USA	Width refers to transition section of spillway tunnel which has the heaviest support steel ever used in a tunnel

Note: The tunnel of largest cross-section driven in ancient times is thought to be the Pausilippo Tunnel. See table 28.

Two steel shells and the catamaran placing craft for an immersed-tube tunnel under the Tama River, Tokyo, 1969

Close-up of the coupling and joint between elements of the BART tunnel; hydraulic cylinders were employed to pull the elements together
Photos: San Francisco Bay Area Rapid Transit District

Immersed-tube tunnels (table 30)

The first tunnels completed by this method were three sewer tunnels in Boston, USA, in 1893, but the technique really belongs to the modern era. The effective prototype was the Detroit River Tunnel, and the first to show the potential in cross-section of the technique (multiple-lane roadways plus railway tracks can be accommodated with relative ease) was the Maas Tunnel in Rotterdam. Eighty-five tunnels of this kind have been constructed, of which at least ten are sewage-pipes rather than true tunnels.

The longest, and the one built in deepest water, is the Hyperion sewage outfall at Los Angeles. The largest cross-section and the biggest prefabricated element are claimed by the John F Kennedy Tunnel at Antwerp. Neglecting sewage outfalls, which are of comparatively small cross-section and are a specialised application, the longest immersed-tube tunnel and the one built in the deepest water is the BART (Bay Area Rapid Transit) Tunnel in San Francisco.

Table 28. Chronological list of the world's longest tunnel.

Name	Length between portals km	miles	Year of completion	Country and location	Other details
Delaware Aqueduct	168·9	105	1937–44	USA, New York City	See table 25
Shandaken	29·1	18·1	1923	USA	See table 25
Simplon 2	19·5	12·4	1918–22	Switzerland–Italy	See table 25
Simplon 1	19·5	12·4	1898–1906		

Table 28. cont.

Name	Length between portals km	miles	Year of completion	Country and location	Other details
St Gotthard	15	9·3	1872–82	Switzerland	See table 25
Mont Cenis	12·79	7·79	1857–71	France–Italy	See table 25
Noirieu	12	7·45	1822	France, Saint-Quentin Canal	Three tunnels on this canal – also Tronquoy (1·1 km (0·68 mile)), see text, Riqueval (5·5 km (3·4 miles)) gave a total length in tunnel of 18·5 km (11·5 miles)
Standedge*	5	3·1	1794–1811	England (Huddersfield Narrow Canal)	Britain's longest canal tunnel. The canal was authorised in 1794, and completed except for the tunnel in 1798. Tunnel has passing place in centre. Railway tunnel built beside it in 1849
Sapperton	3·5	2·2	1789	England (Thames and Severn Canal)	Canal tunnel under Cotswolds
Norwood	2·9	1·8	1771–77	England. (Chesterfield Canal)	Canal tunnel. Originally 80 yards (73 m) longer: eastern end fell and was opened out. Tunnel collapsed in 1908
Harecastle	2·6	1·6	1766–77	England, near Stoke-on-Trent (Trent and Mersey Canal)	First major British canal tunnel. Closed early this century after subsidence. Second canal tunnel built 1824–27, and later a railway tunnel
Nochistongo	6·4	4	1608	Mexico, north west of Mexico City	Intended for drainage of Valley of Mexico. It was not lined and, when water was diverted through it, it collapsed. Opened out to a cutting 160 years later
Lake Fucinus†	5·6	3·5	c. AD 40–50	Italy, near Rome	Built to drain the lake at great cost in the lives of the slave labour employed (30 000 men for ten years). Constructed from shafts 37 m (120 ft) apart and up to 120 m (400 ft) deep. This was the greatest of the Roman tunnels
Lake Alban	1·8	1·1	–	Italy, near Rome	Driven through hard lava from 50 shafts, high enough for a man to walk and 1 m (3·5 ft) wide
Pausilippo	1·4	0·9	36 BC	Italy, between Naples and Pozzuoli	Road tunnel, 7·5 m (25 ft) wide by 9 m (30 ft) high

* Width 2·7 m (9 ft); height 5·2 m (17 ft) including a 2·4 m (8 ft) depth of water. Sapperton was wider – about 4 m (12–15 ft) wide, the other canal tunnels similar. † See also the note on Roman aqueducts in chapter 3.

Table 28. cont.

Name	Length between portals km	miles	Year of completion	Country and location	Other details
Baebolo	2·2 (or 1500 paces)	1·4	3rd century BC	Cazlona, Spain	Tunnel forming a silver-mine. A wet tunnel kept open by a human chain of bailers. Yielded 300 lb (135 kg) of silver per day
Isle of Samos	1·0	0·6	6th century BC	Greece	Water tunnel through limestone
Babylon	0·9	0·6	22nd century BC	Babylon	Under the Euphrates River linking royal palace and temple. Built by diverting the river in the dry season‡

‡ 'If it ever was executed, it did not even amount to a tunnel at all but was nothing more than an arched passage built in a large trench or ditch, excavated across the bed of what had been a river' – Henry Law, *Memoir of the Thames Tunnel*.

Table 29. Chronological list of the world's longest subaqueous tunnel.

Name	Length beneath water km	miles	Length between portals km	miles	Year of completion or years of construction	Country and location	Other details
Seikan	23·3	14·48	53·85	33·47	u.c.	Japan	See table 25
BART	5·8	3·6	19·1	11·9	1970	USA	See table 30
Hyperion	10·5	6·5	10·5	6·5	1960	USA	See table 30: sewer outfall, so perhaps does not qualify
Kanmon	0·75	0·47	9·9	6·2	1958	Japan	Described as the world's longest subaqueous road tunnel. Strictly true, but the length under water is short
Severn	3·7 (high tide) 1·3 (low tide)	2·3 0·8	7·01	4·35	1873–86	England, Gloucestershire–Monmouthshire	Opened December 1886. Lined with 76 million bricks
Mersey (rail)	1·2	0·75	*c.* 3·7	*c.* 2·3	1879–86	England, Liverpool	Opened February 1886, for double-track railway. Lined with 38 million bricks. Stations at the low level
Wapping–Rotherhithe	0·35	0·2	0·37	0·22	1825–43	England, London	Brunel's Thames Tunnel. First use of tunnel shield
Babylon	–	–	0·9	0·6	22nd century BC	Babylon	See entry and note in table 28

Table 30. Chronological list of the largest immersed-tube tunnels. (Criteria: length of prefabricated construction; length, and/or cross-section of prefabricated element; depth of water over tunnel; sewage outfalls included, but the list is progressive for other types.)

Name of tunnel	Location	Purpose	Year of completion	Length of prefabricated part of tunnel m	ft	Prefabricated elements Number	Length m	ft	Cross-section m	ft	Length between portals m	ft	Other data
Rotherhithe	Thames, London	Trial length	1811 (trial)	15	50	2	7·6	25	2·7 (diameter)	9	–	–	Brick cylinders: see page 122. Trial length to test feasibility of the technique
Shirley Gut Siphon	Boston, Mass., USA	Sewer beneath river	1894	96	315	–	16/21	48/68	2·6 (diameter)	8·5	457	1500	Cylinders of steel plate lined with brick. Three such tunnels were built
Detroit River	USA/Canada border	Rail (Detroit–Windsor)	1910	813	2668	10	80	262·5	17 × 9·4	55·5 × 31	2552	8373	Two steel tubes joined into a rectangular section by extensive concreting. Navigation depth of 12·5 m (41 ft) over tunnel
Harlem River	New York City, USA	New York subway	1914	329	1080	–	61 & 79	200 & 260	23 × 7·5	76 × 24·5			Prefabricated in steel and concrete, now the standard US form of construction
Posey	Alameda County, California, USA	First road tunnel of this type	1928	743	2436	12	62	203	11·3 (diameter)	37	1080	3545	Named after its engineer. Navigation depth of 18 m (60 ft) over tunnel
Bankhead	Alabama, USA	Road; Mobile–Blakely Island	1940	152	500	1	91	300	11 (octagon across flats)	36	951	3119	Steel shell 9 m (30 ft) diameter, concreted to 11 m (36 ft) octagon
Maas	Rotterdam, Holland	Road	1941	587	1926	6	61	201	25 × 9	81·25 × 28·75	1075	3528	Two 6 m (20 ft) roadways, 5 m (16·5 ft) cycle track above 4 m (13 ft) footway
Baltimore (Patapsco)	Baltimore Harbour, USA	Road	1957	1920	6300	21	91	300	21 × 10·7	70 × 35	2332	7650	Navigable depth of 15 m (50 ft) over tunnel. Two two-lane apertures

Table 30. cont.

Hampton Roads	Virginia, USA	Road, Hampton–Norfolk	1957	2091	6860	27	91	300	11·3 (octagon)	37	2280	7480	One two-lane road. Maximum navigable depth over tunnel is 21 m (70 ft)
Hyperion	California, USA	Sewer outfall, Los Angeles	1960	10 500 (approx.)	34 450	many	58·5	192	3·7 (diameter)	12	10 500	34 450	7·3 m (24 ft) sections of concrete tube built up into 58·5 m (192 ft) elements. Outfall goes about 8 km (5 miles) out to sea in maximum depth of water 61 m (200 ft)
Rendsberg	Germany	Road under Kiel Canal	1961	140	459	1	140	460	19·5 × 7·6	64 × 25	640	2100	Two two-lane road apertures plus central ducts in a binocular or lunette-shaped cross-section
Coen	North Sea Canal, Amsterdam, Holland	Road	1966	450	1476	5	90	295	23·3 × 7·7	76 × 25	540	1772	Two two-lane apertures plus service tunnel
Louis Hippolyte Lafontaine (Boucherville)	Montreal, Canada	Road	1967	768	2520	7	110	361	37 × 7·7	121 × 25	1390	4560	Two three-lane apertures plus ventilation aperture
John F Kennedy	Antwerp, Belgium	Road and rail, beneath River Scheldt	1969	512 (approx.)	1680	4 at 1 at	99 115	325 377	48 × 10	157 × 33	689 (road) 1690 (rail)	2260 5545	Two three-lane roads, one two-track rail and three subsidiary apertures. Navigable depth of 12 m (39 ft) over tunnel (see illustrations)
Hernandarias	Santa Fé, Argentina	Road tunnel, under River Parana	1969	2900	9514	–	–	–	–	–	–	–	Connects Santa Fé with Parana. 32 m (105 ft) deep below river
BART (Bay Area Rapid Transit)	San Francisco, USA	Rail (métro)	1970	5826	19 114	57	83 to 115	272 to 377	14·6 × 6·5	48 × 21	5826	19 114	Binocular steel tubes. Two rail apertures plus service aperture. Navigable depth over tunnel 30 m (98 ft)

EARLY TUNNELS

In the years 1766–77, a canal tunnel 2926 yd (2675 m) in length was built under the direction of James Brindley, to carry the Trent and Mersey Canal under Harecastle Hill, near Stoke-on-Trent. The tunnel was 16 ft (5 m) high, wide enough to take a narrow boat, but without a walkway or tow-path, and the narrow boats were 'legged' through the tunnel. That is a board was laid across the boat and two men, one for each side of the boat, lay on the board and 'walked' along the side or roof of the tunnel to propel the boat. If that tunnel was the longest of its day, it was a typical feat of the canal era. It was one of four tunnels on the Trent and Mersey Canal, which was built for the transport of goods from Wedgwood's Etruria Pottery Works. About 70 canal tunnels were built in Britain, and rather more than that number in France.

The intriguing feat of which visionaries dreamed, however, was to pierce a natural barrier – to tunnel under a river or through a mountain range. The dream became a reality respectively with the Thames Tunnel, and the Mont Cenis Tunnel. How should the cost of a dream be counted? In both cases the cost in lives was heavy, the risks enormous, and the disappointments numerous. The cost in money and resources was very high. Tunnelling only gradually evolved away from those conditions. The hazards remained high, and the loss of life utterly intolerable by modern standards. For example, in building the East River tunnels in New York, 'blows' into the river, when great quantities of compressed air were lost, were frequent. The introduction of a medical air-lock is said to have reduced the death-rate from 26 per cent to 1·5 per cent per year. On one occasion a tunneller was blown out of the tunnel into the river, and was picked up apparently none the worse! Hazards will never be eliminated from tunnelling, but today they can generally be predicted and managed by better exploration and hence understanding of ground conditions, by techniques of treating bad ground *in situ*, ahead of the tunnel drive, and by better plant and machinery.

SOFT GROUND TUNNELS

The Tronquoy Tunnel on the Saint-Quentin Canal, France (length 1098 m (3603 ft)), completed in 1803 is claimed to be the first tunnel driven through soft, sandy ground. The tunnel was 8 m (26 ft) wide and consisted structurally of a semicircular arch of brickwork. A small pilot heading – well timbered – was first excavated along the line of the abutment of the arch, and the brickwork of the abutment constructed within it. The heading was then backfilled. A second, and subsequently a third, heading were opened above it, to permit construction of the next section of the arch; hard-packed sand was shaped to form the soffit of the arch; a fourth heading extended across to the crown of the arch. The same sequence was necessary on the other side. Having built the arch and backfilled the headings, the soft ground was excavated from within the tunnel.

The brick immersed-tube tunnel of 1810

In 1802–8 an attempt to drive a tunnel under the Thames was made by Vazie ('the Mole') and Trevithick, between Rotherhithe and Wapping. A shaft was sunk at Rotherhithe, and a heading – 5 ft (1·5 m) high and 3 ft (90 cm) wide at the base and 2 ft (60 cm) wide at the top – was driven for a distance of 315 ft (96 m) from the shaft to the river, and altogether 1046 ft (318 m), that is most of the way to the Wapping side. Quicksand was then found, the Thames broke in, and the heading was lost. Notable experiments were then started to find out whether it would be possible to build brick cylinders, float them out and sink them end to end on the line of the tunnel, in a trench in the quicksand, so that their ends could then be removed and the cylinders would then form the tunnel in the difficult region. Two cylinders were built and sunk to the river-bed experimentally in 1810 and 1811, at a site nearby. The experiments were completely successful, thus initiating and proving the concept of the immersed-tube tunnel.* But they had been costly, and there was no money available to complete the tunnel.

Marc Isambard Brunel's tunnel at a site close by also suffered many vicissitudes and was abandoned for seven years, and then restarted and completed. It was noteworthy for the use of the first tunnelling shield. The shield is a rigid shell, the size of the tunnel, that holds the ground at the tunnel face. Behind it, the tunnel lining is built, progressively in rings, right up to the shield. The face is excavated at the front of the shield; the shield is then pushed forward to expose a band of ground behind it, where another lining ring is added. Its use proved to be the secret of success.

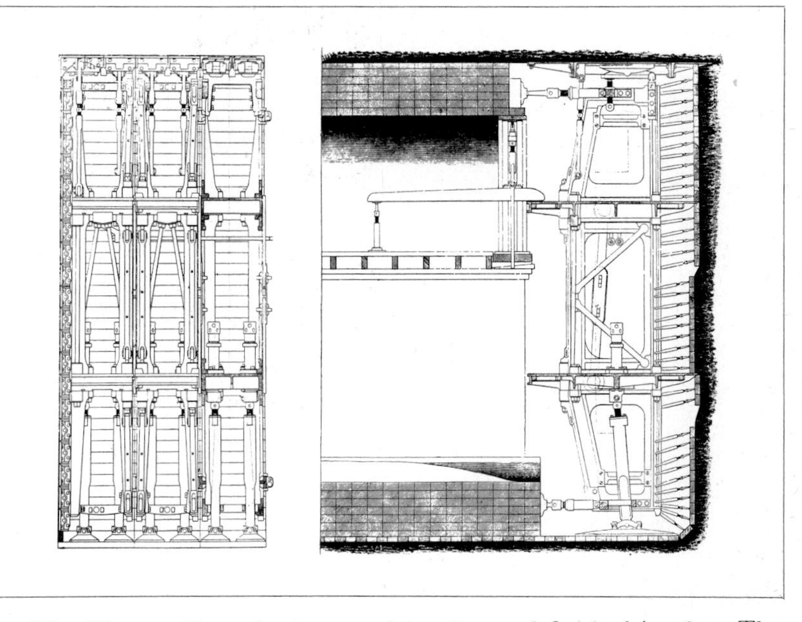

Brunel's shield. Its overall size was:

width	11·43 m	37 ft 6 in
height	6·78 m	22 ft 3 in
length	2·74 m	9 ft

It consisted of twelve cast-iron frames, each 0·99 m (3 ft 3 in) wide and each of three 'storeys'; that is 36 cells for excavation at the face. Each frame, and each cell could be moved forward independently. Chisel-shaped roof plates (the forerunners of the cutting edge of subsequent shields) supported the muck at the roof and served as floor plates in the intermediate positions. Poling boards – about 15 for each cell – were held by screw stretchers. Screw jacks bore against the finished masonry lining to shove the shield. Iron tail plates bridged between the tail of the shield and the masonry. Alternative frames were moved forward about 0·15 m (6 in) at a time. The greatest length built in one week was 4·27 m (14 ft)

The Thames Tunnel was started in 1825 and finished in 1843. The shield used in the first phase of the work (1825–8) was removed when work restarted in 1835. The 1835 shield is the one described and illustrated here. Copperthwaite, in his book, quotes and comments on Brunel's patents in detail and concludes: 'The second shield described and figured in the patent specification of 1818 is the one suggested, as described by Brunel himself, by the screwlike action of the Teredo Navalis, a marine worm that can pierce the hardest woods. The shield was never practically tested, and it is one of fame's little ironies that Brunel is popularly supposed to have derived his great invention from his observation of a natural excavating machine, whereas in fact, the actual shield used by him borrowed nothing from any previously known natural or other mechanism.'

Brunel's shield is shown in the illustration. Greathead (1844–96) improved it, and the shield he used for a little tunnel under the Thames at

*C A Pequignot, 'The first immersed-tube tunnel', *The Consulting Engineer*, February 1974

Rear view of the shield for the second Dartford–Purfleet Tunnel under the Thames. The shield is shown in the shield chamber where it was erected, and the two arms that are used to place lining segments in position behind the shield are prominent. The external diameter of this tunnel is 10·3 m, the same as the second Mersey tube. However, this tunnel is being driven under compressed air, the drive being in progress during 1975. *Photo: Mott Hay and Anderson*

Tower Bridge in 1869 was the prototype of the modern shield. This tunnel was the second one to be constructed using a shield, but in the same year an American engineer, Beach, was using a shield in New York.

Greathead's Tower Subway was the first tunnel with a cast-iron lining with grout filling behind it, and the first tunnel built with a shield movable in one piece – the model for soft-ground tunnelling for the best part of a century. However, Greathead's apparatus for injecting grout was not patented until 1886. The grout at Tower Subway was injected with hand-syringes and was not satisfactory – it had to be too wet and there was not enough pressure. Both shields were based on a design by Barlow. Beach's shield was the first one to be shoved forward with hydraulic cylinders.

Hydraulically operated shutter for lining a hard-ground tunnel

Another basic technique complements the tunnelling shield, when bad ground is encountered. It is the use of compressed air. A pressure greater than that of the water is maintained in the tunnel. This technique is also used to construct deep foundations in water-bearing ground. At first the use of compressed air proved hazardous. If air is lost suddenly through a weak face for example, or a cave-in from the bed of a river, the enclosure behaves rather like a champagne bottle when the cork is removed, and then, when the pressure has been lost, rapidly fills with water and soil. Those who work in compressed air also get caisson sickness – the bends – which can be fatal. It is caused by too-rapid de-compression; nitrogen dissolved in the blood at the high pressure forms bubbles within the circulatory system, rather than being ejected through the lungs. The remedy is to go back into compressed air, and then reduce the pressure slowly. The medical air-lock does this. Yet another hazard of prolonged exposure to compressed air is that it makes possible a form of cancer known as bone necrosis.

In 1886 compressed air was used for the first time with a shield to build a tunnel for the City and South London Railway in water-bearing ground.

Greathead carried out this work. He had built a shield for use with compressed air in 1874, to build the Woolwich Tunnel, but the scheme had been abandoned.

Greathead's shield has been superseded in the modern era by shields that permit completely mechanised excavation, removal of spoil, and emplacement of the tunnel lining, and hence rapid progress. A revolutionary innovation that can make compressed air unnecessary in some cases is the bentonite shield.

COMPRESSED AIR IN CONSTRUCTION

The diving bell was in use in the 16th century, but at that time, air inside the bell could not be renewed. In 1778, Smeaton used a diving bell to repair the foundation of a bridge in the River Tyne at Hexham and for the first time, pumped air into the bell. This was the first use of compressed air in the modern sense, i.e. depending on the use of a pump or compressor.

In 1830 a patent specification was published by Sir Thomas Cochrane for building shafts and tunnels with compressed air in water-bearing ground. Previous applications of compressed air had been in open water. The specification contained all the essential elements, as subsequently developed, but the method was not used until 1839. In that year a shaft was sunk for a colliery at Châlonnes-sur-Loire, in France, using compressed air.

The first bridge foundations to be constructed using compressed air were at Rochester (Kent) and Chepstow (Monmouthshire, now Gwent). In 1851, the foundations of Rochester Bridge had been constructed by sinking cylinders under compressed air and fitted with air locks. Fourteen cylinders were sunk at Chepstow Bridge, each 2 m (7 ft) in diameter, down to chalk at a depth of 12 to 18 m (40 to 60 ft) below high water. The cylinders were of cast iron and were filled with concrete or brickwork. The method became established for bridge foundations in the next two decades.

The first use of compressed air in tunnelling was in 1879, almost simultaneously in New York and Antwerp. In New York, Haskin used the technique for a tunnel under the Hudson River but work was suspended

Patterns made by a mole tunnelling machine

Hydraulic-boom drilling rigs

in 1880, when twenty men were drowned in the tunnel. In Antwerp, Hersent, a few months ahead of Haskin, built a cast-iron-lined tunnel 1·5 m (5 ft) in diameter without a shield.

Re-compression to cure 'the bends' (compressed air sickness) was recommended by M Foley in 1861 at Argenteuil, where a compressed-air job was in progress. In 1871–2 Dr Smith published a book in which he set out a physiological explanation, and the reasons why re-compression effected a cure. His work was ignored and was apparently unknown to Moir, the Chief Engineer of the Hudson Tunnel when he installed the first medical air-lock in 1890.

The maximum air pressure that can be used is about 0·34 N/mm² (50 lb/in²) corresponding to a head of water of 33·5 m (110 ft). On occasion, the maximum has been exceeded, and medical experimenters have repeatedly experienced 0·52 N/mm² (75 lb/in²), and on one occasion 0·63 N/mm² (92 lb/in²).

ROCK TUNNELLING

The traditional sequence of operations in rock tunnelling is to drill, charge the holes with explosives, blast the face, clear the rubble, make the new length of tunnel roof and walls safe if necessary, and then repeat the cycle. If the quality of the rock is poor, apart from temporary support the tunnel will have to be lined; the ubiquitous lining material is concrete. Unlike soft-ground tunnelling, the lining is a separate operation, and in some cases can be done after the drive has been completed. Just as the shield was the threshold invention for soft-ground tunnelling, so was the invention of a practical, mechanically powered drill for tunnelling in rock. Drilling was done by hand in the early stages of driving the Mont Cenis Tunnel,

The most ambitious tunnel under construction in the world today, and one without rival now that the Channel Tunnel has been cancelled, is the Seikan Tunnel that will link Honshu with Hokkaido (table 25). This constructional view shows, on the left, a connection to the main tunnel, in the centre the chamber for the tunnel-boring machine, and on the right, the service tunnel. *Photo: Japan Railway Construction Public Corporation*

An early stage in the construction of the Keban Dam in Turkey, showing excavation for the concrete control and power structure. The remarkable feature of this view is that it gives an idea of the enormous size of the solution cavities in the karstic rock at the site, one of which can be seen in the picture. See also page 170

Rig for compressed-air-operated drills at Mont Cenis Tunnel in 1871

and the pilot tunnel (in many tunnelling jobs it is expedient to drive a small tunnel first and then enlarge it to full size) advanced at a rate of only 0·3 m (9 in) a day. Steam-operated drills could not be used because ventilation was inadequate for them. The Italian engineer Sommeiller invented a drill operated by compressed air, which was used on the Italian face in 1861 and the French face a year later. Christmas Day 1870 was the climax – the two pilot tunnels met. The art of rock tunnelling then rose to a peak with the completion of the Simplon Tunnel, the most famous tunnel of all.

The greatest need for rock tunnels arises in the exploitation of hydro-electric power. Table 26 shows that the Lochaber Tunnel completed in 1930 is still the longest in this category, but since its day the volume of work has grown and techniques, based on the cycle of operations just described, improved immensely. The practical result has been greater confidence and more rapid and more economical tunnelling through rock (table 31).

Construction of Brooklyn Battery Tunnel, New York City. This picture, taken in June 1948, in the East Tunnel shows the front of the Brooklyn shield just after it entered the length of tunnel in rock that had been driven from Manhattan. *Photo: Triborough Bridge and Tunnel Authority*

For example, in Norway, 100 km (62 miles) of tunnels are driven each year for the State Power Board, mostly unlined and with cross-sections from 4 to 150 m² (43 to 1615 ft²). Where the rock is sound, tunnels of great size can be built and left unlined (table 27). According to some engineers, excavating a tunnel like Stornoforrs is more like quarrying than tunnelling. The ability to drive such tunnels with relative ease has influenced the design and economics of hydro-electric schemes significantly.

Table 26 lists about a score of tunnels over 16 km (10 miles) in length. When the Innertkirchen Tunnel of the Oberhasli hydro-electric scheme in Switzerland was completed in 1942, only two longer hydro-electric tunnels existed – Lochaber (in the table) and Orco (13·8 km (8·6 miles)) completed in 1929 in north-western Italy, and one longer, Malgovert in France, was under construction. The Innertkirchen Tunnel is 10 km (6·2 miles) in length and serves one of the early underground power stations.

Recent advances have also tended to blur the distinction between rock and soft-ground tunnelling. Tunnelling shields evolved with excavating devices, such as rotary cutters at the face, to replace hand-digging. From the hard-rock end of the scale – or rather for rock on the soft side – came the mole, as exemplified by the Robbins mole that drove the tunnels at Oahe Dam. This family of machines can be described as 'hard-ground shields', fully equipped mechanically to excavate and remove the spoil. Mechanically operated picks scour the rock face, and the grooves made by the picks are then attacked by cutting-wheels that flake off the rock. The two machines delivered to dig the preliminary lengths of the Channel Tunnel illustrate the point. The English Priestley machine is a development of the strong native tradition of clay tunnelling, adapted to the similar, but harder, chalk. The US-made Robbins machine on the French side is a development of the rock-cutting mole, equally adapted for the softer chalk.

This family of machines has as its prototype Beaumont's tunnelling machine which was used in 1880–3 to drive two trial lengths of the Channel Tunnel, one in England and one in France; it was then used in 1883–6 to drive the ventilation tunnel of the Mersey Rail Tunnel. Beaumont's machine consisted of a rotary cutting head 2·1 m (7 ft) in diameter, driven through reduction gearing by a steam-engine, and with a muck-conveyor beneath the drive.

ROCK-TUNNELLING TRIBULATIONS

The Simplon Tunnel gave experience of the great heat that is encountered inside a mountain, and of the tremendous thrusts exerted by the rock. The rock itself may be under stress. The tunneller may thus have to face unexpectedly large crushing forces; he may also run into decomposed rock (one of the worst sorts of soft ground) or water which may be perhaps hot or under pressure.

At Malgovert Tunnel in the French Alps (constructed in 1938–51, 15·3 km (9·5 miles) long) such conditions were met. In one place the line of the tunnel had to be diverted round a particularly bad region which it was not possible to excavate through. Decomposed rock of silt-like consistence submerged the face, travelling into the tunnel at a speed as fast as a man could run.

High temperatures and extreme pressures are expected in driving the Isère–Arc Tunnel, again in France, listed in table 26. The overburden in this case is about equal to that at Simplon.

Far right: Tunnelling mole at Oahe Dam, Missouri in 1959. *Photo: US Corps of Engineers*

Below: Constructing the Victoria underground railway line in London, in 1965. A drum digger shield is being hauled through the station tunnel at Green Park after having driven the running tunnel from Victoria, and prior to driving the running tunnel to Oxford Circus. *Photo: London Transport*

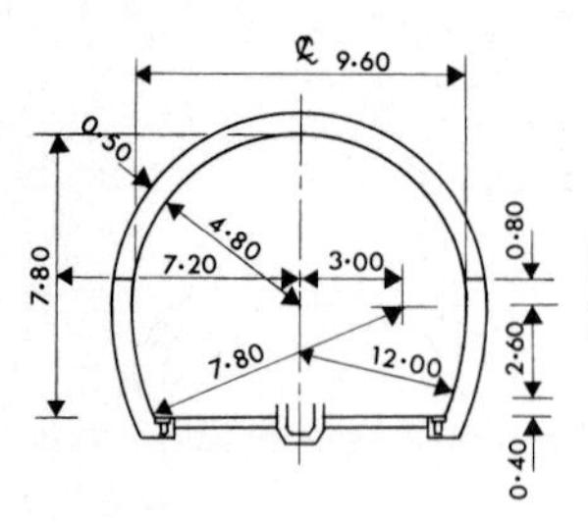

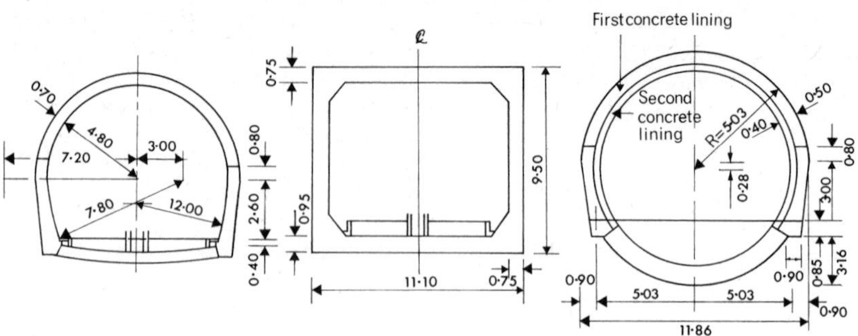

The Shin–Kanmon Tunnel forms part of an extension of the Shinkansen (Japan's very-high-speed railway) 400 km (250 miles) in length from Okayama to Hakata, remarkable for the fact that for 221 km (137 miles) – that is 56 per cent of its length – it is in tunnel. The various types of construction are shown; on the right is the 'problem' length, where the tunnel passed through a fault zone under the sea

The Shin–Kanmon Rail Tunnel crosses a fault zone of clayey, shattered rock about 30 m (100 ft) in width, at a point where the cover of ground between the tunnel and the sea-bed is a minimum – 30 m (100 ft). The tunnel was stopped with 5 m (16·5 ft) of sound rock separating it from the fault zone, and widened out to form a working chamber. Pipes were inserted across the zone and grout was injected. A small pilot tunnel was cautiously excavated, and enlarged, bit by bit. Three months' work was needed, from installing the first grout pipe to completing the tunnel across the fault.

During exploratory works for the Seikan Tunnel, several major faults were identified and studied, and regions of heavy inflow of water, some salt, some fresh, were discovered. In one case, 90 m (295 ft) up from the bottom of the Honshu shaft, a fault designated 'F15' discharged water at a pressure of 26 kg/cm² (nearly 350 lb/in²). This happened in 1969; water flowed in at a rate of 11 t/min (nearly 2 m³/s or about 65 ft³/s) and flooded the shaft to a depth of about 150 m (nearly 500 ft). Work at the face was stopped almost 50 times when conditions of that kind were encountered, and grouting had to be done. Igneous dikes proved to be the most difficult regions to excavate, and as a result of the exploratory work, regions with many dikes were avoided. Advance borings locate inflows of water; grouting in advance then seals the surrounding rock before the tunnel is excavated.

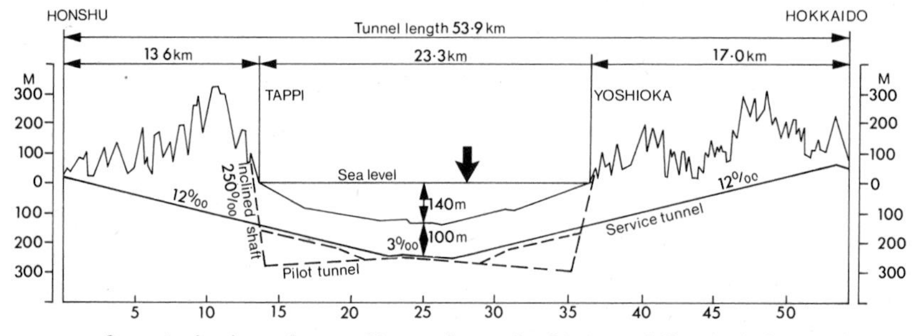

Seikan Tunnel. Work on the main tunnel started in 1972, but was preceded by elaborate exploration which started in 1946. A double-track tunnel, with a service tunnel running parallel to it, are being constructed. The external diameter of the main tunnel is 11·3 m (37 ft). Slopes have been kept flat, to give a track of the standard required for Japan's very-high-speed trunk railway. The route of the tunnel lies mainly through igneous rocks, especially on the Honshu side, but also through sedimentary rocks of the Miocene period

Construction of the Seikan Tunnel.
The upper half-section and a heading in the bottom half

Far right: The upper half-section of the main tunnel constructed as the first stage

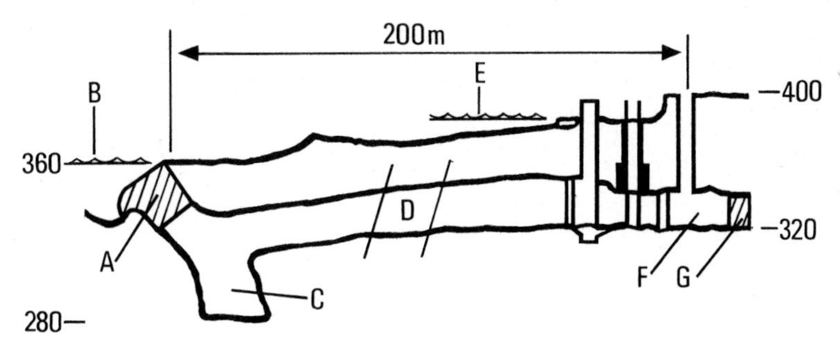

The plug at Chute des Passes.

A Plug
B Reservoir level at blast
C Sump
D Lined part of tunnel
E Maximum reservoir level
F Main tunnel
G Temporary bulkhead

Permanent control is by the gates to the left of G. After the blast, the gates can be lowered and tested, the tunnel downstream of the gates dewatered, and the bulkhead removed. The portion to the left of the gates remains permanently submerged

OPERATION BATH-PLUG

An existing lake may be used as the reservoir of a hydro-electric scheme. If a tunnel is driven to lead to or from it, a subaqueous connection has to be made. This has been rather a rare feat, and depends on very precise work. A sump is excavated at the end of the tunnel, as close as possible to the bottom of the lake. The plug of rock between tunnel and lake is then blown; the sump is made big enough to take the debris from the plug.

At the Folgefonna project in Norway, construction of which started in 1973, there will be 40 km (25 miles) of tunnels, and ten subaqueous connections into lakes will be made. The most difficult one will be made at a depth of water of 60 m (197 ft). The number and depths of these operations are unprecedented.

The largest connection of this kind was made at the reservoir of the Chute des Passes scheme in the Province of Quebec, Canada, in 1962. It was complicated by the fact that the dam was only 200 m (650 ft) from the plug. The sketch shows the layout of the tunnel. The plug had a volume of about 10 000 m³ (13 000 yd³) and was cylindrical, 21 m (69 ft) long and 18 m (59 ft) in diameter. The intake tunnel is 25 m (82 ft) high in the central section. The plug was successfully blown with 27 t of high explosive.

RECORD RATES OF TUNNELLING IN HARD GROUND

The record length of tunnel driven in one day is **127·7 m (419 ft)** at the **Oso Tunnel** of the San Juan–Chama irrigation scheme in Colorado, USA, in 1967.

Second is the rate of advance claimed at **Baidayevskiye Colliery,** Siberia, USSR, in 1958 – 154 m (404 ft).

Mole used to dig the Oso Tunnel, the record holder for speed of advance. *Photo: The Robbins Company*

Table 31. Weekly rates of advance for tunnels in rock.
This table is based on a table in W T Halcrow's (later Sir William Halcrow) Thomas Hawksley Lecture given in 1941, with additions for later dates.

Tunnel	Year of completion	Length km	Length miles	Approx. area m²	Approx. area ft²	No. of working faces	Progress per face per week m	Progress per face per week ft
Average, European double-track railway tunnels driven with hand-drills and black powder	pre-1870	–	–	–	–	–	2–3·8	6·5–12·5
Average, US tunnels hand-drilled	pre-1870	–	–	–	–	–	2·7–5	9–16
Mont Cenis, hand-drilling	1870	12·9	8	49	530	2	3·7	12
Mont Cenis, power-drilling	1870	12·9	8	49	530	2	9·8	32
Hoosac, USA	1873	7·2	4·5	24	259	4	5–9	16–30
Simplon, Switzerland	1906	20	12·5	23	250	2	29	96
Shandaken, USA	1922	29	18	10	110	12	28	91
Florence Lake, USA	1925	21	13	20	211	6	38	125
Lochaber, Scotland	1930	24	15	17	180	22	28 (max.)	91
Garry, Scotland	1937	8	5	6	62	2	37	121
Yewbarrow, Yorkshire	1931	1·3	0·8	7·5	81	2	39	127
Colorado Aqueduct tunnels	1938	109 (12 tunnels)	68	23	250	–	36–53 (dry) 12·5–46 (wet)	118–173 41–150
Delaware Aqueduct	1940	–	–	25	270	–	36·6 (hard rock) 76 (shale)	120 250
Cripple Creek, Colorado, Carlton	1939	9·7	6	7	72	1	104 133*	340 437*
Kemano, BC, Canada	1953	16	10	46	490	4	69†	226†
Oahe, USA (mole)	1954	–	–	78	840	1	193 (max.)	635
Allt na Lairige, Scotland	1955	2	1·25	4	45	1	135 (max.) 63 (average)	444 207
Eucumbene–Tumut, Australia	1955	22·5	14	42	450	–	147 (max.) 70 (average)	484 231

Table 31. cont.

Tunnel	Year of completion	Length km	Length miles	Approx. area m²	Approx. area ft²	No. of working faces	Progress per face per week m	Progress per face per week ft
Breadalbane, Scotland	1956	7·2	4·5	5·5	60	2	170 (max.)	557
Saskatchewan River Dam Canada (mole)	1962	–	–	78	840	1	203 (max.)	666
Poatina, Tasmania (mole)	1963	–	–	19	200	1	229 (max.)	751
Awe, Scotland	1965	–	–	4·5	50	–	171 (max.)	560
Azotea, New Mexico, USA (mole)	1965	–	–	11	120	1	286‡ (max.)	940‡
Navajo No. 1, USA (mole)	1967	3	1·9	23	250	1	85§	280§
Blanco, Colorado, USA (mole)	1967	13·4	8·3	5·5	57	1	533 (max.)	1748
Oso, Colorado, USA (mole)	1969	8	5	5·5	57	1	580·6 (max.) 147§	1905 482§

* Average from 1879 ft (573 m) in one month.
† Average over twelve weeks.
‡ Average from 4077 ft (1243 m) in best month.
§ Average from start to finish of complete tunnel drive.

Oso Tunnel also holds the record weekly rate of progress (see table 31) and the record monthly rate – 2087 m (6849 ft).

Blanco Tunnel (also in table 31) is second, its monthly maximum being 2046 m (6713 ft).

The fastest rate of advance sustained for the entire drive of a long tunnel is 21 m (69 ft) per day for the Oso Tunnel. Other notable examples are 11·7 m (38·5 ft) per day over 22·3 km (13·9 miles) by a US contractor on the Eucumbene–Tumut Tunnel of the Snowy Mountains scheme in Australia, completed in 1959. This rate was exceeded in the much shorter Number One Tunnel of the Navajo irrigation scheme in New Mexico, USA – 12·2 m (40 ft) per day over 3072 m (10079 ft), i.e. driving completed in nine months. Over an eight-month period a daily rate of 12·9 m (42·5 ft) was maintained in 1959, at Bingham Canyon, Utah; and 13·5 m (44·5 ft) per day for four months for the 1629 m (5345 ft) Starvation Tunnel, Utah, in 1967.

The fastest rate of excavation – that is, cubature of rock removed – in a hard-ground tunnel, was achieved at **Mangla Dam** in Pakistan in 1963 and 1964, when 2982 m³ (3900 yd³) was removed in one day when excavating the diversion tunnels. This quantity compares with 1079 m³ (1411 yd³) excavated in one day at one face of the great tunnel at Stornoforrs, Sweden.

The record British advance in hard ground was achieved at **Sherwood Colliery,** Mansfield, 207·3 m (226·7 yd) of access tunnel to a coal face driven in a week (6–11 October 1975).

Rotary head and cutters of the mole at Mangla Dam, Pakistan. This is the mole of largest diameter so far, but a mole for the Paris Métro, of 10·4 m (34 ft) diameter, is the heaviest, weighing 500 short tons. *Photo: The Robbins Company*

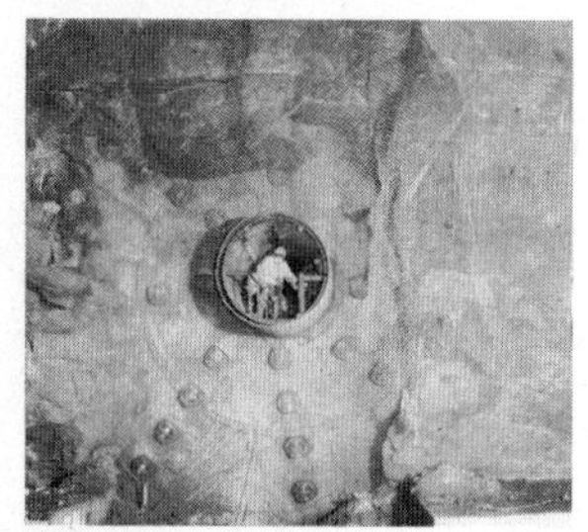

The same mole, after modification, shown at the breakthrough of the second Mersey Tunnel (the first of twin 10·4 m – 34 ft – tubes) in March 1970. The snout of the mole projected into the pilot tunnel; note the changed disposition of the cutters. The two blocks of concrete will keep the mole upright when it has cleared the rock

RECORD RATE IN SOFT GROUND

The maximum length of tunnel driven in soft ground (i.e. shield-driven and lined) in one week is 434·4 m (1425 ft) on the Ely Ouse Tunnel in England in 1969.

PIPE-JACKING

Pipe-jacking is a technique that was evolved to avoid the expense and hazard of driving a small tunnel. For example, if a sewer has to be built beneath a railway embankment, jacks are installed on one side, and a pipe with a cutting edge is pushed into the bank by the jacks; pipes are added progressively at the jacking chamber, until it has been pushed through to the other side. This technique – also known as thrust boring – has been developed enormously. At one extreme it reduces to a torpedo-like device that punches itself through the ground, pulling a cable behind it. In its normal mode it is used to build sewers in busy streets, simply from two small pits, one at each end, that are hardly noticed. At its largest, a complete road tunnel is pushed through the ground.

The largest cross-section thrust by this technique is the North Road Tunnel at Brent Cross Interchange in north London, built in 1974. Three precast concrete elements were cast in line, on a substantial reinforced concrete jacking base. Their cross-section (width × height) was 9·5 × 7 m (31 × 23 ft) and their lengths respectively 11, 16·8 and 18·3 m (36, 55 and 60 ft).

A steel shield was built at the front of the leading element, within which excavation was carried out. Drag sheets were paid out along the roof of the tunnel and bentonite was injected between the roof and the drag sheet to lubricate the slide. Twenty-eight jacks, each of 100 t capacity, shoved the elements into place. The longest thrust for a large cross-section was at Durban, South Africa, in the Albert Park area. A rectangular culvert about 8 m (26 ft) wide and 4·5 m (15 ft) high, is being jacked over a total length of 160 m (525 ft) in three separate drives of 51, 52 and 57 m (167, 171 and 187 ft) respectively. The first drive was beneath railway tracks with only 0·6 m (2 ft) of cover; the railway track was supported on rollers on the roof of the culvert.

Large-diameter pipe can be successfully jacked. For example, a pipe 3·66 m (12 ft) in diameter was jacked 35 m (114 ft) at Cumberland Basin, Bristol, in 1966.

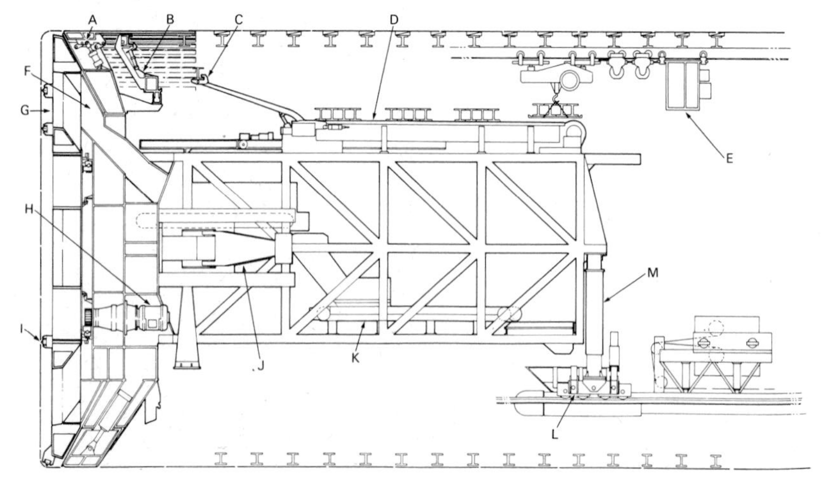

The mole used to drive the tunnels at Mangla Dam, Pakistan. Its diameter is 11·2 m (36 ft 8 in). It was later modified to drive the second Mersey Tunnel.

A Roof support shield
B Ring beam jig
C Ring beam positioner
D Ring beam conveyor
E Monorail trolley (for operator handling ring beams)
F Bucket
G Cutter head
H Gear motor
I Cutter
J Propulsion unit
K Jumbo conveyor
L Tunnel conveyor
M Vertical steering

The maker's own (US) terminology has been used

The storm-water culvert at Durban, described in the text. The first element has been jacked under the railway a distance of 25 m (82 ft) – note that trains are running – and the second has been cast, but has still to be jacked sideways into line on the thrust base, before the main thrust is continued. *Photo: Cementation Group*

The longest distance that has been constructed by thrusting from a single anchorage point appears to be 487 m (1600 ft) achieved with 0·9 m (3 ft) diameter pipe beneath a main road (I 35W) near Burnsville, 25 miles (40 km) south of Minneapolis in Minnesota, USA. The claim has to be qualified because intermediate jacking points (seven to thirteen of them) were inserted in the pipeline and it was moved forward in the manner that an earthworm moves, i.e. the front section was pushed forward; then the next section while the front section remained stationary and the jacks between the two took up the movement; then the next section.

This pipe 3·66 m (12 ft) in diameter was jacked 35 m (114 ft) at Cumberland Basin, Bristol, in 1966. *Photo: Cementation Group*

Southern entrance to the Grand St Bernhard Road tunnel between Switzerland and Italy

ODDEST TUNNELLING STORY

Erich von Däniken, in his book *Gold of the Gods*, refers to various extensive tunnels and underground excavations of mysterious origin, but he does not say very much about them in a factual, descriptive way. His theory is that they were built by a prehistoric race of astronauts who had lost a cosmic war and so needed secret shelter. He gives these facts:

● Tunnels in Ecuador. 'A gigantic system of tunnels thousands of miles in length. Hundreds of miles of underground passages have already been explored and measured in Ecuador and Peru.' Von Däniken visited these tunnels and descended by rope 230 m (750 ft) below the surface into passages with smooth walls and flat ceilings, perfectly rectangular, sometimes with seemingly polished walls or glazed ceilings. A hall measures 153 × 164 yd (140 × 150 m) in plan.

● Tunnels in Peru. An expedition in 1971 explored caves 6700 m (22 000 ft) up in the Andes discovered by Pizarro. At the end of the caves they found watertight doors made of gigantic slabs of rock. The doors pivoted on water-lubricated stone balls, to give access to vast tunnels leading straight to the coast, with a slope of 14 per cent and with anti-slip floors. These tunnels are 90 to 105 km (55 to 65 miles) long (exceeded in length, therefore, only by the Delaware Aqueduct Tunnel in table 25) and descended to a point 24 m (80 ft) below sea-level, or perhaps extend to the island of Guanape.

From the Alps to New York City. Entrance to the Holland Tunnel. *Photo: Port of New York Authority*

New Jersey approach to the Lincoln Tunnel, New York City. This photo was taken in 1970, on the first day when some lanes were used exclusively for buses. *Photo: Port of New York Authority*

● India, Kanheri, near the Malabar coast, Bombay: 87 caves, 15 m (48 ft) high 'blasted out of natural stone, mostly granite, ... this titanic architectural feat'. Also 150 caves at Junnar on the Deccan plateau and other examples.*

● Caroline Islands, Temuen. The ruins of Nan Madol. An extensive city with buildings and walls, made largely of octagonal columns of basalt laid on their sides, 3–9 m (10–29 ft) long and weighing up to 10 t each. Von Däniken estimates 4 million of these columns in 80 minor buildings alone. The city extends below the sea. It has a regular plan, which is centred on a 'well', but von Däniken insists it is the end of a tunnel.

KANATS

Tunnels driven for water-supply in Iran and other Middle Eastern countries are called kanats. They consist of a primary shaft, sunk into a hillside aquifer (to confirm the presence of water in the quantity required) then shafts sunk at intervals of 20–50 m (65–165 ft), from which a tunnel is constructed on a falling gradient to terminate as a well in a village or town. If the supply needs to be increased, another shaft is sunk in the aquifer, and connected to the tunnel. It was estimated that 100 000 miles (160 000 km) of kanats were in use in Iran in 1962. Tehran relied on them for its water-supply – eventually it was supplied by 36 kanats – until its population surpassed 300 000. Their construction is an ancient art, and dates back to 2800 BC. A kanat supplying water to Aleppo in Syria is 12 km (7·5 miles) long.

VOLUME OF TUNNELLING

The greatest amount of tunnelling work in any country during the 1960s was carried out in Japan. Next came the United States and Italy. Japan's tunnels are expected to be tripled during the 1970s, and the United States' tunnels to be quadrupled.

*Compare this comment with the Abu Simbel temples – now reconstructed on a new site – which were excavated out of the sandstone rock in 1250 BC and the following from Sandström:

'A thousand years later (i.e. after Abu Simbel) a similar craze for building rock temples began in India, where Brahmins and Buddhists vied with each other in carving out temples and monasteries, until in the end there were about a thousand such rock structures throughout India, mostly in the state of Bombay'.

The greatest amount of tunnelling per head of population is in Norway, Sweden being in second place. In Sweden, 5 m³ (6·5 yd³) of hard rock is fragmented per year per head of its population.

BIG HOLES UNDERGROUND

The rapidity and sureness with which a big tunnel, like the one at Stornoforrs, can be driven – provided the rock is sound – has also encouraged the widespread construction of underground power stations. The largest one is at **Churchill Falls** in the Province of Labrador in Canada.

	Length m	Width m	Height m	Length ft	Width ft	Height ft
the machine hall measures	296	25	47	972	81	154
the transformer hall (separate because of fire risk) measures	261	15	12	856	50	39
the surge chamber (for water to swash when the flow changes abruptly) measures	234	12–19	45	763	40–64	148

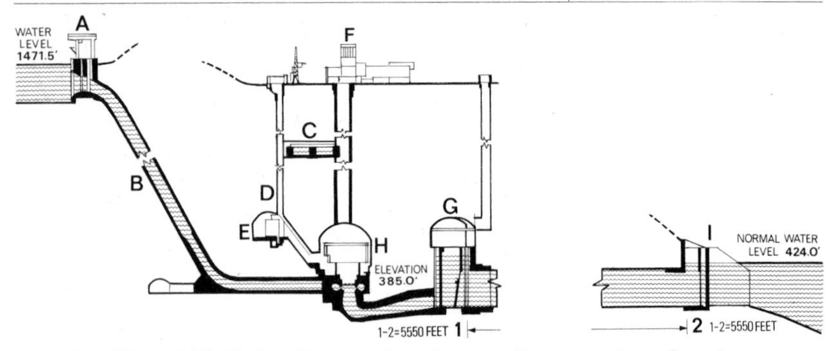

Section through Churchill Falls Power Station, Labrador

At Churchill Falls the total volume of excavation for the power station is 1·76 million m³ (2·3 million yd³) not including the main tunnels. The power station is served by two tail-race tunnels, each 1692 m (5550 ft) long (18 × 14 m (60 × 45 ft) in cross-section) and carrying a flow of nearly 1400 m³/s (49 000 ft³/s); eleven penstocks; seven cable shafts; a surge shaft; a ventilation shaft; and an access shaft. Excavation extends to 300 m (984 ft) below ground level; 5 million lb of explosives were used to excavate 5 million short tons (4·5 million t) of rock. Churchill Falls was completed in 1975, but the excavation was finished in 1970. For comparison, Stornoforrs tail-race tunnel has a total volume about equal to the figure quoted above for the Churchill Falls; its dimensions are in table 27.

Again for comparison, the world's largest natural cavern at Carlsbad, New Mexico, USA, is 400 m (1320 ft) deep and has over-all dimensions (length × width × height) of 1300 × 200 × 100 m (4270 × 656 × 328 ft) and if one assumes its volume to be one-third of the rectangular volume given by these dimensions, it is roughly three times bigger than Churchill Falls.

Cabora Bassa Power Station, under construction in Mozambique, will have a machine hall and transformer hall each about three-quarters the size of those at Churchill Falls. Kemano Power Station, British Columbia (the previous record-holder, and completed in 1954) has a machine hall that is longer than the one at Churchill Falls – 344 × 25 × 42 m (1128 × 82 × 138 ft).

The largest underground power station in Europe is **Santa Massenza** Power Station in Italy. The largest in Britain is the pumped storage station at **Foyers,** completed in 1974, which is marginally larger than the **Cruachan** pumped storage station. All three will be overtaken by the much larger underground pumped storage power station now under construction at **Dinorwic** in North Wales.

The first underground power station was built at **Snoqualme Falls** near Seattle, USA, in 1898. The second was built at Fairfax Falls, Vermont, USA, in 1904. A notable prototype in Europe was the Porjus Power Station, built in Sweden, north of the Arctic Circle, in 1910.

Large underground chambers have also been excavated as reservoirs for water or oil, or for reasons of defence. Little information is available in the latter case. Probably the largest underground excavations of all have been created by nuclear explosions. This has been done in Project Gnome and Project Plowshare in the United States, and undoubtedly in similar projects in Russia.

Underground Storage

It seems likely that excavation of cavities for the storage of oil will become an accepted practice. In some cases a storage can be utilised without physically excavating a hole. If a hole is needed, however, a volume in sound rock of about 2·5 million m^3 (3 million yd^3) – rather larger than the total volume of Churchill Falls – is considered a reasonable maximum. At Gothenburg, Sweden, four chambers each measuring 550 × 18 × 30 m (1800 × 60 × 100 ft) are to be excavated to store oil (total volume about 1·2 million m^3 (1·5 million yd^3)), but the total volume of excavation for the scheme is apparently to be over 2 million m^3 (2·5 million yd^3). Construction started early in 1975 and is expected to take three years.

The largest underground factory was at Mittelwerk near Nordhausen, Germany, in which V2 rockets were made during the war. Its floor area was 118 000 m^2 (1·27 million ft^2). That area is exceeded by about ten times by an underground commercial storage area near Kansas City, USA. A limestone stratum 6·7 m (22 ft) thick is exposed in a river valley and is mined with horizontal access, as a source of aggregate. A thickness of 3·9 m (12·67 ft) is extracted, leaving 7·62 m (25 ft) square pillars of limestone about 16 m (50 ft) apart on a square grid to support the roof. As the mine extends, so the area left is taken over for commercial storage. The present gross area is over 120 ha (300 acres) and is growing at about 16 ha (40 acres) per year. These figures correspond to a storage area of over 1 million m^2 (11 million ft^2). The total excavated volume is about 4 million m^3 (5·25 million yd^3).

Excavation by Leaching

An unusual underground excavation is in progress near Hornsea on Humberside on the north-east coast of England. Its purpose is to create storage underground for North Sea gas – enough for 24 hours' consumption for the United Kingdom.

Sea-water is being pumped into a stratum of rock-salt 1650 m (5400 ft) below ground level, to leach out the salt and leave six gas-tight cavities. The cavities will be constructed in pairs, each pair taking about two years to leach out. This is the third project of its type; a similar scheme exists in Canada and one in France. Each cavity will be leached to a height of about 150 m (500 ft) and a maximum diameter of 100 m (330 ft) and its volume will be about 230 000 m^3 (300 000 yd^3). Leaching of the first pair has started.

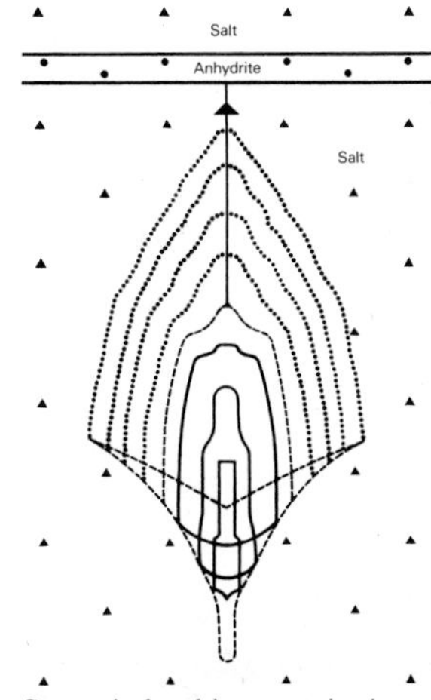

Stages in leaching a cavity in salt at Hornsea, Humberside

Table 32. Tunnels and shafts under the highest pressures.

Name	Head		Cross-sectional area	
	m	ft	m²	ft²
For unlined rock				
Hovatn, Norway	485	1591	12	129
Tafiord 4 shaft, Norway	450	1476	–	–
For lined rock (steel lining with concrete backing) **Shafts**				
La Bathie (Roselend), France 1960	1203	3946·8	7	75
Albigna, Switzerland	1046	3432	1·8	19
Fionnay (Grande Dixence), Switzerland	872	2861	7	75
Tunnels				
Kemano, Canada 1954	841	2760	8·8	95
Lyse, Norway 1954	630	2067	2·8	30
Reisach, Germany 1955 (prestressed concrete)	176	578	18·7	201

Hydro-electric tunnels may, when in service, contain water under pressure. The highest pressures are taken by steel pipes (penstocks) above ground, but high pressures are also sustained by steel-lined shafts and tunnels underground, and in some cases fairly considerable pressures are taken by unlined rock.

MINING TUNNELS

The largest gold-mine measured by the volume of spoil extracted is Randfontein Estates Gold Mine Company Limited's mine in South Africa. The volume extracted is $129 \times 10^6/m^3$ ($170 \times 10^6/yd^3$) – nearly enough to build Tarbela Dam. The total length of the tunnels in this mine is 4184 km (2600 miles).

The largest underground copper-mine, El Teniente, Chile, has about 320 km (200 miles) of tunnels. The largest underground mining operation is claimed to be the Rustenburg platinum-mines in South Africa.

Sinking the shaft at Kloof Gold Mine, about 70 km (45 miles) from Johannesburg, South Africa. Its excavated diameter was 10·36 m (34 ft) and its finished, lined diameter 9·6 m (31 ft 6 in) and it is the world's largest shaft. It is 2042 m (6700 ft) deep. *Photo: Cementation Group*

DEEPEST MINE

The world's deepest mine is the Western Deep Levels gold-mine at Carltonville, South Africa. The depth reached was 3840 m (12 600 ft or 2·38 miles) in May 1975.

The world's longest vertical shaft is at this mine. No. 3 ventilation shaft is 2948 m (9673 ft) long.

SHAFT-SINKING

The fastest rate of shaft-sinking was at Buffelsfontein Eastern upcast shaft (South Africa), which, in March 1962, was sunk 381 m (1250 ft); the shaft has a diameter of 6·7 m (22 ft).

The deepest shaft in Britain is at Wolstanton Colliery and its depth is 1070 m (3434 ft).

excavation and dredging

Aerial view of Bingham Mine from directly overhead

LARGEST EXCAVATIONS

Muck-shifting is the starting-point of civil engineering. Yet records tend to be held in mining.

The world's largest excavation is Bingham Canyon copper-mine, near Salt Lake City, Utah, USA. Since 1904 when the mine was opened, over 3500 million short tons (equal to about 1200 million m³ or 1570 million yd³ assuming a specific gravity of 2·65) has been excavated from it.

The life of the mine is confidential, but over 50 years of working apparently remains, so excavation could continue to perhaps three times the present figure. The mine is 3·6 km (2·25 miles) across and 759 m (2490 ft) deep. Its area is 729 ha (1800 acres) and it has 55 benches.

Ore and overburden are excavated by face shovels and hauled by dump trucks or by rail. There are 39 face shovels at work, with bucket sizes from 11·5 to 4·5 m³ (15 to 6 yd³); 97 dump trucks of 65 to 150 short tons capacity; 18 rotary drills; 62 locomotives working on 160 km (100 miles) of standard-gauge track including four entries (three in tunnel) to the mine. All this plant is to give an output, on average, of 433 000 short tons (150 000 m³ or 200 000 yd³) per day, according to the 1975 production plan.

The greatest quantity that has been handled at the mine in a single 24-hour period is 504 167 short tons (say 173 000 m³ or 226 000 yd³) on 13 October 1974.

Bingham Canyon will soon be overtaken by the largest opencast mine in West Germany, where lignite (brown coal) is won. This mine is the Fortuna–Garsdorf mine near Bergheim. The first of the really large bucket-

Bingham Canyon Mine in the Oquirrh Mountains, west of Salt Lake City, Utah

Prins der Nederlanden, the world's most productive muck-shifting tool. Her installed power is 16 MW (21 500 hp), she is 143 m (470 ft) long, and can dredge to a depth of 35 m (115 ft) through twin 1·2 m (4 ft) diameter suction tubes. She is the flagship of the fleet of the Westminster Dredging Group

wheel excavators was put to work at this mine in 1955; it has become the largest and most modern opencast mine in the world, and the proving ground for the dramatic success of the large bucket-wheel machines. In 1974, the output of the Fortuna–Garsdorf mine was 41·5 million t of brown coal (say 18 million m^3 – 23·5 million yd^3) and 76·17 million m^3 (99·6 million yd^3) of overburden, a total quantity of 94 million m^3 (about 123 million yd^3). These figures give an average output per working day of 289 750 m^3 (378 960 yd^3) – around double the average planned for Bingham Canyon for 1975. The excavated cut is 4 km wide by 3 km long (2·5 by 2 miles) by 250 m (820 ft) deep at its deepest part. This one mine produced in 1974 about one-third of the output of the West German brown coal industry.

MOST PRODUCTIVE TOOLS

The world's most productive muck-shifting tool is the trailer-suction dredger, *Prins der Nederlanden*. She can dredge about 20 000 t of sand (9000 m^3 or 11 770 yd^3) from the sea-bed in 30–60 mins., and, after travelling to the dumping-ground, deposit the same quantity in 5 minutes. The rated hourly capacity of the ship's two dredging pumps is 22 000 m^3 (28 775 yd^3) against a head of 22 m (72 ft).

During the first year of its life this vessel dredged, on average, 1 million m^3 (1 300 000 yd^3) of sand every week.

Below: Bucket-wheel excavator with a daily output of 110 000 bank m^3 (i.e. volume measured *in situ* – 144 000 yd^3) mining lignite in West Germany. Its maximum digging height is 44 m (145 ft); digging depth 12 m (40 ft); the total length of the machine is about 200 m (say 650 ft); its operating weight is 7760 t and total power installed 8·6 MW. Ten machines of more than 5000 m^3 per hour output have been built and are in use

Far right: Bucket wheel of the excavator shown in the previous illustration. Its diameter is 17·2 m (56·4 ft) and it has ten buckets each of 4·5 m^3 (5·9 yd^3) capacity; it turns at 3·45 revolutions per minute and is driven by three 600 kW motors

The largest bucket-wheel excavator in use has an output of 110 000 m^3 (144 000 yd^3) per day. In this case, the volume of material is measured *in situ*, and a day means twenty hours at work.

The mean hourly output of this machine is thus 5500 m^3 (7193 yd^3) and its weekly output, assuming six working days, is 660 000 m^3 (863 000 yd^3), compared with the 1 million m^3 of the *Prins der Nederlanden*.

The measured output of the 4250 W dragline – the world's largest land machine – is, at peak, 6782 m^3 (8870 yd^3) per hour, on average 4130–4600 m^3 (5400–6020 yd^3) per operating hour. The output of the 4250 W over a period of days is thus probably a little less than that of the bucket-wheel machine. In practice, it is used to rehandle the muck excavated by other machines, so no exact measurements are made.

Base section of very large bucket-wheel excavator shown under construction. This particular machine will have an output of 200 000 m^3 per day. *Bucket-wheel excavator photos: Orenstein und Koppel AG*

A bucket-wheel excavator is under construction that will have a guaranteed output of 240 000 m^3 (almost 314 000 yd^3) per day, more than double the figures just discussed; its productivity will comfortably surpass that of the big dredger.

The new bucket-wheel excavator will have a length of about 210 m (about 690 ft), a height of about 82 m (about 270 ft) and the total weight of 13 000 t. Surprisingly enough the weight of this great machine will be only about 10 per cent or so more than that of the 4250 W. It will be used at the proposed opencast lignite-mine of Hambach, near Jülich in West Germany. The intention is to remove some 200–300 m (650–1000 ft) of overburden to expose a stratum of brown coal 100 m (330 ft) thick. A 200 000 m^3/day (260 000 yd^3/day) machine is to be delivered in 1976. Its wheel will be 21·6 m (71 ft) in diameter with 5 m^3 (6·5 yd^3) buckets and driving power of 2·5 MW.

GREATEST QUANTITY FOR A SINGLE PROJECT

The greatest quantity of excavation for a single civil engineering project is that for the flood-protection works, mainly embankments or levees as they are called, on the Lower Mississippi River. The most recent figure (1975) for the total quantity is 1150 million m^3 (1500 million yd^3).

GREATEST QUANTITY IN A SINGLE STRUCTURE

The greatest quantity of material in a single structure of the modern epoch is 142 million m^3 (186 million yd^3) – the bulk of the Tarbela Dam in Pakistan. Tarbela Dam's volume probably exceeds that of the Great Wall of China. Assuming that the wall is – or was – 2000 miles (3200 km) long, 10 yd (9 m) high and 7 yd (6·4 m) wide, its volume would be about 250 million yd^3 (200 million m^3), but it never existed apparently with that cross-section for its entire length.

New Cornelia mine tailings dam near Ajo, Arizona, USA, when completed in 1973 had a volume of 209·5 million m^3 (274 million yd^3). 'Dam' is a courtesy title, and 'spoil heap' might be more appropriate.

FASTEST RATES

The fastest rate of excavation **on record** is 873 750 m^3 (1 142 777 yd^3) per day, achieved by manual labour and without mechanical plant, on the North Kiangsu Canal in China. Available details are on page 151, from which it can be inferred that it is likely that this record was surpassed in the 6th or 7th century during construction of the Grand Canal (chapter 4).

The 3850B stripping shovel at work, standing on the coal seam, at the River King Mine, Lenzburg, Illinois, USA, of the Peabody Coal Company. The productivity of this machine is about three-quarters that of the big dragline. Its bucket has a capacity of 107 m³ (140 yd³); its boom is 61 m (200 ft) long; and its crowd arm (also called dipper arm or handle) is 38 m (125 ft) long. It can dump at a radius of 61 m (200 ft), and at a height of 41 m (135 ft) at a slightly smaller radius. It travels on four sets of crawler tracks, and its revolving frame stands 21 m (68 ft) above the ground. It is electrically powered, with four 3000 hp (2·25 MW) motor-generators and with eight hoist, six swing and two crowd motors

Top right: Bucket of the 3850B. Present opinion is that, as the unquenchable hunger of today's society for resources of fuels and minerals grows, a vast expansion of opencast mining will be needed. The start of work to exploit the oil shales of the Rocky Mountains in the USA and the Athabasca tar sands in Canada are examples of the trend. It is reasonable, then, to expect that in a few decades time, the enormous excavating machines described here will, in their turn, have been dwarfed by even more gigantic and mightier machines

The fastest **sustained rate** of excavation on record is at the Fortuna–Garsdorf mine – 289 750 m³ (378 960 yd³) per working day average throughout 1974.

The record for daily rate of excavation in the West is held by the Lower Mississippi flood-prevention scheme. When both graders and dredges were working in 1974, 391 000 m³ (512 000 yd³) of excavation was carried out in a day. The work is seasonal, but the August or September figures could easily exceed Tarbela's monthly maximum. The following information from Vicksburg explains the figures:

'The greatest rate of excavation' is divided between the revetment graders and the government owned and leased dredges of The Lower Mississippi Valley Division of the Corps of Engineers. During the work season of 1974, the graders of both Memphis and Vicksburg mobilized on 5 August with the Vicksburg grader completing all grading on 10 December. This approximates 114 working days, two shifts per day, each plant. There were 12 869 539 yd³ excavated which computes to approximately 112 000 yd³ per day. If these figures are exceeded, it can or has only been done by a hydraulic dredge. On any given day, however, a grader can easily excavate 50 000 yd³.

'The bank graders are barge mounted draglines carrying a 15 yd drag bucket on a 165 ft boom. The four machines are either steam or diesel-electric operated.

'On any given day during the low water season (July through October) when the four government dustpan dredges and three privately owned contract dredges are operating between Baton Rouge, La., and Cairo, Ill., 400 000 yd³ of material are moved by hydraulic dredges. The government dustpan dredges are selfpropelled 32 in discharge pipeline dredges. The discharge size of the contract dredges range from 27 to 30 in.'

The fastest rate of muck-shifting **from start to finish of a single project** is 4·56 million m³ (6 million yd³) per month, the average rate for the

Fairland sheaves on Big Muskie, and (*below*) a ninety-piece high-school band just fitting inside the bucket

construction of Tarbela Dam. This is equivalent to about 6000 m³ (8000 yd³) per hour continuously.

The maximum rates of placing material in Tarbela Dam were:

	m³	yd³
peak daily rate	230 000	301 000
peak monthly rate	5 010 000	6 550 000

achieved in April and July 1972.

About two-thirds of the total was borrow material, and about one-third spoil from excavation necessary for construction. This operation is more elaborate than mining. The fill material in the dam is graded, and is placed in layers and compacted, and there is strict quality control. The plant installed must excavate, crush, screen, blend, correct water content, and convey to site. At site, the layers must be spread and compacted. Process plants with a rated capacity of more than 30 000 short tons per hour were installed. Some material was won from a source 5 km (3 miles) from the dam, for which two conveyor-belts were installed.

The fastest rate of progress on a **marine dredging** job is 18 000 t (from 7000 to 8800 m³) per hour continuously achieved during the construction of the new entrance to Europoort, Rotterdam.

This figure compares with 16 000 t per hour, the rated capacity of each of the two loaders at Tubarao Bulk-loading Terminal (chapter 3), but is only about two-thirds of the daily rate quoted for the Mississippi dredges.

LAND-BASED EXCAVATING PLANT

William Otis built the first steam-shovel in the United States in the 1830s. It was a rail-mounted half-circle machine made of iron and wood; the first machine worked on railways in Maryland and in Canada, and was later, in 1905, used as a dredger, so it had an active life of 70 years. Otis's second and third machines were exported to Russia and the fourth to England in 1840.

By the 1880s, steam-navvies were being made by various companies in

The world's largest land machine, Big Muskie, the 4250W dragline, at work at Muskingum Mine, Eastern Ohio, USA, of the Central Ohio Coal Company. In a working life of, say, 25 years, this machine will move about 1000 million m³ (say 1350 million yd³) of muck.

Principal statistics are:

Bucket size	168 m³	220 yd³
Working weight	11 933 t	13 150 short tons
Length of boom	94·5 m	310 ft
Maximum digging depth	56·4 m	185 ft
Maximum suspended load	499 t	550 short tons
Operating radius	92 m	302 ft
Maximum dumping height	39·3 m	129 ft
Diameter of base	32 m	105 ft
Length of walking shoes	39·6 m	130 ft
Length of step	4·3 m	14 ft

Six 5000 hp (3·7 MW, i.e. about 22 MW total) motor-generator sets convert the incoming ac power to dc to supply 10 hoist, 8 drag and 10 swing motors.

Undoubtedly representative of one of the greatest feats – yet attracting little recognition. The spillway at Mangla Dam in Pakistan where a million cubic feet of water (the conversions and the exact figures are in table 43) can be discharged per second, dissipating about 40 million hp as it flows down the spillway. Even more remarkable, it is built on soft silt, where a fraction of such a titanic discharge flowing freely would happily excavate enormous craters in next to no time. The spillway is in two stages, elaborately contrived to dissipate energy in the discharge. *Photo: Harza Engineering*

Britain and the United States. The tool came of age with the construction of the Manchester Ship Canal in England, and the Chicago Drainage Canal and Panama Canal for the United States. Ninety-seven machines were in use at one time on the project, according to the records of the Manchester Ship Canal Company. One of them filled 640 wagons in twelve hours – that is shifted 2500 yd³ (1900 m³) of sand in one working day.

A similar concentration of machines excavated the Panama Canal, but they included machines with 5 yd³ (4 m³) buckets, an unprecedented size (highest output per machine per day: 4465 yd³ (3414 m³); per year: 543 481 yd³ (415 500 m³)). These rail-mounted half-circle machines were so robust that they survived burial in rock-slides. The last of them was retired from a Spanish mine in the 1950s.

Twin-powered scrapers and bulldozers at Oahe

Face shovels loading big bottom-dump trucks at Oahe, dragline in the background

The first dragline was introduced in the United States in 1884. Improvement was continuous – full-circle machines, crawler tracks, diesel power, walking machines, and always increasing size.

In 1944 a new record of size was set with the 1150B Bucyrus-Erie dragline with a bucket of 19 m³ (25 yd³) capacity and a jib 55 m (180 ft) long. This machine excavated at rates up to a maximum of about 1000 m³ (1300 yd³) per working hour, having a dumping height of 21 m (70 ft) and an operating radius of 53 m (174 ft). The world's largest machines today are the 4250 W dragline which has a bucket that holds 168 m³ (220 yd³) and the 3850B stripping shovel (bucket – 107 m³ or 140 yd³).

The Clark 675 tractor shovel is the largest wheeled loader in production. It stands 6·5 m (21·3 ft) high and is equipped with an 18·3 m³ (24 yd³) bucket. It is powered by twin turbocharged diesel engines each delivering 593 flywheel hp (444 kW each) to torque converters, and thence to all-wheel-drive axles with reduction gears in each wheel

Three generations of dump trucks:

Top left: 1930s. Excavating gravel for the toe of Fort Peck Dam

Top right: 1950s. Truck of 61 m^3 (80 yd^3) capacity purpose-built for Oahe Dam, and of record-breaking size (1958)

Left: 1970s. The world's largest today, the Terrex Titan. Its carrying capacity is 350 short tons (317 t), the struck volume it will carry being 115 m^3 (150 yd^3) and the heaped volume (heaped at 3:1) 151 m^3 (198 yd^3). Its own, net, weight is 255 short tons (231 t); it is powered by a diesel engine of 3300 gross hp (almost 2·5 MW) at 900 rpm, with electrical transmission, and, according to its makers, it 'operates more like a luxury car than the largest off-highway hauler'. The tyres are 3·35 m (11 ft) in diameter. *Fort Peck and Oahe photos: US Corps of Engineers; Titan: Blackwood Hodge*

A stripping shovel is a face shovel with an extra-long boom and crowd arm, intended for stripping overburden at an opencast mine. It simply digs the muck off the top of the coal-seam and dumps it on the other side of the cut, where the coal has been won, and it must have a reach long enough to do that.

A dragline stands above the excavation and pulls its bucket in and up as it digs, in contrast to the face shovel which digs at its own level. A dragline has a longer reach, other things being equal, than a face shovel. The largest of these machines at work in Britain is the dragline Big Geordie (50 m^3 or 65 yd^3 bucket) at Sisters opencast mine in County Durham. The largest ever built in Britain is the Ransomes and Rapier W1800 dragline – 23 m^3 or 30 yd^3 bucket and 86 m or 282 ft jib – built in 1961. Several similar machines were built, and are at work at the ironstone mines, at Corby, Northants. At that time they were the most advanced machines of their kind. After a take-over bid the design team was disbanded and production ceased. One of the features of the history of excavating machines has been the voracity with which competing makers have excavated out their rivals.

Chain-bucket excavators built at Lübeck – precursors of the bucket-wheel machine – excavated more than half of the Kiel Canal between 1887 and 1895, and the first machine for lignite mining went into service in 1890. They were steam-driven, but the first electrically driven machine was in use just before the turn of the century. The first crawler-tracked

The first 1150B dragline, the largest excavating machine in 1944

Photos, 3850B, 4250W and 1150B: Bucyrus-Erie Company

machine dates from the mid 1920s, and the bucket wheel in its modern form from the mid 1930s. In 1939, bucket-wheel excavators, each with a daily output of 10 000 bank m³ were in service. The great depth of overburden that would have to be removed if further lignite were to be won prompted the enormous jump to two prototype machines with outputs of 60 000 and 100 000 bank m³ respectively; the latter went into service in 1955. A machine with a slightly smaller rate of excavation carries its bucket wheel on a 100 m (328 ft) boom, which gives it greater capability than the other machines. A spreading machine with an output matching the excavator (plus belt conveyors) is also needed with each of these machines.

THE FIRST BULLDOZER

'I have it on good authority that no one knows where the name bulldozer comes from, nor who invented it. But that's its name, and I'll put my name in with many others in claiming to have invented it. Up to now, with minor concessions, I've won all my arguments with other claimants to the title. Plowboards, or what we now call dozer blades, were used by the Mormons in the early days of Salt Lake City, and were later used with varying degrees of success with mule power and tractors. All had the same weakness. The blade couldn't be lifted out of the ground for turning or backing.

'Stuck up there in Crow Canyon, my contribution was the first practical blade that could be raised or lowered at will, with push-button

The world's largest bulldozer, the VCON machine of the Marion Power Shovel Co. It has an engine of 1320 hp (almost 1 MW) with electric motor drive in each wheel. It weighs 136 t (300 000 lb)

controls. . . . I put a steel scraper blade out in front of my new Best tractor, rigged it up with an electric cable winch, and wove the cable through some sheaves. A press of a button would lift the blade out of the ground for easy turning or backing up. Another touch of the button would shut off the power, and the blade would dig into the ground of its own weight. It was to change the course of heavy construction history, but all I knew then was that it was bulldozing me out of a tight spot.'

This quotation is from *Mover of Men and Mountains* the autobiography of R G LeTourneau (published by Prentice-Hall Inc.). It refers to the author's first big job as a muck-shifting contractor, building a highway between Oakland and Stockton, California, in the spring of 1926.

Quoting again from the same source:
'The Philbrook Dam was a milestone in the engineering business, and in my life. It was the first major project in which the new broke entirely away from the old. There was not a mule on the site. We were still using some men with shovels and pick-axes for clean-up work, but the heavy labour was done with power shovels, mechanized dump-trucks, and, in the starring role, my scrapers. From the start, it was clear that nothing short of an earthquake would stop us from getting an all-time record in dam building.'

Philbrook Dam is an embankment on the western branch of Feather River in California; it was completed in 1926. The contractor was Henry Kaiser; LeTourneau supplied the scrapers and the key men to operate and maintain them.

The scrapers that were so successful at Philbrook became obsolete when LeTourneau introduced a cable-controlled machine in 1929; then in 1932 he built the first rubber-tyred earth-moving plant, and in 1939, the first of the Tournapull two-wheeled prime movers.

DREDGING

The earliest records of dredging machines are from China. In AD 1073 attempts were made to clear silt in the Yellow River. A river-deepening harrow – an 8 ft (2·4 m) beam with iron spikes 1 ft (0·3 m) long – was worked with windlasses from two ships. Several thousands of these harrows were put into service, but there is no record of how successful they were. Their use was again recommended in 1595.

The greatest depth dredged by a bucket dredger is believed to be 37·5 m (123 ft) below mean sea-level, by the rock bucket dredger Kil in the Beer Canal at Europoort, Holland in 1973, when a trench was being excavated to accommodate a bundle of 12 pipes (see text below, table 49). A dredging depth of 60 m (nearly 200 ft) can be reached by the suction dredger Sliedrecht 27.

Potential of the Dredger

'There has never been a piece of engineering plant which has given the engineer so much power to change the formation of sea and land to the benefit of society generally. Even mountains can now be cut down to size and the material pumped out to sea to form islands as has been done recently in Japan. Areas were thus reclaimed for heavy and unpleasant industry, leaving a plateau upon which new towns were established. The greatest natural barrier to the engineer in using the dredge with its capacity to remove and deposit material at such enormous volumes seems to be the ability to think in big enough terms.' – Bernard L. Clark.

PRODIGIES OF MANUAL EXCAVATION

In his book, *Dredge, drain, reclaim*, Dr van Veen, a former Chief Engineer of the Rijkwaterstaat, refers to Pliny's description of the Frisians of his day. At storm tide, Pliny said, they resembled groups of miserable shipwrecked sailors, marooned in the midst of a waste of water on mounds they themselves had made for protection when abnormal tides occurred. Today many villages in Holland are centred on such mounds. There are 1260 of them in north-eastern Holland, each one 2–16 ha (5–40 acres) in area, sometimes 9 m (30 ft) above mean sea-level and containing up to 1 million yd^3 (0·75 million m^3) of material. The start of this was probably in about 400 BC and is the earliest example of reclamation from the sea. The mounds were raised continuously through the centuries. By AD 1600–1700, reasonable security against flooding had been achieved. The total quantity of clay in these mounds is about 100 million yd^3 (75 million m^3).

By comparison, the quantity of material in the Pyramid of Cheops is 3500000 yd^3 and of Chephren 3000000 yd^3. By 1860 there were 1750 miles of sea-walls or dikes in Holland – all built by manual labour, barrows and horse-drawn carts. They contained about 200 million yd^3 of material. In addition, abandoned dikes contained about 50 million yd^3 of material. For land drainage only, ditches and canals at that date in Holland had necessitated about 800 million yd^3 of excavation. Canals for ships – 4800 miles – had needed about 200 million yd^3 of excavation. Peat had been excavated extensively for use as a fuel and to create lakes – later drained to become fertile land. Excavation of peat totalled about 10000 million yd^3.

Dr van Veen's estimate of the grand total of manual excavation for reclamation works in Holland up to the beginning of the age of mechanical excavation is thus about 1350 million yd^3 (1032 million m^3) (i.e. excluding excavation of peat). These quantities seem to outpace even digging the Grand Canal (see chapter 4) several times over. The average performance of a man excavating, Dr van Veen quotes as 10 yd^3 (7·5 m^3) per day. He records that the record figure is 76 yd^3 (58 m^3) dug by one man in one day.

The Big Hole at Kimberley, South Africa, was dug manually from 1871 to 1914, to a depth of 365 m (1200 ft) and with a diameter of about 460 m (1500 ft). Three tonnes of diamonds were extracted from it. Its total volume is about 250000 m^3 (320000 yd^3).

The greatest feat of manual excavation of the modern era is claimed to be the construction of the North Kiangsu main irrigation canal. This canal is the largest of the canals of the groups of multiple-purpose schemes for harnessing the Huai River; it has a bed width of 126 m (413 ft) and a capacity of 707 m^3/s (25000 ft^3/s). The first stage of this group of schemes was partly completed by 1969, including the Putzeling multiple-arch dam on the Philo River, some 150 km (93 miles) from Hafe, the capital of Anhwei Province. The earthworks of the canal were completed in 80 days without the use of any mechanical plant. The quantity of earthwork was 69·9 million m^3 (91·4 million yd^3). The exact dates, the length of the canal, and the number of people employed are not known. Applying known rates of manual output, it appears that at very least, 120000 strong men must have been at work continuously if straightforward excavation only were involved, but with transport and placing embankments, four times that number appears to be merely a minimum likely number.* This minimum,

*In 1954, to prevent flooding of the city of Wuhan by the Yangtze River, a hill was levelled in 100 days – output 3·5 million m^3 (4·7 million yd^3); labour force 200000. At this rate of working the labour force on the North Kiangsu Canal would have been about 4 million people.

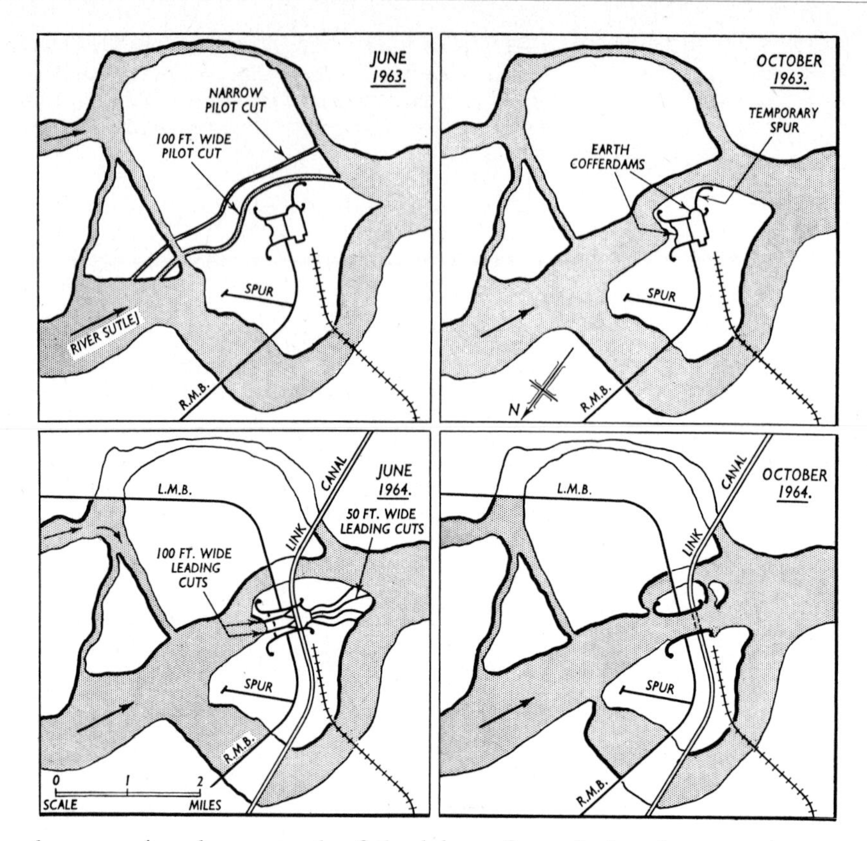

Using the Sutlej River to excavate, by its own power of scouring, in the construction of Mailsi Siphon. This sequence of four diagrams shows the siphon, with its training walls, built in the dry, and how the river excavated the land around it and finally was made to pass through it

however, is only one-tenth of the labour force deployed at one time to build the Grand Canal.

EXCAVATIONS BY EXPLOSIVES

The largest chemical explosion (i.e. conventional, as opposed to nuclear) was for the construction of Baipaza Dam (see chapter 1); 1860 t was used. A much larger one, involving the use of 500 000 t of explosive, is planned for 1980. It will be on the River Naryn in the USSR and will form a dam nearly 300 m (1000 ft) high and nearly 3 km (2 miles) long. An experimental explosion for this project was planned for 1974. Compare this with the annual consumption of explosives at Bingham copper-mine – 15 000 short tons (13 600 t). The greatest quantities of excavation by nuclear explosions are not known.

EXCAVATION USING A RIVER OR TIDE

The natural scour of a river can be a cheap and effective tool for excavation. It is likely that, if this technique is used, the total quantity of excavation will not be measured, because payment to the contractor is unlikely to depend on the amount of muck that the river moves.

The series of four sketches shows how this method was used in the construction of Mailsi siphon in 1963 and 1964. During the 1963 excavation, more than 3 million yd^3 (2·3 million m^3) was excavated inside the upstream channel area. No quantity was calculated for the 1964 excavation, but it was clearly very large.

The technique probably originated in tidal waters. The first recorded use is by the Dutch engineer Christiaan Brunings (1736–1805), who made a deep harbour at Den Helder in 1782 by building a mole on a sandbank which formed a funnel-shape with the shore. The ebb tide, converging through this funnel, scoured out the harbour so that it was deep enough for the largest ships. The function of the long mole at Escravos Bar (see table 36) is similar.

In a more usual early application, a dam would form a pond which filled at high tide. The impounded water was released when the tide was low, to scour out a channel.

DEEPEST DRILL-HOLES

At sea: A new epoch in exploration of the sea-bed has been initiated by the US vessel *Glomar Challenger*. Originally, the idea was that the earth's crust would be penetrated by a very deep hole (the Mohole) drilled at sea in a region where the crust is at its thinnest. The project never materialised. Instead, *Glomar Challenger* has voyaged round the world, drilling holes of unprecedented depths in the sea-bed for scientific exploration. This work is in progress. Reported so far are:

- A hole drilled in the sea-bed beneath 6096 m (20000 ft) of water. Depth of hole not stated.

- A hole 1220 m (4000 ft) deep beneath 3353 m (11000 ft) of water.

- Deepest penetration of the sediment on the sea-bed (i.e. the deepest hole) 1300 m (4265 ft).

- Deepest penetration of basalt (i.e. crustal rock beneath the sediment) 80 m (262 ft).

- Deepest penetration of the oceanic crust (total thickness about 5 km (3 miles)) expected in the programme: about 1 km (0·621 miles).

On land: A wildcat well drilled in 1974 in Washita County, Oklahoma, USA, penetrated to a depth of 9583 m (31441 ft). This is the world's deepest hole.

The deepest wells in Europe are:

6691·6 m (21954 ft) – Nassiet, France
6173·5 m (20254 ft) – Ernesto Nord, Italy (offshore – Adriatic)
7025·4 m (23049 ft) – reported from East Germany

The deepest well in the Middle East is a test well of 5827·5 m (19119 ft) drilled in 1970 in northern Iraq. Wells less than 3050 m (10000 ft) account for over 90 per cent of Middle Eastern production.

The average depth of 4462 wildcat wells drilled in the United States in 1971 was 1806 m (5924 ft), and of 18929 development wells 1350 m (4429 ft).

The cost of drilling to 15000 ft (4572 m) is nine times greater, and to 12000 ft (3660 m) five or six times greater than the cost of drilling to 7000 ft (2134 m). (Source: E N Tiratsoo, *Oilfields of the world*, Scientific Press Ltd.)

The world's deepest well for water was drilled in 1961 in Rosebud County, Montana, USA. It is 2231 m (7320 ft) deep.

Table 33. Feats of excavation in construction.

Quantity $m^3 \times 10^6$	$yd^3 \times 10^6$	Date	Details
1150	1500	1927–75	Mississippi flood protection (see chapter 4)
74	97	1859–75	Suez Canal: during construction
550	719	1859–1967	grand total up to 1967
341	446	1932–59	St Lawrence Seaway, made up of Beauharnois Canal 250 × $10^6 yd^3$ Welland Canal 62 Seaway 54–59 69 Welland By-pass 65 446
261	342	1882–1914	Panama Canal
251	328	*c.* 1970	Indus link canals
164	215	1952	Volga–Don Canal
142	186	1973	Tarbela Dam
70	91	1971–75	Ogoshima Island – 515 ha (1272 acres) created in Tokyo Bay. Mountain sand excavated and transported 34 km (21 miles) across the bay. Average depth of fill 13·6 m (45 ft)
70	91	recent	North Kiangsu Canal. Excavated manually in 80 days
41	54	1894	Manchester Ship Canal. Apparently the largest in Britain, with Vermuyden's Great Ouse scheme second
11 000	14387	1974–75	1975 output on capital projects for irrigation and agricultural improvement in China
Dredging feats in construction			
240	314	u.c.	Antwerp: fill for reclamation
54	71	1967–75	Rotterdam: improving and deepening entrance of port
Mining			
1200	1570	1904–74	Bingham Canyon copper-mine; total to date
3600	4700		Possible ultimate total
1487	1945	1893–1970	West German opencast lignite-mines. Production – lignite only (overburden probably a quantity of similar magnitude)
94	123	1974	for the year of lignite plus overburden at Fortuna–Garsdorf mine
350	457	1974	for the year of lignite plus overburden for the West German industry

Table 33. cont.

Quantity m³ × 10⁶	yd³ × 10⁶	Date	Details
2717	3554	1942–74	British opencast coal-mines. Total overburden moved*
82	107·7	–	Average, per annum (overburden)
124	162	1955–59	Highest, per annum (overburden)
			Average overburden moved per annum at two largest mines
7·6	10		Maesgwyn Cap, South Wales (12·25 years)
3·8	5		Westfield, Fife (14 years)

* *Note:* Area restored 52 600 ha (130 000 acres) out of 62 600 ha (154 586 acres).

foundations

Foundations of structures may be classified thus:

Direct bearing	pad and strip footings
	trench or pier foundations
	(more massive but essentially similar)
	rafts
	buoyant foundations
	caissons, cylinders and monoliths
Piled	direct-bearing piles
	friction piles

Distinction between the various types is not necessarily clear cut. Pad and strip footings are associated with commonplace construction rather than records; similarly so with rafts, except perhaps buoyant rafts. CN

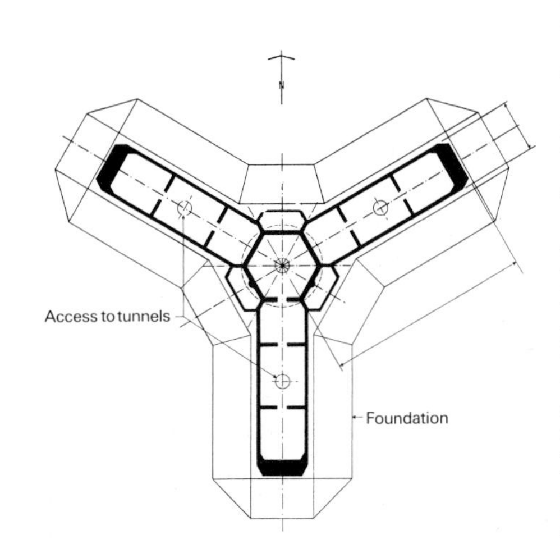

The foundation of CN Tower consists of a Y-shaped slab of concrete, 5·5 m (18 ft) deep. The two dimensioning arrows pick out the length and breadth of one of the buttresses of the tower, about 27·8 m (91 ft) and 7 m (23 ft) respectively. The direct load on the foundation (i.e. the weight of the tower) is 117 000 t. The overturning moment on the foundation caused by the 'design' wind acting on the tower is 443×10^6 kg m (3200×10^6 lb ft). Each limb of the Y supports one buttress of the tower. The foundation is prestressed by cables that help to redistribute the loads caused by the wind. There are tunnels in the foundation pad; the tower is prestressed (i.e. held down) by cables, and the anchorages of these cables are in the tunnels. The shale on which the foundation rests has a bearing capacity of 7 MN/m² (1000 lb/in² or 65 t/ft²). Maximum pressure on it is 1·2 MN/m² (175 lb/in²)

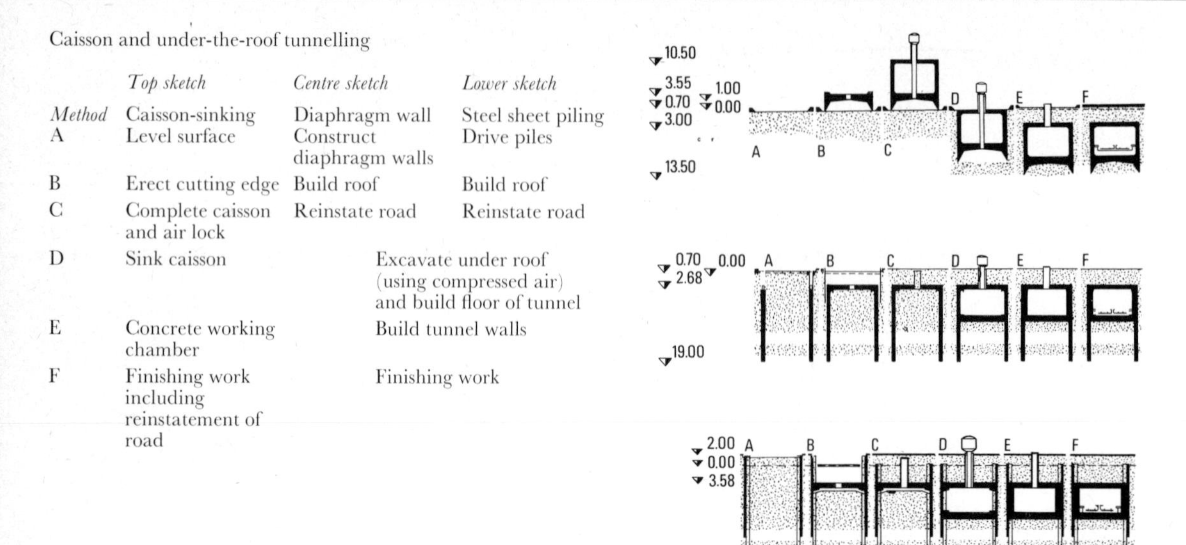

Caisson and under-the-roof tunnelling

Method	*Top sketch*	*Centre sketch*	*Lower sketch*
A	Caisson-sinking Level surface	Diaphragm wall Construct diaphragm walls	Steel sheet piling Drive piles
B	Erect cutting edge	Build roof	Build roof
C	Complete caisson and air lock	Reinstate road	Reinstate road
D	Sink caisson	Excavate under roof (using compressed air) and build floor of tunnel	
E	Concrete working chamber	Build tunnel walls	
F	Finishing work including reinstatement of road	Finishing work	

Tower, Toronto, is the record example of the pad or trench type. The term 'buoyant' is used in a literal sense, the principle of a fully buoyant foundation being that the weight of excavated material equals the weight imposed by the structure. No settlement will take place on compressible soils if this principle is followed; or, if the extra weight imposed by the structure is limited to a modest surcharge, settlement is kept within acceptable limits.

The buoyant principle was introduced in the United States in 1930 in the construction of a telephone building at Albany, New York, the foundation of which was designed as a floating box. The first bridge to use the principle was Tappan Zee Bridge.

CAISSONS, CYLINDERS AND MONOLITHS

A whole variety of foundation techniques (and also some rather inconsistent use of terminology) are grouped under this heading. Basically, a vertically walled enclosure (i.e. a box shape or a cylinder) is positioned where the foundation is to be built, and made to sink into the ground, by excavating inside it, and/or by resting weights on it, and/or by twisting it, until a hard bottom or a satisfactory founding level has been reached.

In the pneumatic method, the caisson has a cutting edge generally made of steel, with a working chamber above it which is airtight and which is entered through air-locks. Compressed air keeps the working chamber dry, and excavation is carried out within it as the caisson sinks. At the same time, the walls at the top of the caisson are extended by further construction to keep pace with the sinking. When down to founding level, the working chamber is concreted solid. The maximum depth to which a pneumatic caisson can be sunk is about 35 m (115 ft) below water-level. That head corresponds to about the maximum air pressure that can be used. Bentonite has also been used to make a seal, so obviating the use of compressed air. This description is somewhat simplified.

An alternative, and in some respects only, more primitive technique, is open grabbing. The construction, which may be like an egg-box or simply a cylinder, is open from end to end in the vertical direction. The open wells become partly full of water; sinking proceeds by excavating with grabs in the wells. When down to level, the wells are filled with concrete, placed under

Table 34. Chronological list of the deepest caisson or monolith.

Site	Size: length × breadth m	ft	Sunk to depth m	ft	Year of completion	Country and location	Other details
South Bisanseto Bridge	75 × 64	246 × 210	129	423	Project	Japan	One of the bridges in the programme described in chapter 1
Salazar Bridge, South Pier	41 × 24	134 × 78	82	270	1966	Portugal, Lisbon	Steel monolith with 28 open wells. Pier designed to resist earthquakes
San Francisco–Oakland Bay Bridge, central anchorage	60 × 28	197 × 92	65·8	216	1933	USA	Depths are below mean low-water datum. Steel monolith with 55 open wells sunk through 22 m (71 ft) of water in current of 7·5 knots (13·9 km/h) to form world's largest bridge pier
pier W3 (west of central anchorage)	38 × 23	127 × 75	73·4	240·7			Twenty-eight open wells; selected wells capped to control sinking; piers W4 and W5 similar but not so deep
pier E3			70	230			Reinforced concrete open-well monolith
Huey P Long Bridge			52	170	1933	USA, New Orleans	Water-jets used to overcome skin friction; monoliths placed on sand islands
Suisun Bay Bridge			44	143	*c.*1930	USA, California	Railway bridge; first use of sand island method
Carquinez Strait Bridge			41	135	1925	USA, California	Monoliths lowered through 27 m (90 ft) of water with swift currents
Poughkeepsie Bridge	31 × 18	100 × 60	41	134	1888	USA, New York City, over Hudson River	Railway bridge, first large egg-box monolith in US bridge construction. Open wells concreted under water with automatically opening skips
Eads Bridge	25 × 22 (six-sided)	82 × 72	26 (St Louis Pier) 37·5 (Illinois Pier)	86 123	1867–73	USA, St Louis, Missouri	Iron-sheathed timber pneumatic caissons. First major use in US and deepest for many years. Depths are below mean water-level (variation 25 m (82 ft)). Pressure 40 lb/in² (0·28 N/mm²); twelve lives were lost
Saltash Railway Bridge	11 (diameter)	15	26	86	1855–59	England, Devon/Cornwall, River Tamar	First big bridge using pneumatic caissons. Wrought-iron cylinder with inclined cutting edge to fit rock surface; pressure 35 lb/in² (0·24N/mm²)
Rochester			18·6	61	1851	England, Kent	First use of pneumatic caisson in bridge construction (shared with Chepstow Bridge 1843–51, 25 m (70 ft) deep)

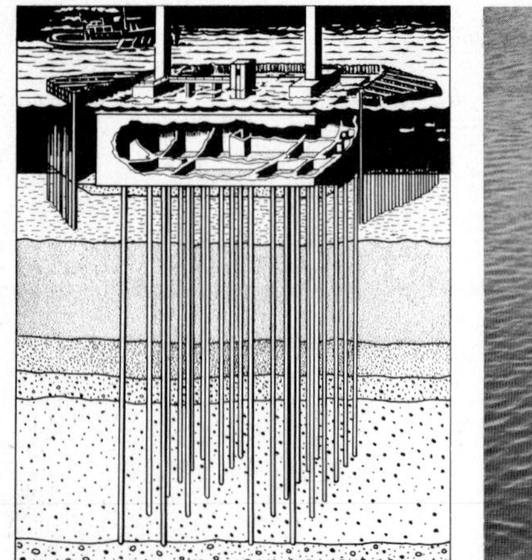

Foundations of Tappan Zee Bridge. *Top left:* Cutaway sketch of the foundation of a main pier at Tappan Zee Bridge. The caisson was floated into position, and then filled with water to sink it on its prepared base. Steel piles up to 82 m (270 ft) long were then driven to bed rock through holes left within the walls of the caisson, and bonded to the caisson by concreting. As construction of the superstructure added weight to the pier, so water was pumped out of the caisson to give the required amount of buoyancy. In the completed bridge, with the caisson pumped dry, 20 per cent of the dead load of the bridge is carried by the piles, and the remainder by the buoyancy of the caisson. There is a comprehensive control system with pumps, access, and lighting, inside the caissons
Top right: Constructing the main pier
Left: Foundations under construction for approach spans. These are not buoyant foundations, but the picture, taken in 1954, gives a good idea of the difficulties and the size of the job
Photos and sketch: URS Madigan Praeger Inc.

water and the 'bottom' is never seen. This method is obligatory when the head exceeds 35 m by a substantial margin. It can be improved by various expedients, one of which is to cap selected wells and dewater them with compressed air (there is no limit to pressure because men do not enter), to exercise control over the sinking.

Strictly, in English usage, the term caisson refers to the pneumatic device and monolith to the open well. The word 'caisson' is now used rather indiscriminately however. Large-diameter piles are called 'caisson piles' or simply 'caissons'. In any case they have become so big in recent practice that a distinction between 'pile' and 'cylinder' is no longer possible.

There is yet another variety of caisson: a large box built in a dry-dock, floated into position and then sunk on to a prepared base to form a permanent structure such as a breakwater or a quay wall, or to make a tidal closure.

Largest in Britain. Uskmouth Power Station in Wales has a cooling water pump house, built in the tidal estuary of the Usk, that sits on a single caisson. The caisson was sunk in 1951. It measured 50 × 33·5 m (164 × 110 ft) and was sunk to a depth of 24·5 m (80 ft) below the level of the pump-house floor which is 8·8 m (28·75 ft) above mean sea-level. The welded steel shoe of the caisson was 3·7 m (12 ft) high and contained 510 t of steel.

Pier 57 of the Grace Line, New York City, is an interesting example of the semi-buoyant foundation principle, and of large floating 'caissons' of concrete of the last variety enumerated in the text. The sketch shows the construction of the pier. Buoyancy is provided by three similar caissons, that were cast 61 km (38 miles) up the Hudson River, and towed down the river and into position, an operation shown in the photograph, that took place in 1951. Each caisson measured 110 × 38 m (360 × 125 ft) length × breadth, and was 11·6 m (38 ft) in height; its displacement was 80000 short tons. This weight makes each of the three caissons the largest ever constructed, until the concrete tank for the Ekofisk field (displacement 215000 t – see chapter 4) was built more than 20 years later. For comparison, the nearest rival was one of the five prefabricated elements for the Kennedy tunnel at Antwerp, which weighed nearly 50000 t. The big dock gate at Nigg Bay displaces 16000 t, the Tappan Zee caissons each about 12000 t, and the Phoenix AX, the largest caissons for the wartime Mulberry harbours, 7000 t. *Photo and sketch: URS Madigan Praeger Inc.*

Early US caissons. The main piers of the Brooklyn Bridge rest on pneumatic caissons. The caissons were of timber, up to 6·7 m (22 ft) thick, with a cast-iron cutting edge. Excavated material was dredged from a vertical shaft, the bottom edge of which was sealed within the caisson by keeping it under water. Maximum pressure was 36 lb/in^2 (0·25 N/mm^2). The Brooklyn caisson measured 31 × 51 m (102 × 168 ft) and the Manhattan one was slightly longer; they were sunk to depths of 13·5 m (44·5 ft) and 24 m (78 ft) respectively. They were built in 1869–72.

The first use of pneumatic caissons in building construction in the United States was for the Manhattan Life Building in New York City in 1893.

The following quotation is from Sir Hubert Shirley Smith's *The world's great bridges*:

'During the sinking of a caisson in the bed of the river Neva at St Petersburg in 1876–78, the caisson, which was in very soft ground, suddenly plunged deeply down. Of twenty-eight men in the working chamber nineteen had time to escape up the shaft as the mud and water flowed in, but nine were imprisoned. It was twenty-eight hours before the rescue parties succeeded in reaching them and by that time only two remained alive. A year passed before work was renewed, and then there was another disaster. Under the high pressure one of the locks gave way. Nine men were blown bodily up the shaft out of the caisson and killed; twenty others were smothered by the inrushing mud and water in the working chamber. Now that engineers have had many years' experience of compressed air work, however, there is little possibility of such an occurrence to-day.'

PILING

Piles of record size or load-carrying capacity today are either:

- steel cylinders, driven or keyed into a foundation stratum; typically, a deep-water jetty may be carried on such piles, standing, unsupported laterally, in the deep water.
- bored piles ('caisson' piles) consisting of a shaft, sometimes lined with steel, excavated in the ground and then filled with concrete. The typical example has a belled-out base that is excavated with a special tool.

The longest piles that have ever been used are the steel cylinders, with a diameter of 1·37 m (4·5 ft) and with a length, as driven, of 270 m

The monoliths for the piers of the Verrazano Narrows Bridge, New York City, are nearly as large in plan area as the Rive caisson, being 40 × 70 m (130 × 230 ft) but this is perhaps the only similarity. The Staten Island one, shown here, was built within the shelter of a cofferdam formed of circular steel-sheet-piled cells. It was constructed of steel, with 66 wells for open grabbing, each 5 m (17 ft) in diameter. It was sunk to a depth of 32 m (105 ft) below mean high water. The Brooklyn monolith was similar but was sunk to a depth of 52 m (170 ft). This work was done in 1959

(886 ft) that pin the four steel production platforms of the Forties Field to the bed of the North Sea. Initially, three piles were driven at each corner of each platform, down internal guides. There is provision for nine piles in external guides at each corner, but it was found necessary to drive only six, making a total of 36 piles for one platform.

These piles were driven with a hammer weighing 330 t (250 t piston). They penetrate up to about 100 m (330 ft) into the sea-bed (i.e. up to about 230 m (750 ft) below mean sea-level). When driven they were cut off above the guides in the jacket structure.

Other notable steel piling. The steel-cylinder piles carrying the greatest load are at West Side sewage treatment plant, New York City, where a 12 ha (30 acre) platform is carried over the Hudson River on 2400 piles. The greatest load carried by one pile is 1800 short tons (1633 t); the pile is 1·1 m (3·5 ft) in diameter and the maximum length of these piles is 79 m (260 ft); they are socketed into rock and filled with concrete.

The steel-cylinder piles of greatest diameter are the cylinders – they have a dual function and are more a complete structure than a pile – that have been constructed to carry the cableway at Eastern Scheldt Dam in Holland. They have a diameter of 15 m (49 ft) and are 77·25 m (253 ft) long.

BP's tanker terminal in the Firth of Forth, Scotland (chapter 3) has piles weighing 100 to 137 t each, that stand in 28 m (92 ft) of water at low tide. Maximum length is 73 m (239 ft), diameter 2·0 or 2·2 m (6·6 or 7·2 ft) and the greatest load carried is 1500 t. In some cases the pile is secured by an anchor penetrating 15 m (49 ft) beyond the toe of the pile.

Apart from the Forties Field platforms, the longest steel piles are those shown in the cutaway sketch of the Tappan Zee Bridge. These piles are steel cylinders filled with concrete, with a maximum length of 82 m (270 ft); the total length of these piles was just over 10 km (33 400 ft).

The world's largest caisson (it is generally so-called, but monolith would perhaps be a better term) has a diameter of 61 m (200 ft). It is of concrete and forms an underground car park, seven storeys in height, at Rive, Geneva, Switzerland. Such a great diameter is a practical proposition for sinking if it is lubricated evenly round its circumference by a layer of bentonite mud; this was done for the Rive caisson, which was sunk without using compressed air, by the Lorentz-Fehlmann method. The first picture shows the sinking in progress, construction of the upper part keeping pace with it. The second shows the completed caisson in its final position. The third shows excavation at the cutting edge of the caisson in what appears to be a stiff clay, giving an indication of the control of the cutting edge

The record length in 1940 was at a bridge over the Potomac River at Ludlow's Ferry, Maryland, USA – they were 59 m (194 ft) long.

Bored piles. The longest bored piles were built for the river crossing of the Parana Guazu Bridge, Argentina (table 9). They are 70 m (229 ft) long, fully cased (with steel) and carrying a load of 1500 t per pile. Similar piles 50 m (164 ft) long were used for the Parana Las Palmas crossing. Total number 200.

The greatest diameter of a bored pile is 6·8 m (22·5 ft) for four piles that support the 50-storey tower of the Development Bank, Singapore; the longest pile of these four is 67 m (220 ft) in length. These piles carry a load of 12 590 t each, the greatest load carried by any pile. According to the firm that built them, these foundations are cylinders not piles; they were constructed by shaft-sinking methods, not by methods normally used for bored piles.

Looking down a bored pile. This one is 2·1 m (7 ft) in diameter and 36 m (118 ft) deep and was for a building in the City of London. *Photo: Taylor Woodrow*

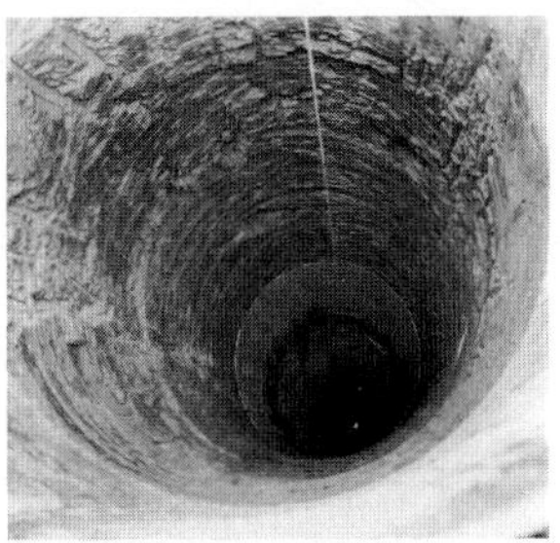

The load-carrying record in Britain is held by the piles (fifteen plus one test pile) that carry an office building in Cannon Street in the City of London. The larger-sized piles of this group each carry a load of 3000 t. They have a diameter of 2·6 m (8·5 ft) for their shafts, and the base of each of them was under-reamed to give a diameter of 5·5 m (18 ft). Their length is 47 m (154 ft). They were constructed in 1974, and the record pour of concrete mentioned in chapter 4 was for the same building. An interesting point about these piles is that they had to be installed very accurately, because a station of the Fleet Line Underground will be built, so that the piles will straddle it (see sketch, page 162).

The longest bored piles in Britain were constructed in 1969, to carry viaducts leading to the second Mersey Tunnel over Bidston Moss, Cheshire. Bedrock lies up to 66 m (215 ft) below the surface, but for economy, a shallow cut was made on the surface, and the maximum length of the piles was limited to 61 m (200 ft). These piles are 1·2 m (4 ft) in diameter and each one carries a load of 550 t.

A notable early record-breaking construction was the Church Street Post Office, New York City, built in 1935. The piles each carry a load of just over 3500 short tons (3000 t) and their toothed cutting edges were rotated and jetted through clay and boulders.

Greatest numbers of piles. Construction of Littlebrook D Power Station, Dartford, Kent, which started in 1974 has required the largest piling contract ever let in Britain. Its value is about £15 million and it involves about 24 000 piles, mainly Raymond step-taper piles about 30 m (nearly 100 ft) in length but also including 1800 bored cased piles. At Grain Power Station, also in Kent, all major structures are carried on friction piles extending into the clay at depths of 19–38 m (60–125 ft), in all about 21 000 piles. Drax Power Station in Yorkshire also has over 20 000 piles.

The steelworks being built at Ogoshima Island, Japan, is using 340 000 t of steel cylinder piles and 25 000 concrete piles.

Tappan Zee Bridge, in addition to the 10 km (6·3 miles) of steel-cylinder piles noted above required 100 km (62·5 miles) of steel H piles and almost 490 km (303 miles) of timber piles, a total length of 598·2 km (371·8 miles) of piling.

The Haringvliet Barrage required about 22 000 concrete piles from 8 to 24 m (26 to 80 ft) in length. In addition, about 24 000 t of steel-sheet piling in lengths from 3 to 21·3 m (10 to 70 ft) was needed.

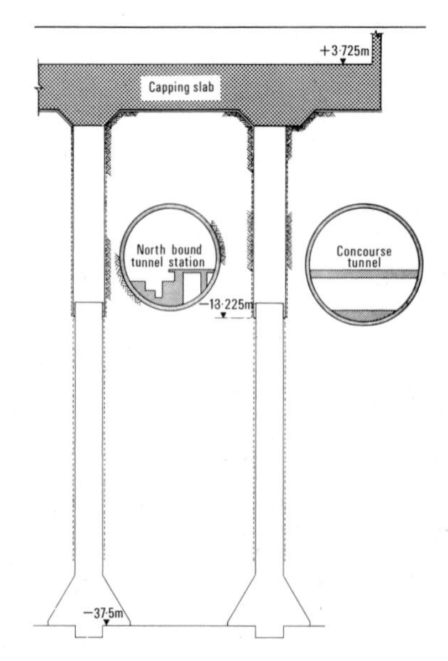

Piles carrying an office building at Cannon Street, London, and straddling a future underground station. The 3000 t working load on a pile imparted a pressure of about 9 t/ft^2 (nearly 1 MN/m^2) giving an immediate settlement of about 12 mm ($\frac{1}{2}$ inch). Where the piles pass close to the future tunnels they were coated with bitumen so that they can slip, relative to the ground; this will prevent load being transferred to the tunnels. The piles were driven very accurately, and loads and movements are being accurately monitored. *Sketch: Cementation Group*

Building the great steel bridge at Niteroi (tables 5, 9 and 54), Brazil. The first picture shows a steel span being floated out to site, floating on another steel span of the bridge; the second picture shows the same span in position at the piers. When a similar span was in position on the other side of the piers, the two were jacked up to the tops of the piers. The same process was then repeated at the two piers seen in the background. Then, the suspended span which joined them together was lifted, slung from the cantilevered ends of the elements already erected. *Photo: Joint venture, Redpath Dorman Long – Cleveland Bridge and Engineering Co Ltd*

Historical facts. Vauban recorded that 38 men were needed to raise a piling hammer weighing 587 kg (1290 lb). It was raised 1·45 m (nearly 5 ft) and then allowed to fall on the pile. Thirty blows were completed, then the team of men rested for a time equal to the time they had worked. This routine gave, over six hours, an effective continuous rate of twelve blows per minute on the pile.

Labelye described, in his book published in 1751, an exceptionally powerful pile-driving apparatus he used to build Westminster Bridge in London. It was mounted on a floating stage and worked by two or three horses, which gave 48 and 70 strokes per hour respectively. The horses turned a capstan, the stroke was 6 m (20 ft) and the weight of the hammer was 770 kg (1700 lb).

Amsterdam Town Hall, built in 1660, rests on 13659 timber piles.

When constructing Dover Harbour, the contractor, Pearson, used a gantry comprising piles of Australian blue gum, each 33 m (100 ft) long and 0·6 m (2 ft) square and weighing 10 t. These piles were used repeatedly as the work progressed.

FOUNDATIONS FOR HORIZONTAL LOADS

An arch, whether bridge or dam, exerts a thrust sideways on its foundations. Usually, sound rock is available to take the horizontal load. Other examples of horizontal loading are, with live loads, the deceleration of traffic on a bridge, or the impact of ships against a quay or jetty. But the greatest loads in this category are those resisted at the anchorage of a suspension bridge. Further, the anchorage must not move under load – that is critical.

A massive concrete pier, keyed into bedrock, is the favoured anchorage. An alternative is wedge-shaped tunnels. The anchorage can be prestressed, so that deflection does not build up as load is hung on the cables. In poor ground, the problem may be so severe as to rule out the construction of a suspension bridge. The Barton anchorage of the Humber Suspension Bridge is an example of an elegant engineering solution to such a problem. The anchorage is a wedge-shaped box in the Kimmeridge Clay 72×44×35 m (236×144×115 ft) deep. Its peripheral walls and four longitudinal walls inside it were built in slurry trenches as diaphragm walls. Then excavation was started, but was confined at any one time to two of the five trenches defined by the walls, so that no great area of clay at the bottom was ever unloaded. Concrete was gradually placed in these trenches to form the permanent structure of the anchorage – the diaphragm walls were not relied on as permanent construction – and as this work proceeded some of the spaces in the box were filled with sand or water to retain roughly unaltered loading on the clay.

When the cables have been spun and the bridge is being hung on them, so load on the anchorage will build up. The reaction of the anchorage box on the Kimmeridge Clay will then be adjusted by adding concrete as structure (for example the deck slab) or filling, so that the load on the clay remains approximately uniform, thus preventing rotation of the anchorage.

MOST ACCURATE MEASUREMENT OF SETTLEMENT

Loads up to 180 kN/m^2 (26 lb/in^2 or about 1·7 t/ft^2) were applied to the chalk ground at Mundford, Norfolk, experimentally, and settlements measured at five places down to depths of about 25 m (80 ft) below the surface. Deflection of the chalk *in situ* was measured to an accuracy of about ±0·0025 mm or ±0·05 mm (±0·0001 in or ±0·002 in) depending on the type of instrument used. This is the most accurate measurement ever made of the deflection of the ground under load. The reason for the work (carried out by the Building Research Establishment in 1967–8) was to test the site to the very strict specification of CERN, the European nuclear research body, for a proposed particle accelerator. A settlement of not more than 0·5 mm (0·02 in), 50 m (164 ft) away from a 10 000 t load, and alignment of the foundation within 0·15 mm (0·006 in) over 50 to 100 m and over a number of years were specified. In fact the device was never constructed.

GREATEST SETTLEMENTS

The Palace of Fine Arts, Mexico City, has experienced settlement that is measured in metres. Settlement of street levels of more than 8 m (30 ft) has occurred in Mexico City. A grain silo built in Canada in 1913 was filled with 20 000 t of grain – a weight about 10 per cent greater than the weight of the structure itself – on one day in 1914. It tipped over until it was over 10 m (34 ft) out of level. The grain was removed and the structure restored to the vertical by careful jacking and underpinning.

ground engineering

COFFERDAMS

Cofferdams are temporary structures that keep water out of a region of deep construction. They can be literally a dam, more often an enclosure formed by embankments or by steel-sheet piling. In small cofferdams the piling can be strutted like timbering in a trench. Otherwise, sheet piling can be anchored in some way, or a double wall filled in to form a gravity dam; or, basically the same idea but more elaborate, circular cells can be linked together. Generally, a system of pumps or well-points is a necessary complement to the physical barrier, to keep the enclosure dry.

The world's deepest cofferdam – measured by the head of water resisted by it – was built for the construction of Akosombo Dam, Ghana. The depth there was over 60 m (200 ft). 'The original site of the dam was several miles upstream where the Volta River was wider and shallower, however, the costs were exorbitant. A study showed that a dam could be built at the present site. The axis of the dam would not be so long, the river (to rock) would be deeper, and the costs would be substantially less. However, at the site selected we had a depth of water of 100 ft to sand and another 100 ft of sand to bedrock. This meant a 200 ft excavation and suction dredges did not dredge that deep. This was 1959–60, and the only way in which to get a dredge to work this deep was to design and build one ourselves. The prototype was shipped to us in Ghana to assemble and work the bugs out. I don't remember too many of the details, but it was a 6″ suction dredge. It worked so well that we had two more made. The cofferdam and dam were designed to be built as a first stage to go from the right abutment to the center of the river and to be overtopped during the first flood season. Although it was anticipated that there would be several millions of dollars worth of damage,

Table 35. Cofferdams.

Site	Head of water		Area in plan		Year of construction	Country and location	Other details
	m	ft	m	ft			
Akosombo Dam	61	200	–	–	1962	Ghana, Volta River	Sealed rock-fill (460 000 m^3 (600 000 yd^3)) upstream and downstream cofferdams eventually incorporated in the dam (see text)
San Francisco–Oakland Bay Bridge, Pier W2	30·9	101·4	16 × 37	52 × 121	1933	USA, California	Conventional steel-sheet-piled. Piling from 10 ft (3 m) above to 88·5 ft (27 m) below datum level; rock from 78·4 to 101·4 ft (24 m to 30·9 m) below datum
Champlain Bridge	29·6	97	–	–	–	–	Similar cofferdams to that above and at George Washington Bridge
Mackinac Bridge	27·4	90	35 × 41	115 × 135	1954–57	USA	Cofferdams for anchorages
Haringvliet	18·5 (generally) 28·5 (max.)	61 93	1050 × 420	3445 × 1378	1957–67	Holland	Ring embankment in open sea, with elaborate well-point dewatering
Rance tidal power station	28·5	93	600 × 250	1968 × 820	1962–63	France, near Saint-Malo	Most extreme conditions of tidal closure ever encountered. Cofferdam consisted of 19 m (62 ft) diameter circular cells of steel-sheet piling (135 kg/m^2 (28 lb/ft^2)) plus strong points – 9 m (30 ft) diameter concrete cylinders (maximum height 27 m (88 ft)) on prepared foundations. A head of 3 m (10 ft) was measured across the gap between the strong points on the spring tide of 22 June. The tidal gap was closed on 4 July
Velsen Tunnel	125	82	63 (diameter)	207	1953–55	Holland, near Amsterdam	First stage. Very heavy steel-sheet piling, 27·5 m (90 ft) long braced by five rings of reinforced concrete (see sketches, page 168)
George Washington Bridge	24	78	30 × 33 (clear)	99 × 108	1927–28	USA	Two similar cofferdams one for each leg of the New Jersey tower. Double skin of steel-sheet piling 8 ft (2·4 m) wide in deeper regions
Golden Gate Bridge	11	35	47 × 91	155 × 300	1933–34	USA	100 ft (30 m) depth to rock; 65 ft (20 m) of concrete placed before dewatering cofferdam

it was a calculated risk and well worth it, if it got us out of the wet. The damage was negligible, clean-up was made, and the dam was ahead of floods from there-on.' (The quotation is from a letter from Mr R W Sheridan, formerly Project Manager at Akosombo.)

A depth of about 40 m (120 ft) is apparently going to be reached inside the cofferdams for the Thames Barrier, under construction as this book goes to press.

CONTROL OF GROUND WATER

Those bone-dry, deep holes one sees when the foundations of a big building are being constructed are highly unnatural, and much care has gone into making sure that they will be dry. In simple cases, pumping from sumps suffices. Otherwise, an excavation can be kept dry by installing well-points – that is a series of suction wells in the ground connected to a suction main – or submersible pumps installed in boreholes, such that water is abstracted from round the perimeter, so depressing the water-table until it is below the lowest level to be excavated. Velsen Tunnel and Haringvliet Barrage cofferdam illustrate two of the largest examples of this principle.

The great cofferdam in which the Haringvliet Barrage sluice structure was built was kept dry for about ten years. Its dimensions are in the table. Ninety well-pumps were installed round the perimeter of the main structure to dewater the deepest region of excavation. Lines of well-points were spaced along the inner slopes of the ring embankment to control the hydraulic gradients within the embankment. Well-points extended down to about 22 m (72 ft) below sea-level, and well-pumps to about 32 m (105 ft). The water-table was maintained at a general level of about 17–18 m (55–60 ft) below sea-level. A power station with seven 770 kVA generators served the installation. Water was pumped out at a rate of about 4500 m^3/h – 1 million Imperial gallons per hour or about 1·25 m^3/s (45 ft^3/s), quantities that can be compared with the pumping outputs listed in table 46, chapter 3. At

The Haringvliet cofferdam, shown nearly at the end of its life. The cofferdam was remarkable for its great size – it was nearly 100 ha (250 acres) in area – that it was virtually in the open sea, and that it was kept pumped dry for about ten years. Here the permanent structure has been completed, and the inner slopes of the ring embankment have been trimmed so that the greatest area possible of the stone aprons that protect the permanent structure from scour could be built in the dry, within the cofferdam

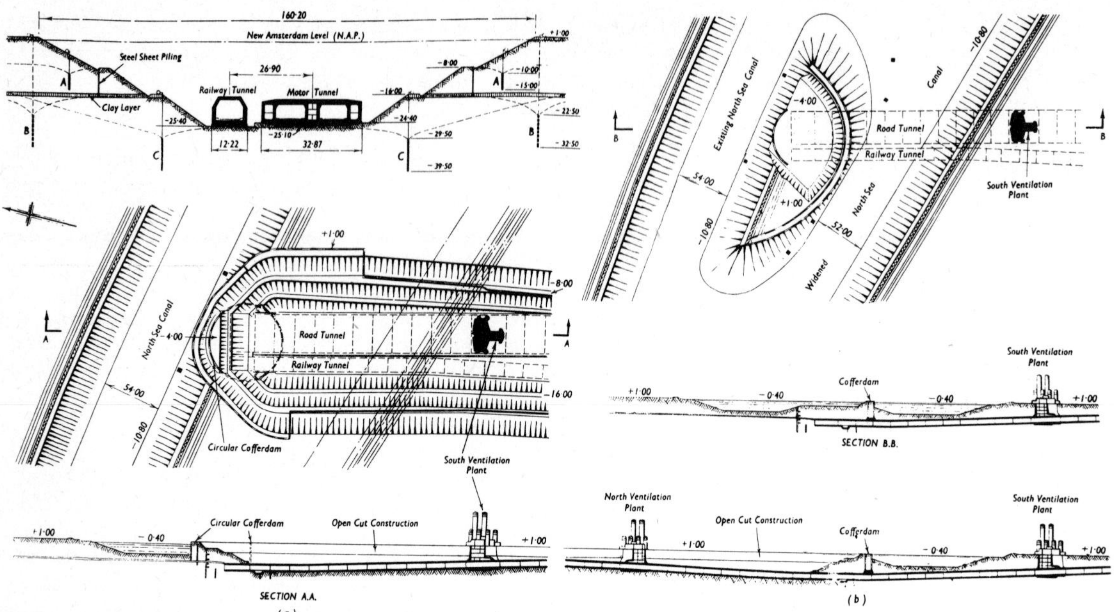

Constructing the Velsen Tunnel beneath the North Sea Canal. The ultimate in cut-and-cover tunnelling, combined with elaborate dewatering, plus a big cofferdam close to the canal.
Stage 1 A circular cofferdam is built beside the canal and a short length of tunnel constructed within its shelter.
Stage 2 Soil is backfilled over the the completed short length of tunnel. It is then possible to remove the steel-sheet piling on the landward side of the cofferdam. The excavation is then opened up as shown in the upper view in three stages. (i) Installing well-points at *A* to lower the water-table as shown by the dotted lines; then steel-sheet piles behind them. (ii) Submersible pumps placed at *B* to reduce the hydrostatic pressure in the fresh water below the clay stratum. With this precaution observed, excavation to the clay stratum (level – 16 m). (iii) Install a further stage of submersible pumps at *C* to lower the level of the fresh ground water below the clay stratum as shown. Then complete the excavation and build the tunnels. (iv) Backfill and reinstate the clay stratum to keep fresh and saline ground water apart.
Stage 3 Excavate a new channel for the North Sea Canal, over the completed section of the tunnel.
Stage 4 Build cofferdams across the existing canal and repeat Stage 2 above on the north side, to complete the tunnel.
Stage 5 Remove the cofferdams and the island site of the circular cofferdam to give a widened canal.
The road tunnel is 768 m (2520 ft) long plus 876 m (2874 ft) of approaches with dual two-lane roads. Total length 1644 m (5394 ft).
The rail tunnel is 2032 m (6667 ft) long plus 1192 m (3911 ft) of approaches and is for two tracks. Total length 3224 m (10577 ft)

Farakka Barrage, on the Ganga River in India, an installation of 12 000 to 15 000 well-points was used to dewater the cofferdams.

Silty soils can be drained by an electrical technique – electro-osmosis – but it is expensive. It was first used about 25 years ago to dewater a railway embankment in Scotland. The most ambitious example was an installation of about 1000 electrodes, used in 1966 to stabilise clay and so stop movement of an embankment dam nearly completed on the Mahoning River, Ohio, USA.

DIAPHRAGM WALLS

A technique that has swept the board in the last fifteen years or so. For years before that, mud was used to stabilise the holes drilled when prospecting for oil. The mud contained bentonite, a substance that is thixotropic (i.e. it becomes thin when stirred, but is jelly-like when undisturbed). The stiffness of the mud in the wall of the drill-hole helped to hold it stable. This same property is made use of in the diaphragm wall. A trench is excavated, and kept filled with bentonite slurry as the excavation proceeds. Without the mud, the trench would collapse, and elaborate shoring and stage-by-stage construction would be necessary. Concrete is then placed with a tremie (a device for inserting wet concrete beneath a liquid) until it fills the trench, displacing the bentonite. Thus a strong wall is inserted in weak ground.

The impermeable cut-off beneath Manicouagan 3 Dam, a hydro-electric dam in the Province of Quebec, Canada (Daniel Johnson Dam listed in chapter 1 was originally known as Manicouagan 5 Dam) is the deepest wall yet built by this technique, its greatest depth being 131 m (430 ft). There are two walls, 191 m (625 ft) long with a total area of 20 718 m^2 (223 000 ft^2).

The greatest quantity of diaphragm walling on a single project is for the Metropolitana Milanese (Milan's underground railway) where 19 km (11·8 miles) of line have side walls built this way. The total area of these walls is 684 000 m^2 (7 362 335 ft^2).

The largest single contract for diaphragm walling was for the construction of the quay walls at Royal Seaforth Dock, Liverpool. The total length of quay walls is 3128 m (10 262 ft) and the volume of concrete they contain is 95 860 m^3 (125 375 yd^3).

The deep-water berth at Redcar Quay, Yorkshire, listed in chapter 3, was constructed of diaphragm walling, and is a record for this class of construction. (Source: ICOS Limited.)

GROUTING AND CHEMICAL TREATMENT

The properties of a soil or rock can be modified *in situ* by injecting grout into it. Grout is a liquid or slurry that sets hard. The commonest one is made with ordinary cement, but many other types are used to make the ground watertight or to give it physical strength. They are important in tunnelling and deep excavation, but the biggest examples are in the construction of dams.

The visible structure of a dam may be like the tip of an iceberg, and only a small part of it. If the rock or soil round the dam is permeable, a watertight screen has to be inserted into it. This may take the form of

- A cut-off. A vertical watertight screen between the foundation of the dam and impermeable rock at some depth below it.
- A watertight curtain extending beside the dam.
- A watertight blanket (i.e. horizontal or sloping layer) in or beneath the reservoir bed on the upstream side of the dam.

Curtains and cut-offs are generally formed by grouting. However, other methods – for example the diaphragm wall at Manicouagan 3 Dam – may be applicable. The Derwent Dam in Yorkshire, built in 1912, was the first dam in Britain to have a foundation that went 30 m (100 ft) below ground.

Camarasa Dam in Spain (about 90 m (300 ft) high) had a seepage problem in 1922, and 186 000 t of grout was injected. This figure is believed to be a record still: it could have been exceeded at Keban Dam, Turkey (see below) but no figure is available in that case.

The largest curtain ever constructed by grouting is at Dokan Dam in Iraq. Limestone or dolomite are the rocks that tend to need extensive treatment of this kind because they form solution cavities. At Dokan,

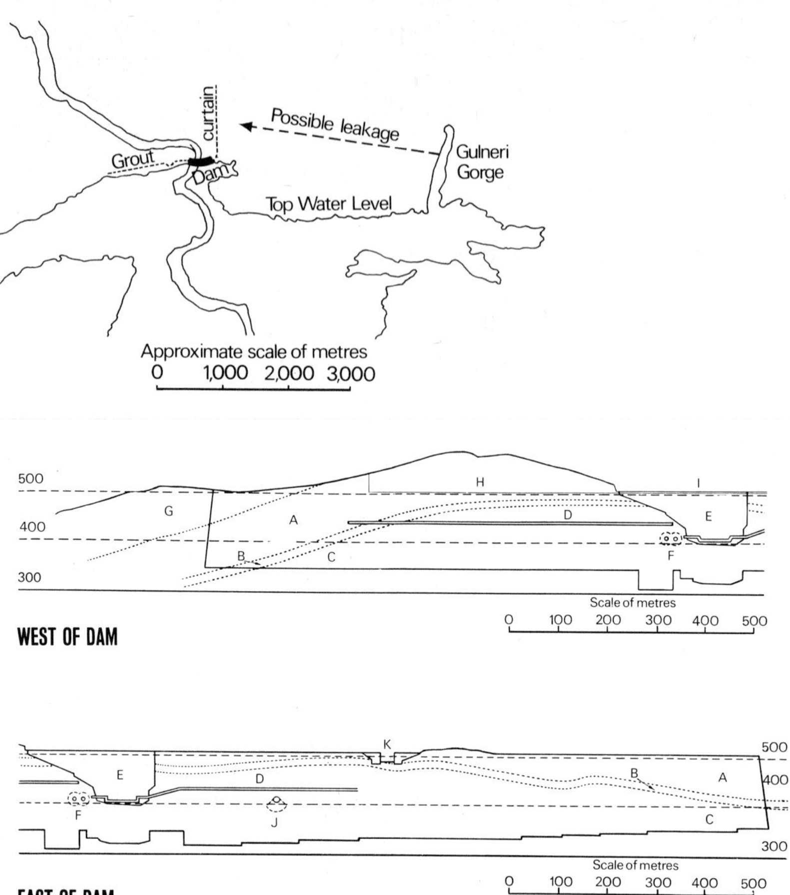

The grout curtain at Dokan Dam, shown in plan, and in section separately for each bank. Most of the grouting was carried out from the two tunnels.
Sketches: Cementation Ground Engineering Ltd

A Limestone
B Contact zone
C Dolomite
D Grouting tunnels
E Dam
F Irrigation tunnels
G Marl
H Top of curtain (511 m)
I Crest of dam (516 m)
J Diversion tunnel
K Gated spillway

where the work was done in the 1960s, the curtain extends over a length of 2877 m (9438 ft) and has an area of 464 500 m² (5 million ft²). Drilling to the extent of 305 000 m (1 million ft) and 114 000 t of cement and sand were necessary. The maximum depth of the curtain below top water-level of the reservoir is 195 m (640 ft). The dam is 116 m (382 ft) high.

Keban Dam (see chapter 1) is built on karstic rock that contains enormous solution cavities. The so-called 'crab cavity' on the left bank of the dam's foundations was 320 m (1050 ft) deep and with maximum dimensions of 100 × 120 × 29 m (330 × 400 × 95 ft) and was filled with ground water different in character from the river water. Its volume was about 100 000 m³ (130 000 yd³) more than half of the filling of which was concrete. Drilling 547 000 m (nearly 1 800 000 ft) for the injection of grout has been carried out, some of the holes being 250 m (820 ft) deep. An incomplete figure of 43 000 t of cement injected has been recorded, and concrete in cavities and diaphragm walls totals 243 000 m³ (318 000 yd³). The success of rock grouting at Vajont Dam was mentioned in chapter 1; in this case its purpose was to increase strength, rather than to make the dam watertight. Another example on a very large scale is at Mauvoisin Dam, where the curtain of grouted rock is 19 000 m² (over 2 million ft²) in area. The Watergrove Dam (1938) in Britain is 26·5 m (87 ft) high above ground level, but it has a foundation that goes 46 m (150 ft) below ground.

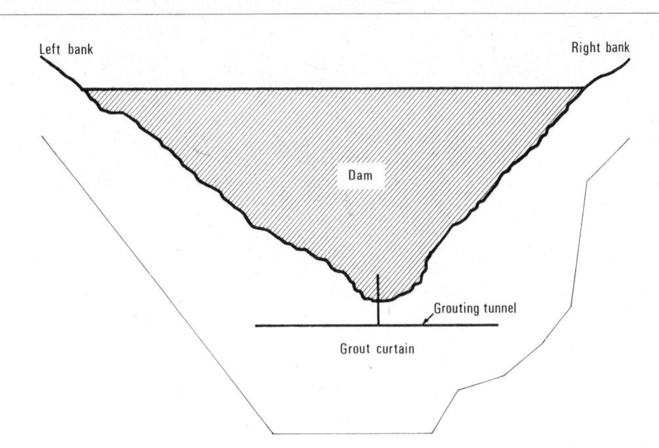

Grouting at Grande Dixence Dam

Alluvial grouting. The grouting described up to now has been in rock, albeit fissured or with cavities. Grouting in alluvial soils is a more recent technique. It depends on a device called a 'tube-a-manchette' – a sleeved tube that can inject grout at any given level down the tube. The tube-a-manchette was invented in 1933, during construction of the Bou-Hanifia Dam in Algeria; but it was not used in a significant manner until the construction of the Serre Ponçon Dam on the Durance River in Provence, France. This dam was completed in 1960, and is the prototype for dams with a grouted cut-off in alluvial material. The Serre Ponçon cut-off is 100 m (over 300 ft) deep and the dam is 120 m (400 ft) high.

The largest cut-off in alluvial soils, and the deepest cut-off at any dam, is at the Aswan High Dam in Egypt. The cut-off is 255 m (836 ft) deep and seals an area of 54 700 m² (nearly 600 000 ft²). The largest single grouting contract ever carried out is claimed to be at Tarbela Dam, Pakistan. Tarbela Dam, however, relies for its watertightness on a blanket of fine-grained soil that extends upstream for 2300 m (7700 ft) across approximately the entire length of the main dam. This blanket is the largest construction of its kind.

Physical treatment. The properties of the ground *in situ* can also be improved by physical treatment – in practice consolidation or removing water. The obvious example of consolidation is compaction on the surface by rolling. Thumping, or dynamic consolidation, consists of dropping weights on the ground.

Weights of tens of tons are dropped from heights of 15–40 m (50–130 ft) repeatedly. At Karlstad in Sweden 110 000 m² (nearly 1 200 000 ft²) of ground was treated by this method. The largest example in Britain is at Teesside, where 50 000 m² (538 000 ft²) was treated prior to the construction of oil storage tanks at the landfall of the Ekofisk pipeline.

Vibration is another technique of consolidation, applicable particularly to sands. At Aswan High Dam, in the first stage of construction, sand was placed under water to form the body of the dam, and consolidated with enormous vibrators. The sand was placed in two 'lifts' each 15 m (50 ft) deep, and covering an area of over 150 000 m² (over 1 600 000 ft²).

An area of ground about 100 000 m² (over 1 million ft²) was treated by Vibroflotation at Port Talbot in Wales (1971–4) at the site of an iron-ore stockyard; its potential settlement was halved.

Freezing. A shaft can be sunk, or a tunnel driven, through unstable

Excavation for a nuclear power station, Colorado USA. The ice wall has plastics sheeting hung on it to protect it from the sun. The pipes for circulating brine can be seen round the perimeter. The heavy plant is standing on frozen ground and so can approach close to the edge of the excavation

water-bearing ground by freezing it. The first recorded example is at a colliery in South Wales in 1862, when an ether engine was used to cool brine that was circulated in tubes round the difficult ground, and the shaft was then constructed satisfactorily, according to the very scant account available.

The method was not used again until 1883, when Poetsch applied it at the Archibald Mine in Saxony. At a depth of 34 m (112 ft) a band of quicksand, 4–5 m deep (13–16 ft) was solidified by freezing; on 8 July the freezing machine started to work; on 31 July the temperature was down to −19 °C (−2 °F); after 83 days the shaft was completed and coal-mining had started.

The first application to tunnelling was in Stockholm in 1884 at Brunkeberg Tunnel – a small cast-iron-lined tunnel for pedestrians. Cold air at a temperature of −50 °C (−58 °F) was pumped into the tunnel at night, and a depth of 2 m (6 ft) of ground was frozen by it. During the day, hand-mining and timbering was carried out. The method was considered to be successful but very costly.

The longest length of tunnel driven in artificially frozen ground is a sewer tunnel 980 m (3215 ft) in length constructed in Hamburg (Hamburg–Wilhelmsburg) in 1967.

The deepest shaft sunk by the freezing process is the Number 1 Shaft at Saskatoon, Saskatchewan, Canada, of the Potash Corporation of America. It was constructed in the late 1950s and frozen from the surface to a depth slightly over 915 m (3000 ft). The record in Britain is the Number 1 Shaft at Boulby Mine of the Yorkshire Potash Company, completed in 1974. Freezing was effective between the levels 610 and 945 m (2000 and 3100 ft) below ground level. More than 350 shafts throughout the world have been sunk using the freezing process.

Tunnel driven through frozen ground for a sewer under Dortmund–Mengede railway station, West Germany

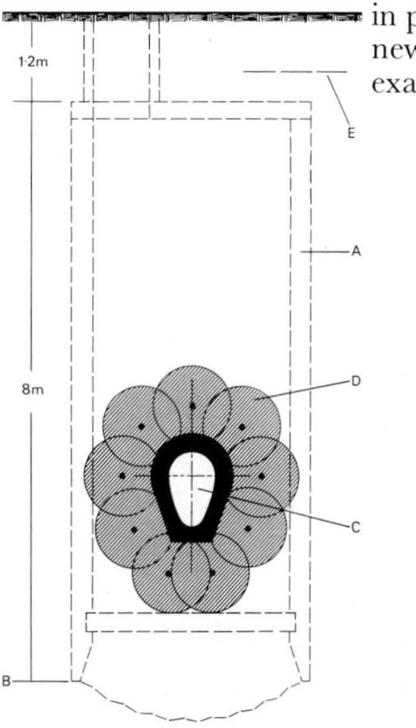

Cross-section through sewer tunnel in Hamburg, described in the text. Shafts were formed by sinking caissons from the surface at intervals of about 40 m (130 ft). Freezing pipes were then inserted horizontally as shown to build up the ice wall.
Freezing photos and sketch: Deilmann-Haniel GmbH

A Caisson
B Cutting edge
C Sewer pipe placed in tunnel and concreted in
D Frozen ground
E Water table

Its most ambitious use to date, however, was for the construction and permanent stability of four underground tanks at Canvey Island, Essex, built in 1968 and 1969. Each of these tanks is 40·6 m (130 ft) in diameter and 40·6 m (130 ft) deep below ground level. They are for the storage of NG (liquefied natural gas) which is at a temperature of −162 °C (−260 °F). An excavation 32×48 m (96×144 ft) in plan by 17 m (51 ft) deep for a nuclear power station at Fort St Vrain, Platteville, Colorado, USA, was stabilised by freezing and the ice wall was maintained for six months. Its only competitor for size in the United States is an excavation 82×49 m (269×161 ft) by 10 m (33 ft) deep in Phoenix, also stabilised by freezing.

The traditional method of freezing since the beginning has been to circulate brine through pipes. An impetus has been given by the recent introduction by Foraky Limited, the British specialists, of liquid nitrogen in place of brine. The ice wall is established in a much shorter time with the new method – less than two weeks instead of about fourteen weeks in one example.

chapter 3
HYDRAULIC WORKS

harbours and canals

By 1976 the world's largest ships were 542,000 t, about a fivefold increase in tonnage over two decades. The 1 million t ship may well be built in the next few years. This rapid increase in size, especially of bulk carriers, together with other changes such as the use of containers, have caused enormous changes in, and keen competition among, ports, ship-repairing and shipbuilding yards.

A 300 000 t ship needs a depth of water of 27–31 m (89–102 ft), deeper than is available at the berths of established ports. Single-point moorings (for example a single buoy) seem to be the favoured installation for very large tankers but the swinging area is, of course, huge. One suggestion put forward for the 1 million t ship is to construct a vertical steel tube, 30 m (100 ft) or so in diameter, in 180 m (600 ft) of water with a tunnel below it connecting it with the shore. Ships would moor at a pontoon, freely attached to the tube, so that it could follow the movement of the ship.

The berths with deepest water alongside are:

Hunterston, Scotland, 39·5 m (130 ft) u.c. 1978. Piled jetty 1·6 km (1 mile) from the shore for iron ore and coal.

Heianza, Okinawa, 32 m (105 ft) 1970. Similar to Bantry Bay and operated by the same company, Gulf, together with **Point Tupper**, on the strait between Cape Breton Island and Nova Scotia, where the depth at low tide is 30·5 m (100 ft).

Kharg Island Terminal, 31 m (102 ft) 1972, about 1450 m (4700 ft) offshore in the Persian Gulf. Kharg 4 is 549 m (1800 ft) long by 84 m (275 ft) wide and consists of a central platform and twelve dolphins connected by walkways and pipe bridges. The depth quoted is the minimum at lowest tide. The breasting dolphins (groups of three steel piles) at this terminal, can deflect 3 m (10 ft) when the tanker is berthing. Kharg 5 is in use.

Bantry Bay, County Cork, Ireland, 30·5 m (100 ft) 1968. Piled jetty 366 m (1200 ft) offshore and 488 m (1600 ft) long. Depth given is for outer berth at low water. Tankers operate from Mina al Ahmadi, Kuwait, to this terminal.

Bantry Bay tanker terminal, Whiddy Island, County Cork. *Photo: Gulf Oil*

Firth of Forth, Scotland, 28 m (92 ft) 1974, for tankers up to 300 000 t. Depth at high tide is 34 m (111·5 ft). There is no link with the shore except a pipeline on the sea-bed.

Antifer, France, 25 m (82 ft) 1976. The new tanker terminal at the port of Le Havre. The depth given is below sea-level datum (mean sea-level) and corresponds to about 29 m (95 ft) at high water. The port will be deepened, and the depth at full development will be 32 m (105 ft) below mean sea-level, or about 36 m (118 ft) at high water.

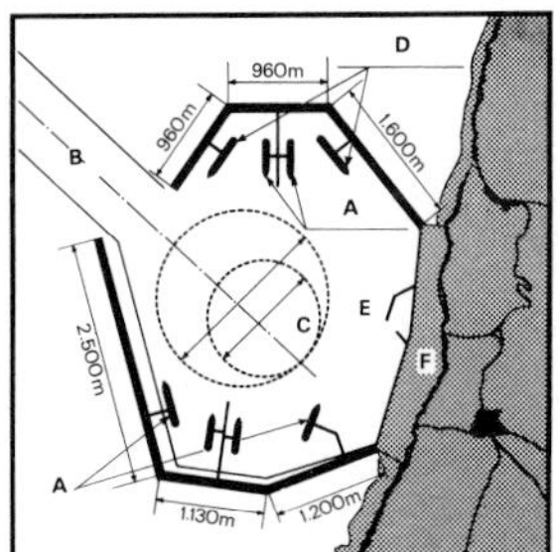

The oil port of Antifer on the north coast of France. The north breakwater is complete. The south one has still to be constructed. The port owes its existence to a geological fault; there exists a bed of sand in this fault, 3 km (1·9 miles) wide and 34 m (112 ft) deep on the sea-bed, and the approach channel and turning circle are dredged out of the sand

- A 500 000 t berths
- B Dredged channel 700 m wide
- C Turning circles of 1200 and 1800 m diameters
- D 1 000 000 t berths
- E Service harbour
- F 28 ha flat area for eight tanks, each 24 m high, 92 m diameter

Fos, Marseilles, France, 23·5 m (77 ft) 1968. Depth available at the outermost berth. Entrance channel is 24 m (79 ft) deep, and the second berth 21·5 m (70·7 ft). The main basin of the port is 4·4 km (2·7 miles) long by 650 m (2133 ft) wide. The date given is for the opening of the port, but the depth given is current.

Rotterdam, Holland, 23 m (75 ft) 1971. The deepened entrance to Europoort. At high tide, and with certain precautions, 23·5 m (77 ft) is available. The 20 m (65 ft) ship is normally accommodated; future development is possible, to handle any ship that can sail in the North Sea. But the problem 'is now shifting seawards' and depends primarily on economic feasibility.

Ogishima Island, Tokyo, 23 m (75 ft) 1976. Berth on east side of artificial island 515 ha (1272 acres) in area, i.e. about 3 × 2 km (2 × 1·5 miles) – in Tokyo Bay.

Redcar Ore Terminal, River Tees, England, 22·86 m (75 ft) 1973. Depth given is below mean sea-level datum, tidal variation being about ±2·5 m (8·2 ft). The terminal has a wharf 518 m (1700 ft) long, which is a solid wall of 28·36 m (93 ft) exposed height and 42 m (137·8 ft) total

Deep-water tanker berth at Fos, Marseilles

Constructional view of the dry-dock at Fos, Marseilles, showing the cofferdam within which the dry-dock was built. The dock is now completed

Lighthouses – each a floating caisson – were positioned at the ends of each of the new breakwaters at Rotterdam. One of them is seen here being manœuvred into position using tugs, two block-laying vessels, and a stone-dumping vessel. The breakwater can be seen in the top right-hand corner of the picture

height. The only solid berth of comparable dimensions appears to be the one illustrated at Fos, Marseilles.

World's largest ports (by length of quay or berth in the port)

Ports are generally classified by tonnages of cargoes handled, but this classification is concerned with construction. Some ships, especially tankers, can be moored at buoys without using a berth, or they can berth at T-head jetties, the length of the jetty being less than the length of the ship.

Port	Length of quay km	miles	Other details
Rotterdam	122	76	Actually 60 km (37 miles), plus 62 km (38·5 miles) of stone revetments along the shoreline serving T-head jetties
Hamburg	110	68	Comprising 63 km (39 miles) of quay (41 for seagoing vessels) and 47 of dolphin berths (20 for seagoing vessels)
Antwerp	99	61	
New York	75	47	Estimate of total for the port
	25·7	16	Marine facilities of the Port Authority of New York and New Jersey
London	35	22	Length in 3 enclosed dock systems only. Additions include 42 berths at tanker terminals
Amsterdam	29	18	

Rotterdam port has a water area of about 2000 ha (5000 acres) for seagoing vessels, in addition to 750 ha (1850 acres) for inland shipping.

Largest impounded ports (Areas of water at high level)

Port	Area of water ha	acres
Antwerp	1306	3226
Le Havre	378	934
Liverpool: Princes to Seaforth	174	430
Birkenhead	73	180

No figure is available for Amsterdam. The North Sea Canal and the port together form a system enclosed by locks, in which the level is maintained about 0·5 m (1·8 ft) below mean sea-level. The area of water enclosed is probably about 700 or 800 ha (say 1500 to 2000 acres). Pumping is necessary to maintain the level.

The first lock was brought into use at Antwerp in 1811.

Impounding was introduced at Liverpool in 1886, and at Birkenhead in 1889, but the term 'impounding' is here used to mean that the high level is maintained by pumping. The impounded level is 4·77 km (15·65 ft) above mean sea-level at Liverpool, and 5·07 m (16·63 ft) at Birkenhead. The quantity of water pumped is on page 202.

Japanese harbour construction.
The most ambitious programme of harbour construction in progress today is in Japan. Modernisation and enlargement of more than 1000 ports of all sizes are involved. Key projects include the construction of islands, each around 400 ha (1000 acres) in area, for industrial use as well as directly for the port, at Yokohama, Kawasaki and Kobe. At Osaka, a reclamation of about 900 ha (2200 acres) is being constructed, plus breakwaters with a total length of 14 km (nearly 9 miles) in depths of water of 12 m (40 ft). Tomakomai, on Hokkaido Island, will be able to berth 200000 t tankers and will have 13·5 km (about 8·5 miles) of breakwaters.

Single-point moorings. There were over 100 single-point moorings for tankers in use throughout the world at the end of 1974. Mostly they were of the type called single-buoy moorings, in which a buoy is anchored to the sea-bed by hanging chains, and the tanker can swing round the buoy when it is moored and unload its oil through floating hoses, connected to the buoy and thence to a sea-bed pipeline. The largest buoy of this type is to be in use in 1976 at Amlwch, Anglesey. The buoy will be in 42 m (138 ft) of water and will be nearly 21 m (69 ft) in diameter and will weigh 500 t. It will be connected to pipelines on the sea-bed, and will moor 500000 t tankers. A similar, but much larger, single-point mooring able to load tankers has been designed.

Breakwaters. A massive mound of rubble is the favoured construction for a breakwater, which is rather surprising, since to stop waves by presenting to them a large, solid obstacle, is neither subtle, scientific nor cheap. Subtlety is introduced by punching the solid full of holes. The Jarlan breakwater dissipates part of the kinetic energy of waves and attenuates their movement at a perforated wall. This device is used for the Ninian Field oil platform (see chapter 4). Another approach is a floating device

The ore-loading terminal at Tubarao

Slewing-bridge shiploaders, each of 16000 t/h capacity

that absorbs or dissipates the energy of the waves. The first such in practice was the Bombardon used for the Mulberry harbours off Normandy in June 1944. A recent development is the Harris floating breakwater. The most intangible breakwater, however, is a curtain of bubbles. Bubbles of air, released on the sea-bed, have an attenuating effect on waves. The prototype was the Brasher air breakwater of 1916. It saved a jetty at El Segundo in California, when waves 4–5 m (12–15 ft) high were 'robbed of their harmful surge' for 23 hours by an installation of 37 m (120 ft) of bubble-generating air main positioned across the head of the jetty and 44 m (145 ft) away from it, plus two 30 m (100 ft) lengths making a T on each side of the jetty. The drawback is today considered to be an excessive consumption of power, if the device is in continuous use.

Table 36. The world's longest breakwaters and jetties.

Name	Length km	Length miles	Year of completion	Country and location	Other details
Breakwaters, moles, or causeways over 4 km (about 2·5 miles) long					
Port of Galveston	10·85	6·74	–	USA, Texas	South breakwater. North breakwater is 7·9 km (4·9 miles) long. Both are of granite, and are 3·2 km (2 miles) apart
Genoa	9·97	6·20	–	Italy	Main breakwater of port, in maximum depth of water of 20 m (66 ft)
Escravos Bar	9·01	5·6	1963	Nigeria	Mole to arrest littoral drift and scour the shipping channel
Dammam Pier	7·79	4·84	1950	Saudi Arabia	Length of causeway only (see below)
Kobe	6·56	4·08	u.c.	Japan, Honshu	For a new port
Europoort	6·00	3·75	1974	Holland, Rotterdam, Hook of Holland	South breakwater. North breakwater is 3 km (1·9 miles) long, depth of water 18 m (59 ft)
Ishakari–Wan	5·47	3·40	u.c.	Japan, Hokkaido	For a new port
Antifer	4·83	3	Project	France, Le Havre	South breakwater. The North breakwater 3·5 km (2·2 miles) long, has been built
Kawasaki	4·20	2·60	u.c.	Japan, Honshu	For construction of artificial island. Consists of 15 m (50 ft) diameter caissons, 9 m (30 ft) high on submarine embankment in 18 m (60 ft) of water
Tuticorin	4·14	2·57	u.c. in 1968	India, Madras	North breakwater
Long Beach	4·07	2·53	1949	USA, California ports of Long Beach and Los Angeles	Longest of a system of four breakwaters. Total length 13·39 km (8·26 miles)
Longest breakwater in Britain					
Holyhead	3	1·86	–	Wales, Anglesey	North breakwater
Longest jetties					
Dammam Pier	10·93	6·79	1950	Saudi Arabia	Causeway, 7·79 km (4·84 miles); steel trestle pier 2·90 km (1·80 miles); main pier 0·226 km (744 ft)
Port Said	7	4·33	–	Egypt	Protects west side of dredged channel into Suez Canal
Longest jetty in Britain					
Bee Ness	2·5	1·55	1930	England, near Rochester	On River Medway at Kingsnorth
World's longest pleasure pier					
Southend	2·16	1·34	1889	England, Essex (Thames Estuary)	Stands on cast-iron screw piles

Conneaut coal and ore installations, Lake Erie, USA. Coal is loaded at a rate of 11 000 short tons per hour, and is unloaded at 10 000 short tons per hour, a record rate. This largest ore unloading plant was completed in 1973. *Conneaut and Tubarao photos: Sorros Associates*

Floating docks. A lifting capacity of about 60 000 t corresponds to a deadweight tonnage of about 250 000 t, because a ship is at its lightest when using a floating dock. In 1971, a very large floating dock was brought into use for shipbuilding at Pascagoula, Mississippi, USA. Its lifting capacity is 60 000 short tons (about 52 700 t) and it has a clear width of 54·8 m (180 ft) and an over-all length of 292·6 m (960 ft).

In 1975, a floating dock made in Poland was due to be brought into use at Gothenburg in Sweden, as No. 5 Dock of Gotaverken Repairs Limited. This dock has a clear width of 55 m (180·4 ft), an over-all length of 303 m (994 ft) and a draught over the keel blocks of 9·2 m (30 ft); it is rated at 55 000 t and will be equipped with a 150 t gantry crane. The Blohm and Voss floating dock planned at Hamburg should be nearly as large – 300 m (984 ft) long by 52 m (171 ft) wide.

An unusual floating dock is under construction at Genoa, Italy. Its structure is of prestressed concrete and steel; its clear width is 65·7 m (215·5 ft), over-all length 350·37 m (1149·5 ft) and a draught of about 15 m (49 ft). Its lifting capacity will be 100 000 t.

The bulk-loading terminal of greatest capacity is the one at Tubarao, Brazil, shown here. It was completed in 1973 and has a throughput of 61×10^6 t of iron ore per year, increasing to 100×10^6 t when 350000 t carriers are used. An access channel 3 km (2 miles) long dredged to 22 m (72 ft) deep, will be deepened to 27 m (88 ft) in the second stage. The two slewing-bridge shiploaders are each of 16000 t/hour capacity; ore is reclaimed from stock by two bucket-wheel excavators. *Photo: Sorros Associates*

The Harris floating breakwater presents a horizontal barrier to the waves and destroys the vertical component of the waves' motion. Mooring forces are surprisingly low. The breakwater is a hollow prestressed concrete structure filled with polystyrene foam; it is similar in shape to a comb, but joined together at both sides. It can reduce wave height by up to 80 per cent for wave lengths that are equal to the width of the breakwater, and waves of shorter length are virtually eliminated. Sea trials at Stokes Bay, Solent show an area of tranquil water behind the breakwater to the right of the photograph. *Photo: Taylor Woodrow*

Artificial island, 14 km (8 miles) off the coast of Brazil in 7 m (23 ft) of water, on which salt, delivered by small vessels, is stored and then loaded in ocean-going ships at a nearby, natural deep-water channel. The island is 100 × 150 m (about 330 × 500 ft) in size. *Photo: Sorros Associates*

There are 29 floating docks in the port of Rotterdam.

When the Panama Canal was opened in 1914, the only craft afloat that was too wide to go through the canal was the US floating dock *Dewey*.

The floating dock to be constructed for use on the River Neva at Leningrad, USSR, will have a unique purpose. When a ship has entered the dock, and its gates have been closed, water will be pumped from the chamber holding the ship, into ballast tanks within the double walls of the dock. The ship will thus be lowered by about 5 m (16 ft). The dock will then be moved up the river beneath seven bridges that would otherwise be too low for the ship to pass beneath. The dock will make four double journeys per day.

The US Navy's advanced base sectional docks have lifting capacities of 71 000 short tons, and in the case of the one at Green Cove Springs, Florida, of 80 000 short tons (72 500 t).

Admiralty Floating Dock 35, built in Bombay in 1947, was towed to Malta and subsequently to Genoa. It has a lifting capacity of 65 000 t and an over-all length of 261 m (857·67 ft).

Table 37. The world's largest dry-docks.

Name	Length × breadth		Depth over sill		Year of completion	Country and location	Other details
	m	ft	m	ft			
Dry-docks for North Sea platforms							
Nigg	305 × 183	1000 × 600	13	43	1973	Scotland, Cromarty Firth	Classified as dry-docks because they have reinforced concrete caisson gates. Other building sites have steel-sheet piling which is removed and an embankment which is dug away to flood the dock
Graythorp	305 × 152	1000 × 500	12	39	1973	England, River Tees	
			(depths at high water)				
Dry-docks over 350 × 50 m (1150 × 165 ft)							
Port Rashid	525 × 100 418 × 80 370 × 66	1722 × 328 1371 × 262 1214 × 216	12·3	40	u.c. 1978	Dubai, Persian Gulf	All three will have swinging caisson gates opening outwards. Largest is for 1 million t ships
Cadiz	525 × 100	1722 × 328	–	–	u.c.	Spain	
Shipbuilding dock	556 × 93	1824 × 305	8·3	28	1969	Northern Ireland, Belfast	For Harland and Wolff Ltd
Fos	465 × 85	1525 × 279	11	36	1975	France, Marseille	For 700 000–800 000 t ships
Botlek	410 × 90	1345 × 295	10·5	34	1973	Holland, Rotterdam	Largest of eight dry-docks in the port
Kiel	426 × 88·4	1398 × 290	9·3	30·5	u.c. Feb. 1976	Germany	For use as a building-dock
Sembawang	384 × 64	1260 × 210	–	–	1975	Singapore	For 477 000 t ships; largest outside Europe and Japan (1975)
Hamburg, Elbe 17	350 × 56	1148 × 184	16	52	In use	Germany	
Lisnave	350 × 54	1147 × 177	–	–	1971	Portugal, Lisbon	For 500 000 t ships; dry-dock for 1 million t ships planned
Puget Sound	351 × 55	1152 × 180	–	–	–	USA, Washington	At US Navy Yard, Bremerton
Saint-Nazaire	354 × 50	1160 × 164	13	14	–	France	

Table 38.
Chronological list of the world's largest dry-dock, showing the rapid increase in size in the 1970s.

Name	Length × breadth m	ft	Depth over sill m	ft	Year of completion	Country and location
Port Rashid	525 × 100	1722 × 328	12·3	40	u.c. 1978	Dubai, Persian Gulf
Fos	465 × 85	1525 × 279	11	36	1975	France, Marseille
Botlek	410 × 90	1345 × 295	10·5	34	1973	Holland, Rotterdam
Lisnave	350 × 54	1148 × 177	–	–	1971	Portugal, Lisbon
East Twin	335 × 51	1100 × 167	11·5	38	1969	Northern Ireland, Belfast
Sturrock	360 × 45 (369 in emergency)	1181 × 148	13·7	45	1939	South Africa, Cape Town
King George V	348 × 41	1142 × 134	11 (below low water)	36	1933	England, Southampton

In 1951 there were 32 major dry-docks in the world, of which 10 were in the British Commonwealth, 10 in the USA, 8 in Europe, and 1 in the Panama Canal Zone.

Invention of the lock. The lock (called 'pound lock' to distinguish it from 'flash lock') was invented in China in AD 984.

Lock working against the highest head. The lock at John Day Dam on the Columbia River, Oregon/Washington, USA, has a lift of 32·9 m (108 ft). The lock is 206 m (675 ft) long by 26 m (86 ft) wide and it was completed in 1968.

The shiplift at Heinrichenburg, near Datteln on the Dortmund–Ems canal in Germany, completed in 1962. The lift consists of a caisson or tank, into which the ship enters like a person having a bath. But the 'bath' is itself floating. There are two shafts, each about 50 m (164 ft) deep, beneath it, in each of which is a great flotation cylinder, 10 m in diameter and 35 m high (33 ft by 115 ft). These cylinders themselves float in water. The 'bath' is supported on columns on the flotation cylinders, and the buoyancy they provide is equal to its weight. The floating assembly is held by four towers, in each of which is a screwed rod. The four rods are turned by an electric motor to raise or lower the lift

Table 39. The world's largest locks (over 305 × 30·5 m (1000 × 100 ft) in size).

Name	Length × breadth		Depth of water		Year of completion	Country and location	Other details
	m	ft	m	ft			
Zandvliet	500 × 57	1639 × 187	18·5	60·7	1973	Belgium, Antwerp	Largest lock serving Antwerp; it increased the locking capacity of the port by 75 per cent
François 1er	400 × 67	1312 × 220	15·5	51	1974	France, Le Havre	The world's largest lock – it will take a heavier vessel than Zandvliet
Ijmuiden	400 × 50	1312 × 164	15	49·3	1930	Holland	Largest of four locks at sea entrance of Amsterdam Ship Canal
West Dock	366 × 42·7	1200 × 140	17·7	58	u.c. 1975	England, Bristol	Entrance lock to new impounded dock: 12 m (40 ft) tidal range
Poe	366 × 33·5	1200 × 110	9	30	1965	USA/Canada, Sault St Marie	Largest lock on 'Soo' Canal
Dunkirk	365 × 50	1197 × 164	13·5	44	–	France	
Boerinne	360 × 50	1181 × 164	15·5	51	u.c.	Belgium, Antwerp	For access to new dock system on south bank of the Scheldt
Baudouin	360 × 45	1181 × 147	15	50	1955	Belgium, Antwerp	
Brunsbüttel Holtenau	310 × 40	1017 × 131	12	39	1914	West Germany, Kiel Canal	
Gladstone River Entrance	326 × 39·6	1070 × 130	14/15	45/50	1927	England, Liverpool	Largest lock serving the port of Liverpool
Iron Gates	310 × 34	1017 × 112	4·5/5·5	15/18	1973	Yugoslavia/ Romania	A hydro-electric dam on Lower Danube. A head of 34 m (112 ft) is navigated in two stages
Panama (twelve locks)	305 × 30·5	1000 × 100	–	–	1914	Panama, US Canal Zone	Head on locks is 8·5 m (28·3 ft) (see chapter 4)
Port Colborne	420 × 24	1380 × 80	9	30	1932	Canada, Ontario	Guard lock of Welland Canal

Shiplift at Lüneburg near Hamburg; the highest vertical lift

Locks 4, 5 and 6, the three twinned flight locks of the Welland Ship Canal, that lift vessels 36·6 m (120 ft) up the Niagara escarpment. *Photo: St Lawrence Seaway Authority*

Highest-head shiplift. The shiplift at Ronquières on the Charleroi–Brussels Canal in Belgium, opened in 1968, overcomes a head of 68·58 m (225 ft). The ramp or inclined plane of the shiplift is 1432 m (4698 ft) long and carries two steel caissons each about 91 × 12 m (298 × 39 ft) in size in which the depth of water is about 3·3 m (11 ft) and in each of which a ship of 1350 t can be carried, or alternatively four 300 t barges. The two caissons are independent of each other. They travel on rail tracks, each with a counterweight, also on rails. The caissons each have 236 wheels (working load on each wheel 22 t) and the counterweights each 192 wheels. It takes 22 minutes to negotiate the lift, which has reduced the number of locks on the canal by 28, and the journey time from 35 to 14 hours.

The highest head surmounted by a vertical shiplift is 38 m (124·6 ft) at Lüneburg in Scharnebeck Shiplift, near Hamburg in West Germany, completed in 1975. Its two caissons are each 100 × 12 m (328 × 39 ft) in size, with a depth of water of 3·5 m (11·5 ft) sufficient for 1200 t ships. They are hung on cables running over sheaves on an endless-chain principle, with counterweights running in shafts beside the caissons. A caisson can be lifted in three minutes, using four electric motors of 150 kW.

Townline road/rail tunnel under the new Welland by-pass section of the Welland Ship Canal. It has been in full use since 1973. *Photo: St Lawrence Seaway Authority*

Zandvliet Lock at the Port of Antwerp. *Photo: Stad Antwerpen*

Gates. The world's largest gate for any lock or dry-dock is the concrete caisson closing the dry-dock used for building production platforms for North Sea oil at Nigg Bay, Cromarty Firth, Scotland. The opening into which the gate fits is 123·7 m (405·8 ft) wide and the depth of water is 13 m (42·7 ft). The caisson is 123 m (403·5 ft) long by 15 m (50 ft) deep and 15 m (50 ft) wide, and it weighs 16 000 t. It has 52 cells, in 4 groups of 13; each group can be pumped separately to trim the caisson. There is a clearance of 350 mm (13·8 in) at each end of the caisson when it is in position. It rests on rubber bearing pads and is sealed by rubber strips.

This size may be compared with the 80 m (262 ft) span of the surge barrier gate on page 197, with the doors 140 m (460 ft) in height of the Saturn Vertical Assembly Building in chapter 1, or with the much larger concrete caissons of the largest immersed-tube tunnels in chapter 2. The swinging caisson at the dry-dock at the Port of Rashid will be a steel gate 100 m (328 ft) in length.

The 34 gates of the Haringvliet Barrage (see chapter 4) span 56·42 m (185 ft) and are either 8·5 or 10·5 m (28 or 34 ft) in depth. The 17 gates on the seaward side (those of the lesser depth) are unique because of the wave forces that they have to resist, being virtually in the open sea.

The gates of the Thames Barrier, which will each span 60 m (197 ft) will also be unique. When closed they lie along the bottom of the structure, leaving the openings of the barrier unimpeded; the gate is rotated to raise and thus close it.

The largest side-hung gate in Britain is the steel guard gate for the Seaforth Passage at Liverpool. It is 40·9 × 16·2 m (134 × 53 ft) with a breadth of 3·6 m (12 ft) and weighs 730 t.

The two sector gates at Seaforth Passage are believed to be the largest of their kind in the world, measuring 28·27 m (92·75 ft) horizontally along the perimeter of the quadrant face, and 16·46 m (54 ft) high.

The largest mitre gates in Britain are at West Dock, Bristol.

Sluice-gates are first mentioned in Chinese literature in the last few decades BC. Dr Needham (see chapter 4) considers they were well known in China by that time, the typical form being the stop-log gate, i.e. a vertical groove in each bank, into which baulks of timber are inserted.

CANALS AND INLAND NAVIGATION

Oldest. Relics of the oldest canal, dated to the fifth millennium BC, were discovered near Mandali, Iraq in 1968.

Longest. The total length of navigable canals in the USSR is about 16000 km (10000 miles). The length of irrigation canals in the Yellow River Plain in China is 8300 km (5160 miles).

Holland has nearly 8000 km (4800 miles) of navigable rivers and canals. When Henry Ford saw them he said 'Fill them in – what roads they'd make!'

Total length of the inland waterway routes in Britain is 3846 km (2390 miles).

Ship Canals

The first ship canal to Amsterdam was built between 1819 and 1825 on the initiative of King William of Orange. It was 79 km (49 miles) long, from Amsterdam to Den Helder, and was 5 m (16·5 ft) deep and large enough for all oceangoing ships. Its locks were 14 m (46 ft) wide and the

Table 40. Inland navigation – the longest routes.

Name	Length of route km	Length of route miles	Year of completion	Country or location	Other details
St Lawrence Seaway	3769	2342	1969	Canada/USA	9 m (30 ft) deep for vessels 222·5 × 23 m (730 × 75·5 ft) for the entire distance – Gulf of St Lawrence to Duluth, 183 m (602 ft) above sea-level
Europe Canal	3500	2175	u.c.	Rhine–Main–Danube	Links the North Sea with the Black Sea. Standardised for 1200 t vessels
Grand Canal	2500	1560	6th century AD	China	Y-shaped route including Yellow River (see chapter 4)
Volga–Baltic	2300 (2960)	1850 (1440)	1965	USSR	Links Baltic and White Seas in the north with Black, Azov and Caspian Seas in the south. Open end of April to mid October
Amazon	*c.* 2900	*c.* 1800	–	Brazil	Substantially natural routes on rivers
Mississippi	2940	1827	–	USA	Length given is from Gulf of Mexico to Minneapolis

Table 41. Ship canals.

Name	Length km	Length miles	Year of completion	Country and location	Other details
Suez	–	–	–	–	See table 53
Panama	–	–	–	–	See table 53
Largest canals on the St Lawrence Seaway:					
Welland Canal	43	27	1932	Canada, Niagara Escarpment	Has eight locks: recently improved (see page 192). Has record for height of lift (head) which is 99 m (326 ft)
Beauharnois Canal	24	15	1944	Canada, Quebec	Built for hydro-electric power. Locks added in 1959. Canal is 1000 m (3300 ft) wide
Manchester Ship Canal	57	35·5	1894	England, Manchester–Mersey	For vessels of 15 000 t to Ince, 12 500 t to Manchester. Five sets of locks
Kiel Canal	98·7	61	1895	West Germany	Enlarged in 1914 and 1966. Now 11 m (36 ft) deep and with a bottom width of 90 m (295 ft). Locks 310 × 40 m (1017 × 131 ft) at each end
Corinth Canal	6·3	3·9	1893	Greece	Depth 8 m (26 ft); bottom width 21 m (69 ft)
North Sea Canal	22·5	14	1867	Holland, Amsterdam	North Sea to eastern part of Amsterdam Harbour. Entrance is through Ijmuiden locks. Depth 15·1 m (49·3 ft). The canal, after recent widening, is 270 m (885 ft) wide at water-level, 170 m (558 ft) wide at the bottom and 15 m (49 ft) deep. It is claimed as the widest and deepest ship canal; recent work at Suez will overtake it. See also note at foot of table 42
Nieuwe Waterweg	27	17	1872	Holland, Rotterdam	For depths today see note on port of Rotterdam in text. There is an approach channel in the North Sea, 12 km (7·5 miles) long, 1200 m (3937 wide and 23·5 m (77 ft) deep

Table 42. The longest canals.

Name	Length km	Length miles	Year of completion	Country and location	Other details
Grand Canal	1700	1060	1327	China	A broad canal with a summit 42 m (138 ft) above sea-level. Summit canal completed in 1283 (see chapter 4). Present length not known.
Kara-Kum	*c.* 1500	*c.* 930	1975	USSR, Turkmenistan	A unique project. The canal takes off from the Amu Darya River and crosses the Kara-Kum Desert (an absolutely waterless sand desert) from east to west. Eventually, its water will irrigate 1 million ha (2·5 million acres) of land. Its capacity is 285 m^3/s (about 10000 ft^3/s) and a length of 450 km (724 miles) is navigable. It also solves problems of industrial and domestic water-supply
California Canai	619	385	1974	USA, California	Canal section of aqueduct listed in table 48
Rajasthan Canal	685·5	426	Second stage u.c.	India, Rajasthan	Carries 524 m^3/s (18500 ft^3) for irrigation. Second stage of canal is 274 km (170 miles) long. Scheme also includes 6308 km (3920 miles) of branches and distributors
Erie Canal	587	365	1825	USA, New York	From Hudson River, Albany to the Great Lakes at Buffalo. Described as the first school of American civil engineering. Appears to be the longest canal of the modern era purely for navigation
Irtysh–Karaganda	450	280	–	USSR	Has 22 pumping stations, 13 earth dams and 14 weirs
V I Lenin Canal	361	224	–	USSR	The Volga–Baltic Canal; has five power stations and seven locks. Note that the Volga–Don Canal – 100 km (62 miles) long – is also called V I Lenin
Nagarjunasagar	378 351	235 218	1974	India	Left-bank feeder canal of irrigation scheme. Right-bank feeder
Lloyd scheme	1670	1040	1932	Pakistan	Total length of seven irrigation canals (see page 206). Largest canal – the Eastern Nara – has a bottom width of 105 m (346 ft). Full supply discharge is 1346 m^3/s (47530 ft^3/s)
Indus scheme	650	400	1971	Pakistan	Total length of the eight link canals of the Indus Basin scheme. The Qadirabad–Balloki Canal, 129 km (80 miles) long, has a bed width of 102 m (335 ft), depth 4 m (13 ft) and flow 527 m^3/s (18600 ft^3/s). The Chasma–Jhelum Canal has a bottom width of 116 m (380 ft)

Largest flow: The North Kiangsu Canal (see chapter 2) is reported to have a bed width of 126 m (413 ft) and a flow at full capacity of 707 m^3/s (24966 ft^3/s). These appear to be record figures for an irrigation canal, but are exceeded by the 25 km (15 mile) head-race canal (also part of the St Lawrence Seaway) of the Beauharnois power station in Quebec; corresponding figures in this case are: width 1000 m (3300 ft); flow (maximum) 9000 m^3/s (320000 ft^3/s).

largest in the world at that time. A sea entrance to Rotterdam was also commanded by the King, from Voorne to Hellevoetsluis with a canal and locks of the same size. It was completed in 1829. These canals became too small. The Nieuwe Waterweg – Rotterdam's present sea entrance canal from the Hook of Holland – was constructed in 1872. It was deepened initially by the use of fifteen steam-dredgers (natural scour proved far too feeble) and it has been dredged deeper and wider ever since. The port is open, without locks.

The North Sea Canal from Ijmuiden to Amsterdam was completed in 1867. Initially, it was 9 m (30 ft) deep and 25 m (82 ft) wide at the bottom; locks have been built successively to give access to this canal as follows:

Date	m width × depth × length	ft width × depth × length	ratio of volumes
1867	18 × 8 × 120	60 × 26 × 392	1
1896	25 × 10 × 229	82 × 33 × 750	3·3
1930	50 × 15 × 402	165 × 50 × 1320	17·7

St Lawrence Seaway

The town of Welland, Ontario, is unique in having been by-passed by a canal. It is claimed that even today, the Panama Canal is a unique feat of construction; in the author's view the St Lawrence Seaway is its peer. The ladder of three pairs of great locks at Gatun is indeed impressive, but the locks at Welland, if smaller, surmount nearly four times the head. The Seaway owes much to the engineers who surmounted the Niagara escarpment, daringly and improbably, with a succession of canals, of which the Welland By-pass is the latest improvement.

The first Welland Canal was started in 1824; it had 40 timber locks. It was used from 1829 to 1845. The second canal, started in 1842, had 27 locks; it was extensively improved in 1871–87. Work on the fourth canal began in 1913, and was completed in 1932.

Welland Canal locks

	Number of locks	m length × breadth × depth	ft length × breadth × depth
First canal	40	33·5 × 6·7 × 2·4	110 × 22 × 8
Second canal	27	46 × 8 × 3	150 × 26·5 × 10
Third canal	26	82 × 13·7 × 4·3	270 × 45 × 14
Fourth canal	8	262 × 24 × 9	859 × 80 × 30

Note: the head surmounted is 99 m (326 ft); the fourth canal has seven locks plus a guard lock at Port Colborne, the dimensions of which are in table 39.

With growth of traffic on the Seaway, the Welland Canal became congested. The Welland By-pass section has been constructed as a remedy. It increased the canal's capacity by about 35 per cent and reduced transit time from 46 to 22 hours. It is 13·4 km (8·3 miles) long with a navigable width of 107 m (350 ft); it is free from sharp bends and bridges – relocation of roads and railways has been an important part of the scheme, which includes a major tunnel under the new canal.

barrages and spillways

Table 43. The world's largest high-head spillways (discharging over 30 000 m^3/s (about 1 million ft^3/s)). (For flood discharges at large barrages see table 44.)

Name	Nominal maximum discharge m^3/s	ft^3/s	Head m	ft	Year of completion	Country and location
Amistad	54 000	1·90 million	–	–	–	Mexico/USA, Bravo River
Guri	47 000	1·66 million	98·5	323	1969	Venezuela, Caroni River
Hirakud	41 626	1·47 million	–	–	1956	India, Mahanadi River
Infernielo	40 200	1·42 million	148	486	–	Mexico, Balsas River
Wanapum	39 660	1·40 million	35	115	1964	USA, Columbia River
Priest Rapids	39 660	1·40 million	29	95	1961	USA, Columbia River
Kadana	32 564	1·15 million	43	141	–	India, Mahi River, Gujarat
Nagarjunasagar	32 560	1·15 million	105	345	1974	India (see table 13)
Mangla	31 160	1·10 million	104	341	1967	Pakistan, Jhelum River

Table 43 can be rearranged according to the rates at which energy is dissipated by the discharge. Assuming the discharges and heads listed in the table, rates of energy dissipation of more than 30 GW are:

Spillway	Power dissipated GW
Infernielo	58
Guri	45
Amistad	42 (assuming 80 m (260 ft) head)
Nagarjunasagar	33
Mangla	32

Intensity of discharge

Spillway	m^3/s per m length
Guri	343·6
Reza Shah Kabir (u.c.)	333·3
Mangla	315·5
Derbendi Khan, Iraq	253·2
Ichair, India	222
Wanapum	216·8

(Source: Mr David S Louie, Harza Engineering.)

The spillway of Nourek Dam should be added; it will discharge 30 000 m^3/s (1 million ft^3/s) and could well head the list, but the author has been unable to ascertain the head.

Table 44. The world's largest barrages over 1000 m in length or passing a flood of over 20000 m³/s (about 3300 ft or 700000 ft³/s).
Note: A broad definition, roughly covering the difference between a dam and a barrage that has become established in English usage is: a barrage is a structure that impounds water over a small range of heads mainly by the use of sluice-gates. Low-head hydro-electric schemes are called barrages, but the distinction disappears as the head increases. For example, Beauharnois on the St Lawrence – see page 200 – is equally dam, barrage or power station.

Name	Length		Maximum head*		Maximum flood passed		Year of completion	Country and location	Gates or openings	Purpose
	m	ft	m	ft	m³/s	millions of ft³/s				
Farakka	2245	7366	13·7	45	70510	2·49	1970	India, Ganga River	108 gated bays of 18 m (60 ft) plus two 8 m (27 ft) fish locks	To maintain navigable depths in the Port of Calcutta (see caption)
Lloyd (Sukkur)	c. 1500	c. 5000	8	25	c. 42000	c. 1·5	1932	Pakistan, Indus River	66 gated bays of 18 m (60 ft)	Irrigates 2 million ha (5 million acres)
Ghulam Mohammed	c. 1000	c. 3300	6	20	c. 42000	c. 1·5	1960	Pakistan, Indus River	44 gated bays of 18 m (60 ft)	Irrigation. The lowest of the Indus barrages, at Kotri
Sone	1407	4616	7	23	40400	1·43	–	India, Sone River	69 gated bays of 18 m (60 ft), 4·3 or 4·9 m (14 or 16 ft) high gates	Irrigation
Kosi	1150	3770	6	20	25750	0·9	1963	India/Nepal, Kosi River	56 gated bays of 18 m (60 ft), 6·4 or 7·9 m (21 or 26 ft) high gates	Irrigation and flood control. Has long earth wing dams
Chasma	1097	3600	12	40	31150	1·1	1971	Pakistan, Indus River	52 gated bays of 18 m (60 ft)	Part of Indus Basin scheme. Has about 20 km (12 miles) of earth dams
Haringvliet	1049 (control structure only)	3440	10·5	34	21000	0·74	1970	Holland, Rhine Delta	17 gated bays each of 56·5 m (185 ft)	Sea defence. Part of Delta scheme. Length given is for control structure only
Qadirabad	1028	3373	10·7	35	25486	0·9	1967	Pakistan, Chenab River	50 radial gates	Part of Indus Basin scheme
Bonnet Carré	2347	7700	–	–	7080	0·25	1930s	USA, Mississippi River	350 gated bays each 6 m (20 ft) wide	Flood protection
Morganza	1191	3906	–	–	16990	0·6	1955	USA, Mississippi River	125 gated bays each 8·6 m (28·3 ft) wide	Flood protection (see chapter 4)

* Maximum upstream pond-level to floor-level downstream.

Farakka Barrage. The flood flow listed in table 44 was passed in 1971; the barrage and its appurtenant works behaved well during the flood. This is the largest flood on record, passed by any civil engineering structure. Farakka Barrage controls the Ganga River. Its purpose is to divert water into the Bhagirathi River (or Hooghly River as it is called in its tidal portion) so promoting scouring of the bed of the Hooghly River and improving navigation depths in the Port of Calcutta. Historically, the Ganga main stream has been capturing an increasing portion of the flow and the Hooghly's flow has decreased. In addition to the main barrage, details of which are in the table, the scheme includes a head regulator, with eleven 12 m (40 ft) gate-controlled openings, a feeder canal 42 km (26 miles) long to the Bhagirathi River, and the Jangipur Barrage on that river. *Photo: D Mookerjea*

Tarbela Dam has two spillways that, together, can discharge 42 720 m^3/s at a head of 101 m (about 1·5 million ft^3/s and 330 ft) and dissipate 42 GW (24 GW for the larger one alone).

The unit here is the gigawatt (GW) which is 1000 million watts (i.e. 1000 megawatts (MW) or 1 million kilowatts (kW)). The world's largest power station, Krasnoyarsk, has an installed capacity of 6 GW, so the spillways listed above are capable of releasing energy five to ten times the rate at which it is generated at Krasnoyarsk when that station is operating at full load.

It is logical to classify Infernielo as the world's largest spillway. On this basis, however, Farakka Barrage cannot be included. At high flood, all the gates of a river barrage are opened and the flood sweeps by, so it is not possible to quote a rate of energy dissipated, only a flow.

Sayany Dam will have two outlets 70 m (230 ft) below normal reservoir level, discharging into a stilling basin 130 m long × 35 m deep (427 × 115 ft). The outlets themselves are lined with steel; air is admitted under the jet to prevent cavitation of the special high-strength concrete. Very accurate work has been specified. Spillway data are: capacity 13 600 m^3/s (0·48 million ft^3/s); head 208 m (682 ft); rate of energy dissipation 28 GW.

When there is a flood at Zeya Dam, it will surely be the most spectacular sight among the world's hydraulic structures. A special form of nose has been devised for the ski-jump, that will throw the jet a distance of 160–180 m (up to almost 600 ft) from the dam. Spillway data are: capacity 28 000 m^3/s (roughly 1 million ft^3/s); head about 90 m (roughly 300 ft); rate of energy dissipation 25 GW.

The figures quoted above are design figures. The extreme flood that will actually be experienced is a matter of probability.

The power in the spillway discharge can be extremely destructive if it

Ski-jump spillway at L'Aigle Dam in the Massif Central, France. An early example of the adventurous use of the ski jump at a constricted site, the spillways being built over the roof of the power station. *Photo: Electricité de France*

is not competently handled. Various devices are used to dissipate energy, of which the most spectacular is the ski-jump, in which the discharge is thrown into the air as a jet, and falls on a deep pool where it can do no harm.

The intensity of discharge has increased sharply in recent schemes. Krasnoyarsk Dam is claimed to be the first example in world practice of a high-head spillway – the head is about 100 m (330 ft) with a discharge of 100 m^3/s for each metre of length of the spillway crest (this is equivalent to a little over 1000 ft^3/s for each foot length of the spillway crest). The list on page 193 also gives some of the highest intensities of discharge, one or two of which predate Krasnoyarsk.

flood control and sea defence

The greatest scheme for protection against river floods is on the Lower Mississippi River. The greatest scheme of sea defence is that which protects the Dutch coast. Both are described in chapter 4. The greatest scheme for protection against river floods in Britain is Vermuyden's Great Ouse scheme, and the largest single work of sea defence is the Thames Barrier.

An area of nearly 1 million ha (2·3 million acres) of the Florida Everglades in the United States is protected from flooding by an extensive system of levees, dams, spillways and canals. Construction started in 1952, the cost of the scheme being about $200 million.

Flood-protection works in India during the period 1954–72 involved the construction of 7300 km (about 4500 miles) of embankments and the installation of 10 000 km (over 6000 miles) of drains. This programme of work has protected some 200 towns, and during the course of it, 4585 villages were raised to a higher level.

The coastal embankment project in Bangladesh involved over 4000 km (2500 miles) of embankment and 770 sluices in its first phase, due to be

The greatest clear span of any sluice-gate is 80 m (262·4 ft). Two gates of this span are in use at the surge barrier on the Hollandse Ijssel River in Holland, one of the structures of the Delta scheme. Each gate is 81·2 m long by 11·5 m high (266 × 38 ft) and is hung at each side on crossheads that carry a roller assembly with tapered bearing surfaces that 'crack open' the gate when it is lifted. The dock gate at Nigg Bay is the only structure of similar kind that has a greater span completed in 1971. The aim was to protect about 1 million ha (2·5 million acres) of land from tidal flooding and salinity in districts close to Chittagong.

Containing the River Great Ouse

Britain's largest flood-protection works are two schemes that contain the River Great Ouse. The earlier and larger scheme was built by Sir Cornelius Vermuyden in 1630–52. It consists of two straight cuts called the Old Bedford and the Hundred Foot rivers each about 32 km (20 miles) long side by side, cutting across a great bend in the river, between Earith and Denver, plus a sluice across the river at Denver. The Ely Ouse was thus turned into a tideless river and storage for river floods provided between the two cuts. The scheme was immediately successful. In time, however, the now well-drained peat soil shrank, until it had to be drained by pumping rather than by gravity.

In 1947 the area experienced its worst flood. A scheme recommended some years earlier was then adopted, and eventually completed in 1964. It consists of: a relief channel – that is a second channel beside the river – for 17·5 km (11 miles) downstream of Denver (i.e. downstream of Vermuyden's two cuts) controlled by head and tail sluices, plus heightening of the tidal embankments; a cut-off channel 44 km (27·5 miles) long that intercepts floods from the hills to the east (Vermuyden also proposed this channel) and takes them to the new relief channel at Denver; and widening and deepening the Ten Mile and Ely Ouse rivers and strengthening their embankments over a length of about 30 km (20 miles) (i.e. increasing the capacity of the river upstream of Denver). So far, the scheme had not been tested by a flood as serious as the one in 1947.

Wadi Tharthar Scheme

The Wadi Tharthar scheme on the Tigris River in Iraq is thought to be unique, excepting only the neighbouring Habbaniyah scheme on the Euphrates River, and apparently the ancient Moeris Reservoir (2300 BC) in the natural depression of Fayum 80 km (50 miles) south-west of Cairo.

A low spillway discharging into a flood storage reservoir is a normal element of a flood-protection scheme, but in the case of Wadi Tharthar, a natural depression larger in area than the Dead Sea provides generous storage for any flood on the Tigris. The scheme consists of a barrage across the river, a head regulator, a protective dike to retain the flood discharge

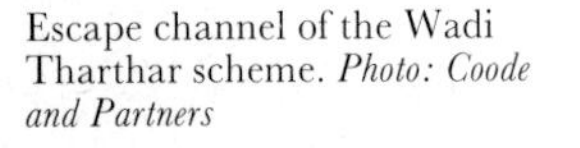

Escape channel of the Wadi Tharthar scheme. *Photo: Coode and Partners*

Artist's view of the Thames Barrier as it will look when completed. Normally the gates will lie flat on the river-bed, and will be rotated to close the barrier. This will only be done if a tidal surge is threatened from the North Sea. Large areas of London are vulnerable to flooding from a tidal surge, though it could be that the barrier will never be required to close which spreads over the desert across a width of 10 km (6 miles) and to a depth of 5 m (16 ft), and finally an escape channel leading to the depression. A flood discharge of 9000 m^3/s (nearly 320000 ft^3/s) can be diverted at the regulator, and the capacity of the relief channel is 7000 m^3/s (250000 ft^3/s). The maximum recorded flood of the river at the barrage is 14000 m^3/s (nearly 0·5 million ft^3/s).

hydro-electric power

LARGEST SITES

The Zaire River (formerly the Congo River) is exceeded in flow only by the Amazon. One site on the Zaire at Inga (near Kinshasa on the lower part of the river) where the river falls 102 m (335 ft) in less than 16 km (10 miles) and turns in a loop, has a potential for development to an installed capacity of 40 GW generating continuously (i.e. an annual output of 370 TWh). The first stage of this scheme, with an installed capacity of 350 MW, less than 1 per cent of the potential, was completed in 1973. There are plans to expand tenfold by 1983.

A site of comparable potential has been identified that could generate between 17 and 38 GW continuously. It is near the border between Tibet and Assam. The Brahmaputra River, as it flows out of the Himalayas, turns in a great loop from east to west, and the site could be exploited by tunnelling across the loop. A third comparable site now being investigated is on the Parana River, about 400 km (250 miles) from Asunción (Paraguay) in Brazil, where it is thought that an installed capacity between 20 and 70 GW would be feasible. An initial development of 12·5 GW known as Itaipu has been started (table 45).

The Gulf Stream has a flow more than 25 times the total average flow of all the world's rivers. At the Straits of Bimini, its narrowest point, the flow is 10 $miles^3$ (40 km^3) per hour and the velocity up to 7 miles (11 km) per hour. Taking two-thirds of this velocity as an average for calculation, the gross power in the stream is about 30 GW.

The runner of the first of the generating sets in the Number 3 power station at Grand Coulee being lowered into position. Some of the 32 wicket gates that control flow into the runner can be seen. The runner weighs 454 t (500 short tons) and had to be fabricated on site because it is so big, its diameter being 10 m (33 ft). This set was due to generate power in August, 1975. *Photo: US Bureau of Reclamation*

Right: Beauharnois power station on the St Lawrence River, Canada. A low-head station, the largest in the world when completed. In the background the locks of the St Lawrence Seaway can be seen. *Photo: Hydro Quebec*

Table 45. The world's largest hydro-electric power stations (over 4 GW installed capacity).

Name	Installed capacity MW	Gross head m	Gross head ft	Average annual output TWh	Year of completion	Country and location	Other details
Itaipu	12600	c.180	c.600	90	u.c.	Brazil/Paraguay	See text
James Bay	10000	–	–	–	u.c. 1985	Canada, Quebec	Will have four major dams
Grand Coulee: ultimate, about	10000	90	300		–	USA, Washington, Columbia River	No. 3 station will have twelve sets. The first six – 3 × 600 MW + 3 × 700 MW – are being installed. A pioneer in great size in all its aspects
present	2300	–	–		1939*		
Krasnoyarsk	6000	101	331	20·4	1968	USSR, Yenisei River	See page 235
Sayany	6400	222	728	22·9	u.c.	USSR, Yenisei River	
Churchill Falls	5225	323	1060	34·5	1975	Canada, Labrador	See chapter 2
Bratsk	4600	106	348	22·6	1964	USSR, Angara River	Largest of the chain of six power stations planned on the Angara River
Ust–Ilim	4320	89	290	21·9	u.c.	USSR, Angara River	

* No. 1 station only.
Note: For definition of GW and MW, see page 195 (1 GW = 1000 MW). 1 TWh signifies 1000 million units. One unit is one kilowatt-hour.
The largest thermal-electric power stations in use are of about 2·5 GW installed capacity; ones of about 4 GW are under construction.
There are about a dozen hydro-electric power stations with installed capacities in the range of 2 to 4 GW.

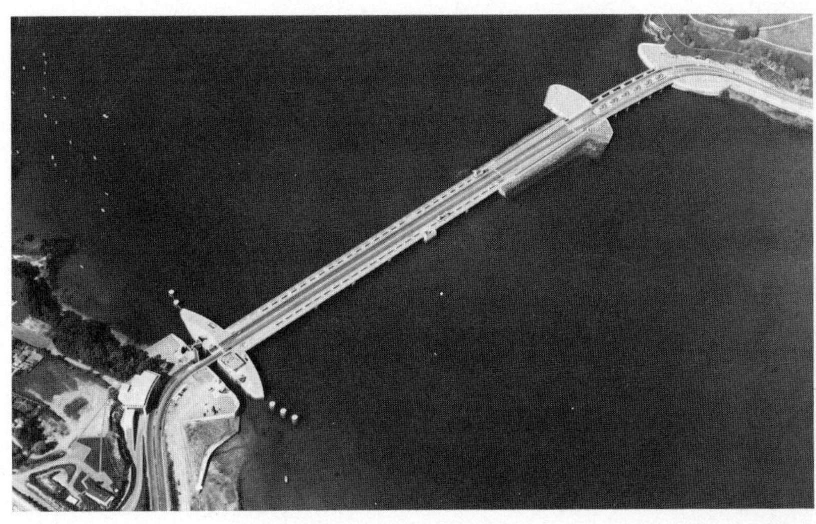

Small – in terms of power output – but unique. The Rance tidal power barrage near St Malo in northern France. *Photo: Electricité de France*

pumping stations

Four-stage high-head centrifugal pump at A D Edmonston plant. The intake pipe is shown at the base of the pump; the motor will be at the top

The largest pump used for water supply in London is at Datchet. It will deliver 5·26 m³/s (100 million Imperial gal/day) against a head of 22 m (72 ft) and its power is 1·6 MW

Archimedian screw pumps for land drainage in the Fylde district in Lancashire. These pumps at Thornton Cleveleys near Blackpool are considered to have the largest output – 4·25 m³/s (150 ft³/s or about 940 Imperial gal/s) – in Britain for pumps of this kind.

As an example of one of the world's major pumping installations for dry-docks, the capacity of the pumping plant at Fos, Marseille (see table 37) may be quoted. Three pumps each have a maximum output of 13 m³/s (459 ft³/s or 2859 Imp gal/s) and can empty the dock in under three and a half hours.

The quantity of water pumped to maintain the impounded level of the port (see page 178) at Liverpool is, on average, 129·6 million Imp gal/day (6·82 m³/s) and at Birkenhead 86·4 million gal/day (4·55 m³/s).

The first pumping station in the modern sense was Morris's London Bridge Waterworks, opened in 1582. This was the City of London's most important water supply until the New River came into use. It continued to operate with several enlargements until old London Bridge was replaced in the 1820s.

London Bridge Waterworks consisted of a floating stage beneath a northern arch of the bridge, on which a water-wheel was turned by the tide. The water-wheel powered, through a crank and connecting-rod, a number of force pumps – the force pump was reinvented and first applied by Morris – and these pumps delivered water to the mains of Morris's supply network. He set the pattern of modern water-supply with this installation – he delivered by pipe to the consumer's premises, and the consumers paid him for the water.

Perhaps the most significant invention on which the Industrial Revolution, and hence our modern age, depended, was a practical heat engine. The first such – Newcomen's – engine was used initially for pumping water from mines. The first Newcomen engine was put to work at Dudley near Birmingham in 1712.

The first steam-powered pumping station in the United Sates started working in 1801. It was at Fairmount Water Works, which supplied Philadelphia with water. The works was converted to water-power in 1822.

Table 46. The world's largest pumping stations.

Name	Quantity pumped m³/s	Quantity pumped ft³/s	Head m	Head ft	Pumping power MW	Year of completion	Other details
Pumped-storage hydro-electric stations							
San Fiorano II, Italy	13·8	487	1438	4718	212	1972	Highest head of any hydro-electric station
Ludington, USA	1870	66 000	90/110	295/362	1900 (max.)	1973	Beside Lake Michigan. The world's largest pumping station in terms of power and flow. Energy stored 15 GWh
Cruachan, Scotland	127·5	4500	333 to 367	1094 to 1205	400	1965	Largest in Britain. Four reversible sets, beside Loch Awe
Hornberg, West Germany	144	5085	630	2067	1000	u.c. 1976	Four similar sets. River Wehra (Upper Rhine catchment)
Dinorwic, Wales	390	13 770	530	1739	1800 (max.)	u.c. 1981	Full power will be reached in ten seconds. Six sets; output 1440/1800 MW generating
Pumping stations for irrigation							
Kokhovskaya, USSR	580	20 483	–	–	–	1975	Serves one of the largest irrigation schemes in the Ukraine taking water from the Kakhovka Reservoir on the Dnieper River
A D Edmonston, USA	116	4095	587	1926	832	1971	State Water Project, California
Delta, USA	292	10 303	74	244	257	1971	State Water Project, California
Grand Coulee, USA:							
present*	368	13 000	89 to	292 to	391	1940	Pumps from Colorado River
ultimate	560	19 800	94	310	591		
Don Amigo, USA	374	13 200	33 to 38	107 to 125	179	1940	California, serving State and Central Valley Projects
San Luis, USA	311	11 000	88	290	376	1973	Pumping/generating plant, Central Valley Project
Pumping stations for land drainage							
De Blocq van Kuffeler, Holland	50 (nominal)	1765	0 to 6	0 to 20	3·6	1972	Serves South Flevoland Polder, Ijsselmeer
	64 (max.)	2260	5·2	17	4·2		
Wortman, Holland	33·3	1176	0 to 6	0 to 20	3	1950s	Four similar pumps, three serve Eastern Flevoland Polder; fourth installed for South Flevoland
Altmouth, England	84	2960	5·3	17·5	6	1972	Britain's largest for land drainage, Formby between Southport and Liverpool, catchment 230 km² (89 miles²)
Pumping station for cooling water for a thermal power station							
Longannet, Scotland	90	3200	9·5	31	2·6	1973	The installed generating capacity at Longannet is 2400 MW. This is exceeded only by Paradise, Kentucky, USA
Largest pumping stations for main drainage in Britain							
Abbey Mills	42·3	1493	–	–	–	–	Sewage pumping station serving an area in east London
West Hull, England	31	1100	17	55	5	1956	The city of Kingston-upon-Hull lies below the high-water level of spring tides in the Humber Estuary
	40	1400			7	(ultimate)	

irrigation

Irrigation is reputed to have started in the Middle East, and maybe was the basic technique that permitted agriculture, and, therefore, a settled, civilised life and the growth of towns. The oldest form of irrigation is basin irrigation. The river, when in flood, overflows its banks and inundates the fields. Regulation of the river plus precise control of water are the basis of modern schemes. Ground water can be utilised also, generally by tube wells, i.e. submersible pumps installed down well-holes. Spray irrigation is widely practised and may involve the use of domestic water-supply, or simply equipment to abstract from a local stream.

Drainage is an essential complement to irrigation. The total area of land in the world artificially drained is about 100 million ha (250 million acres) of which 37 million ha (91 million acres) are in the USA and 2·5 million ha (6·2 million acres) in Britain.

Historical

King Menes (*c.* 3100 BC) first undertook large-scale basin construction in the Nile Valley. Perennial irrigation was introduced in the Nile Valley in 1820. In 1968, with the utilisation of the reservoir of the Aswan High Dam, basin irrigation ceased.

Mohenjo-Daro – a city on the banks of the Indus River in what is today Pakistan – also relied on basin irrigation for its crops over 5000 years ago.

The Nahrawan and Shatt-el-Hai irrigation canals date from about 2200 BC and are the largest of which traces remain in the Euphrates and Tigris valleys. The former is reputed to have been 9–15 m (30–50 ft) deep and about 120 m (400 ft) wide.

Irrigation was practised in the Amu Darya catchment of the USSR in the 6th to 4th centuries BC.

The highest pumping head and the largest pumping station for irrigation are claimed by the A D Edmonston pumping plant of the California State Water project (table 46). This plant was completed in 1971. Its purpose is to provide the main lift for water conveyed over the Techapi Mountains to southern California. The plant has 14 pumps. *Edmonston Plant photos: California Department of Water Resources*

Table 47. Areas of land irrigated.

World totals of land under irrigation

Year	million ha	million acres
1800	8	20
1900	48	120
1949	92	227
1959	149	368
1969	more than 200	more than 494

Totals by country for countries with more than 3 million ha (7·4 million acres) irrigated in 1969

	1900		1969		1969
	millions ha	millions acres	millions ha	millions acres	% of cultivated area irrigated
China	*		74†	183	68
India	11·7	28·8	30	74	22
USA	3	7·4	17	42	10
Pakistan and Bangladesh	3·9	9·5	12	30	42
USSR	3·8	9·4	10	25	4
Iraq	–	–	4	10	53
Indonesia	–	–	3·8	9	27
Japan	2·7	6·7	3·4	8	57
Mexico	–	–	3·3	8	22
Iran	–	–	3·1	7·7	45
Italy	1·3	3·2	3·1	7·7	11
Egypt	2	4·9	3	7·4	100
Britain	–	–	0·1	0·25	1·5

* Area sown with crops in 1900: 57 million ha (140 million acres).
(Acknowledgement is made to K K Framji and I K Mahajan, compilers for the International Commission on Irrigation and Drainage's publication *Irrigation and Drainage in the World.*)
† 1960 figure. The figure for 1949 was 20 million ha (50 million acres). Planned total of irrigated land is 100 million ha (almost 250 million acres) that is about 90 per cent of China's total area of arable land.

The Grand Anicut, a weir on the Cauvery River in southern India, dates from the 2nd century AD. It is of masonry and is 330 m (1080 ft) long, about 4·5 to 5·5 m (15 to 18 ft) in height and 12 to 18 m (40 to 60 ft) wide. It was remodelled and fitted with sluice-gates in 1899–1902. For centuries it irrigated 0·4 million ha (1 million acres) of land.

Some early Chinese examples are in chapter 4, and early Japanese and Sri Lankan dams are discussed in chapter 2. In Sri Lanka, from the 6th century BC to the 12th century AD over 10 000 reservoirs appear to have been constructed, varying in size from about 5 to 123 million m^3 (6·5 to 160 million yd^3).

Irrigation in Iran by kanats (see chapter 2) is very ancient and widespread.

Modern

The Hsinhua News Agency quotes a report from Peking dated 8 September 1975, which states that every winter and spring, 'dozens of millions or even up to one hundred million peasants with tens of thousands or even up to a million rural cadres taking part' undertake capital projects to improve or gain agricultural land. 'Between last winter and the end of June this year alone, a total of 11 000 million $metres^3$ of stone and earthwork was done, twice as much as the average of the previous four years. One and a half million irrigation works of different sizes and types were completed, including 23 000 pump-wells with all necessary accessories. An additional four million hectares of farmland was brought under irrigation, another two million freed from the threat of waterlogging, 5·33 million levelled, 1·13 million hectares of slope land terraced, and 0·4 million hectares of new land created. All these figures exceed comparable ones in any previous year.'

The Sadovia–Corabia irrigation scheme in Romania, although it breaks no records for size (it serves about 80 000 ha (200 000 acres)) is a prototype for a modern concept of irrigation. Water is pumped from the Danube through main feeder canals. The farmers have sprinklers and other controls of piped water that are simple and convenient to use. The main supply system automatically responds to the demand for water; virtually limitless supplies can thus be switched on by an individual farmer. The scheme is economical in capital cost; it has no high-level storage and needs none, because of the responses of its automatic control.

Lloyd Scheme

The Lloyd Barrage at Sukkur in Pakistan (see table 44) completed in 1932 is the key structure in what is still the world's largest irrigation scheme, in the sense of a single, strictly defined project, as opposed to a comprehensive 'project' such as the development of a whole river basin.

It commands almost 3 million ha (7·4 million acres) of which two-thirds (2 million ha; 5·01 million acres) was under cultivation at full development, a stage reached in 1934. Some of this area was irrigated previously, and the new area under crops was 0·8 million ha (2 million acres).

The scheme has seven main canals (total length 1670 km (1040 miles)), 1970 bridges and regulators, and a total length of all canals of 9920 km (6166 miles).

The originator of the scheme was A A Musto, and the Chief Engineer,

Sir Charlton Scott Cholmeley Harrison. Its cost was £15 million, of which the barrage itself accounted for £4·3 million.

Irrigation In California

Eighty per cent of California's farm produce comes from irrigated land. The two greatest schemes that sustain California's irrigation are the multiple-purpose Central Valley project and the State Water project. The latter is claimed as the largest unified water development so far undertaken anywhere. The author thinks that the Indus Basin scheme is the record-holder.

The primary purpose of the State Water project is to conserve, and deliver to areas that need it, 5·2 km^3 (1·25 miles^3, or 4 230 000 acre feet in the official literature) of water annually, for domestic, industrial and agricultural use. But the scheme also has benefits from flood control, hydro-electric power generation and recreation.

Water is distributed along the entire length of the scheme – in the general vicinity of Oroville, the north and south San Francisco Bay regions, the southern San Joaquin Valley, San Luis Obispo and Santa Barbara counties, and about 60 per cent of the total south of the Techapi and San Bernardino mountains. The main supply extends along 1329 km (826 miles) of which 264 km (164 miles) are natural – the Feather and Sacramento rivers and the Sacramento–San Joaquin Delta – and 1065 km (662 miles) consist of canals, lakes, tunnels and pipelines.

The California Aqueduct in the San Joaquin Valley. *Photo: California Department of Water Resources*

Thirty-one agencies purchase water from the project. They serve a gross area of nearly 10 million ha (about 25 million acres), but this is much larger, of course, than the area actually under irrigation. About 320 constructional contracts were needed to complete the project, its cost being about $2300 million. Major installations comprise 22 dams, 23 pumping stations, 8 power stations and the 1065 km (662 miles) of constructed aqueduct mentioned above. The State Water project was substantially completed by 1974 after twelve years of construction.

In a short description it is difficult to separate the State Water project from the Central Valley project, since both serve areas that overlap, and jointly use certain reservoirs and aqueducts – the difference being that the Central Valley project has been the responsibility of the Bureau of Reclamation (i.e. the US Department of the Interior) and not the State of California. Construction of the Central Valley project started in 1937 and water was supplied from it for the first time in 1940. Each year it supplies from 2·5 to 3 million acre feet (say 3 to 3·75 km³) of water for irrigation that is used on 0·8 million ha (2 million acres) of fertile land, domestic and industrial water for 750000 people, and generates 5·5 TWh of electricity. The yield from the irrigated land in 28 years was more than five times the cost of construction.

The Central Valley project is still growing and there are plans to double its size, and then double again. The need for an extra 5 or 6 million acre feet (6 or 7·5 km³) of water annually is foreseen in the next twenty years.

Irrigation in India.
About one-sixth of the potentially useful flow of Indian rivers was utilised for irrigation by 1951: more than one-third by 1969.

Since 1951 34 important irrigation projects have been undertaken. The largest in terms of ultimate irrigated areas are:

	ha × million	**acres × million**
Bhakra–Nangal	1·46	3·61
Gandak	1·46	3·61
Rajasthan	1·27	3·14
Nagarjunasagar	0·83	2·05
Mahanadi Delta	0·68	1·68

At partition, just over two-thirds of the irrigated area of the sub-continent remained in India. Nineteenth-century schemes commanding over 0·5 million ha (1·2 million acres) and their dates of construction are:

	ha × million	**acres × million**	**date in service**
Upper Ganga Canal (Uttar Pradesh)	0·695	1·72	1854
Lower Ganga Canal (Uttar Pradesh)	0·592	1·46	1878
Sirhind Canal, Sutlej River (Punjab)	1·031	2·55	1887
Godavari Delta system (Andhra Pradesh)	0·508	1·26	1890
Western Jamuna Canal (Haryana)	0·547	1·35	1892

aqueducts

The Colorado River Aqueduct was unprecedented at the time of its completion. It takes water from the Colorado River at Parker Dam 249 km (155 miles) downstream from Hoover Dam, and delivers it to a terminal reservoir, Lake Mathews, 16 km (10 miles) south of Riverside, California. It supplies water to Los Angeles and to 66 cities in southern California in which about half of that State's population live. Supply extends by pipeline as far as San Diego.

There are five pumping stations along the line of the aqueduct, giving

Dedication day at Los Angeles Cascades, in 1913. Part of the aqueduct from Owens River to Los Angeles is seen – the first of the major US aqueducts, now a national historic civil engineering landmark. *Photo: American Society of Civil Engineers*

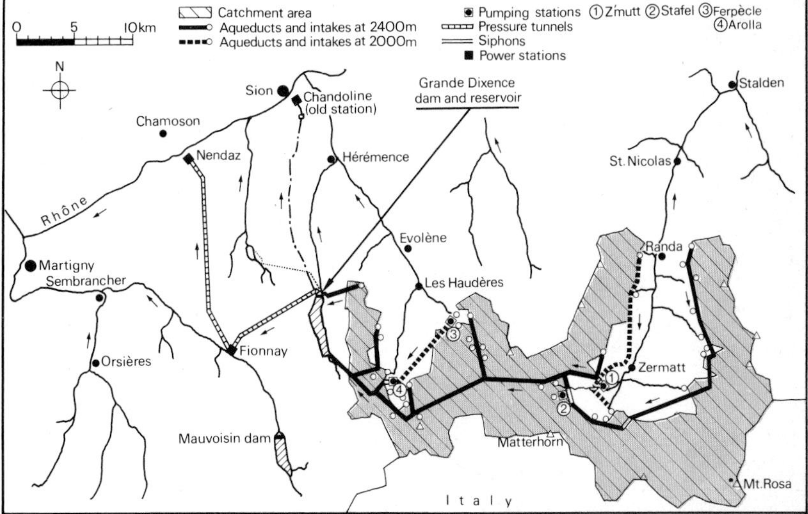

Catchwater aqueducts. Hydroelectric schemes often have reservoirs at a high level, the flow into which is augmented by a diversion from neighbouring catchments, or by collecting the flow from hillside streams by a system of catchwater aqueducts. The most extensive system of aqueducts of this kind appears to be that feeding the Grande Dixence Dam. The idea in this case is to catch the flow when the snow melts, and store it for use the following winter. There are two sets of aqueducts, one at 2400 m (7874 ft) above sea-level – just above the top level of the reservoir – and one at 2000 m (6562 ft), and four pumping stations. The longest aqueduct extends over 50 km (over 30 miles) from the reservoir

Table 48. The world's longest aqueducts (over 300 km (186 miles) in length).

Name	Length km	Length miles	Year of completion	Country and location	Other details
Kara-Kum Canal	*c.* 1500	*c.* 930	u.c. 1975	USSR, Turkmenistan	See table 42
California State Water Project	1329	826	1974	USA, California	1065 km (662 miles) built artificially (see text for description)
Canadian River Project	519	322·7	1967	USA	Prestressed concrete pipeline 2·4–0·4 m (8–1·2 ft) diameter for water-supply: main aqueduct of the project
Colorado River Aqueduct	389	242	1939	USA, California	See text for description
Owens River	357	222	1913	USA, California	For the water-supply of Los Angeles
Hetch Hetchy	305	189	1934	USA, California	For the water-supply of San Francisco
Frying Pan – Arkansas	over 480	300	u.c.	USA, Colorado	85 million m³ (3000 million ft³) is to be diverted each year from the Colorado River through 42 km (26 miles) of tunnels to the Arkansas River for irrigation. The distance quoted is the maximum water is transported, using mainly the natural river
See also Rajasthan Canal and Nagarjunasagar Canals, table 42					
Longest in Britain					
Thirlmere	154	96	–	England	Haweswater and Thirlmere aqueducts carry water from the Lake District to Manchester
Haweswater	132	82	1955	Cumberland/ Lancs	

a total lift of 493 m (1617 ft) the largest single lift being 134 m (441 ft). An interesting point in design was that an optimum was struck between the capital cost of tunnelling and the running costs of pumping. There are 31 tunnels on the aqueduct, the longest being San Jacinto Tunnel.

Historical

Between 312 BC and AD 226, eleven aqueducts were built to supply Rome with water. Their sources were springs, except for the three noted below. Seven of them were routed into Rome across a narrow saddle of land.

The first seven

Appia: 312 BC. Entirely underground, up to the walls of Rome.
Anio Vetus: 269 BC. Source: Aniene River, 64 km (40 miles) long.
Marcia: 144–140 BC. Most celebrated of all, 92 km (57 miles) long with tortuous, contour-hugging route. Direct route is 39 km (24 miles). At Rome, this aqueduct consisted of a colonnade of high arches built of dressed masonry. Marcia had the best drinking-water.
Tepula: 125 BC. Its water was tepid – hence the name. For the last 7 miles (11 km) it was built on top of the Marcia Aqueduct. Later the Julia Aqueduct was built on top of the other two.
Julia: 33 BC.

Virgo: 19 BC. 13 km (8 miles) long, this being the shortest. Water of good quality.
Alsietina: 2 BC. Source: Lake Alsietina (now Lake Martignano).

The new era
Claudio: AD 52. 69 km (43 miles) long.
Anio Novus: AD 52. Its source is also the Aniene River. 87 km (54 miles) long. Close to Rome it was built on the Claudio Aqueduct.
Trajana: AD 109. 56 km (35 miles) long.
Alexandrina: AD 226.

The total length of these eleven aqueducts was about 560 km (350 miles), but they were frequently shortened by modifying their route, or lengthened by adding new construction. Of this total 480 km (300 miles) was underground. Arcades of arches covered a distance of nearly 11 km (7 miles) near Rome: these structures had a maximum height of about 30 m (100 ft). Where the aqueducts were tunnels, shafts were sunk about every 86 m (240 ft).

After the construction of the Marcia Aqueduct, concrete became the universal building material, and was used to encase earlier works.

As Rome decayed, so the aqueducts decayed. By about AD 1000, Rome had no supply of water except the Tiber. Pope Sixtus V built a new aqueduct – Acqua Felice – in 1585, using the arches of the Marcia, and some of Claudio. It is still in use, but much of the Roman masonry was dismantled or embedded in concrete, to make this 'massive and ugly' aqueduct. Later popes did similar work, and an entirely new aqueduct was built in 1870. (Source: E M Winslow, Newcomen Society paper, 1963.)

The aqueduct of Carthage in Tunisia had a length of 141 km (87·6 miles). It was built in AD 117–38.

The first aqueduct in Britain was Drake's Leat, completed in 1591 – little more than a ditch, about 29 km (18 miles) long from a weir on the River Mewe or Meavy to Plymouth. It was named after Sir Francis Drake, and its stated purposes were for the preservation of the harbour at Plymouth, for watering ships and for protecting them from fire. In fact it served as Plymouth's main source of water for nearly 300 years.

The New River, 63 km (39 miles) in length, was opened in 1613. It carried water from the Amwell and Chadwell springs in Hertfordshire to London. It was opened after four and a half years' work and its cost, including some mains in the city, was £18525 0s 1d.

Ganga–Cauvery canal project

Serious consideration is being given in India to an enormous project for water conservation that would involve construction of an aqueduct, mainly a canal, 3300 km (2050 miles) long and the diversion annually of 25 km^3 (6 miles^3) of water. Water would be taken from the Ganga River, southwards for irrigation and for relief in time of drought. Seven major rivers would be linked.

The aqueduct would start from about Patna – a town 500 km (300 miles) north-west of Calcutta – where the Ganga is less than 100 m (300 ft) above sea-level. It would take off west and south, and climb to 460 m (1500 ft) to a reservoir on the watershed between the Ganga and Narmada rivers on the Chotanagpur Plateau. The general route of the canal would then be southerly, linking successively with the Tapti, Godavari, Krishna, Penner and Cauvery rivers. The length of navigable waterway in the country would be more than doubled, and in addition to drought relief, 8–19 million ha (20–47 million acres) of land irrigated.

pipelines

The world's longest pipeline is the Trans Canada natural gas pipeline. By mid 1974 its length was 9099 km (5654 miles) and its diameter up to 1·07 m (42 in).

The longest crude oil pipeline is from Edmonton, Alberta, to Buffalo, NY (Canada/USA), a distance of 2856 km (1775 miles).

The Nadyn–Moscow natural gas pipeline is due to be in use with a length of about 3200 km (2000 miles) in 1975. Its route involves crossing the Ob and the Volga, and great lengths of forest, swamp and permafrost.

Table 49. The longest sea outfalls and pipeline crossings (this table lists pipe-strings that are pulled out from the shore to their final positions on the sea-bed).

Name	Length		Year of completion	Country and location	Other details
	km	miles			
North Wirral	4·8	3	1971	England	0·9 m (3 ft) diameter
Humber oil discharge pipeline to Tetney, Lincs.	4	2·5	1970	England	0·9 m (3 ft) diameter, for off-loading tankers. Marine length is given; total length of pipeline is 8 km (5 miles). Initial weight of 3700 t (i.e. four-fifths of total) needed an initial pull of 80 t
Baglan Bay	4	2·5	1974	Wales (Neath/Port Talbot)	1·2 m (4 ft) diameter for refinery effluent and municipal wastes
Santo and São Vicente sewage outfall	4	2·5	Contract let, 1974	Brazil, for Santo and São Vincente	Thirteen sections each 300 m (984 ft) long. Pipe 1·75 m (5·7 ft) i.d., steel, with epoxy-resin lining and glass-fibre coating and reinforced concrete outer skin. Pulling capacity 800 t, claimed as largest pipe-pull
Humber gas main crossing	3·5	2·2	1973	England	0·6 m (2 ft) diameter pipe pulled across estuary.
Windscale	3·2	2	1950	England, Cumberland	For radio-active effluent from nuclear factory. Prototype of long sea outfalls in Britain

Note: The 1200 m (3937 ft) long sewage outfall at Cannes, France, is in a maximum depth of water of 85 m (279 ft). At Europoort, Holland, a bundle of twelve pipes, each 1100 m (3609 ft) long was floated out and sunk into a prepared trench across the harbour in 1973.

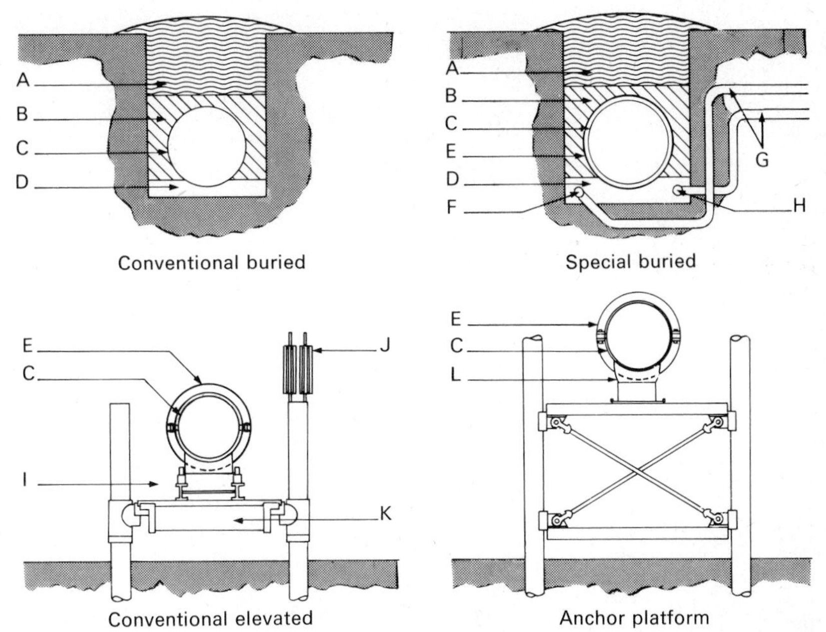

The Alaska pipeline, 1284 km (798 miles) long will carry hot oil across the tundra from Prudhoe Bay oil field on the north coast of Alaska to the port of Valdez on the south coast. Completion of the first phase is expected in mid 1977. A steel pipeline 1·2 m (48 in) in diameter will convey 1·2 million barrels of oil a day in the first phase, and ultimately, with additional pumping stations, 2 million barrels (about 280 000 t) per day. The end-of-1974 estimate of cost was $6000 million. The line is zig-zag, and thermal expansion is taken up by movement sideways. There are three types of construction:
1 Conventional elevated, 615 km (382 miles). Supports 15–21 m (50–70 ft) apart. Thermal device is self-operating and keeps the ground frozen. Anchor platforms every 250 to 550 m (800 to 1800 ft)
2 Conventional buried, 658 km (409 miles). From 0·9 to 3·7 m (3 to 12 ft) deep
3 Special buried, short sections totalling 11 km (7 miles) at places such as caribou crossings. Refrigerated brine is circulated beneath the insulation to keep the ground frozen

A Fill
B Lagging
C Pipe
D Bedding
E Insulation
F Return } of refrigerant lines G
H Supply }
I Sliding shoe
J Thermal device
K Support beam
L Anchor saddle

A project for a natural gas pipeline over 4000 km (about 2600 miles) long is likely to be started in Canada and the USA. It will carry gas from fields at Prudhoe Bay, Alaska, and the Mackenzie River Delta south to Calgary, Alberta, and onward to centres of consumption in the USA and Canada.

It is difficult to state unequivocally what is the longest pipeline in Britain, because major routes tend to be part of a network. British Gas has just completed several lines of pipe, all of 0·9 m (3 ft) diameter, that together form a route from St Fergus in Scotland to Samlesbury, Lancashire, a distance of 1013 km (630 miles). There is a main-line oil pipeline – serving several systems and branches – 458 km (285 miles) long from Milford Haven to Seisden, Birmingham.

A Government oil pipeline laid during the war was rehabilitated in the 70s, and is claimed as the longest multi-product, continuous, underground pipeline in Britain, and the second in Europe. Its length is 573 km (356 miles) and its route is from Humberside, Killingholme, to Aldermaston via Manchester, Chester and Bristol; it is of small diameter.

Submarine Pipelines

Cable-laying in the deep oceans is a long-established technique, but the need for sea-bed pipelines in deep water is new. Those for North Sea oil or gas include the Forties, Piper, Brent, Frigg and Ekofisk lines in depths up to 120 m (400 ft). The last, 354 km (220 miles) long, will bring oil from the Ekofisk Field in the Norwegian sector of the North Sea, to a terminal on Teesside. The Frigg line is to be roughly of similar length, and the other two are shorter. Forties is complete.

The Ekofisk gas pipeline to Emden on the German coast is the longest submarine pipeline to date (excluding submarine cables which are well established and cross the major oceans). It was finished in mid 1975. Its length is 418 km (260 miles) and its diameter 0·9 m (3 ft).

The largest pipe. Penstock of unit 19 at Grand Coulee, looking down its 45-degree slope. The shingle effect is from coal-tar enamel, applied by hand on the interior of the steel pipe

12·19 m (40 ft) length of the steel penstock – its diameter is also 12·19 m (40 ft) – at Grand Coulee, ready to be lowered into position for welding

The pipeline in deepest water so far is at a maximum depth of 350 m (1148 ft), between Italy and Sicily. It is 15 km (9·3 miles) long and 0·275 m (0·9 ft) in diameter. The next stage of this line is to be in a maximum depth of 500 m (1640 ft) of water and 160 km (100 miles) long, between Sicily and Tunisia.

The Brent pipeline, 148 km (92 miles) long from Firths Voe, Shetland, to the Cormorant Field is in a maximum depth of water of 162 m (530 ft) and is 0·9 (3 ft) in diameter.

Largest Pipe

The penstocks of Number 3 Power Station, Grand Coulee Dam, are each 12·19 m (40 ft) in diameter, and each one can handle a flow of 850 m^3/s (30 000 ft^3/s). This flow is equivalent to a power of 700 MW (actually 600 MW or 700 MW depending on the set) transmitted through the pipe.

These penstocks are the world's largest pipes and have the greatest capacity of any pipe in terms of both the quantity of water carried and the energy transmitted by the pipe.

The highest head on any pipe (in civil engineering applications) is at the San Fiorano hydro-electric scheme listed in table 46. Second is Portillon, in the Pyrenees, France, where the head is 1414 m (4639 ft). In 1873, a siphon was built to serve the Marlette Lake water supply scheme, Virginia City, Nevada, USA, with a head of 518 m (1700 ft) claimed to be more than double the head sustained by any other pipe at that time.

Wheel-mounted gate that controls flow into the Grand Coulee penstock. *Grand Coulee photos: US Bureau of Reclamation*

Cast-Iron Pipes

Cast-iron pipes were first used to carry water at the Castle of Dillenburgh in Germany, in 1455. Two other early examples in Germany are at Logensalya, in 1562, and Braunfels, in 1661. In 1932, the Braunfels main was replaced by a main of larger diameter, and was found to be in excellent condition after 271 years.

Mailsi Siphon, Pakistan. The gates controlling flow from the canal into the siphon are shown in the foreground. The River Sutlej flows from left to right through the openings of the barrage, over the top of the siphon. *Photo: Coode and Partners*

In 1664–8, a cast-iron pipeline was laid for the water-supply of Versailles between the reservoirs of Picardy and Monbauron, for a length of almost 8 km (5 miles). The main is still in use.

Largest Inverted Siphons

The four-tube inverted siphon carrying the Welland River under the new Welland By-pass Canal was completed in 1971. It is 194 m (638 ft) long, 28·6 m (94 ft) wide and can carry a flow of 340 m³/s (12 000 ft³/s).

The two inverted siphons at Panoche Creek on the San Luis Canal of the Central Valley project, California, USA, each have a diameter of 8·35 m (28 ft) and together carry a flow of 334 m³/s (11 800 ft³/s). They are 322 m (1056 ft) long.

Mailsi Siphon carries one of the link canals of the Indus Basin scheme in Pakistan across the River Sutlej, at a site where the Sutlej is dry for most of the year, but meanders over a wide flood-plain. It consists of

- an inverted siphon, comprising four reinforced concrete tunnels, each 4·1 m (13·5 ft) square and 573 m (1881 ft) long.
- a barrage on top of the siphons, comprising 24 openings, each of 18·3 m (60 ft), and controlled by a gate 2 m (7 ft) high.
- a road bridge.
- guide banks and about 35 km (22 miles) of marginal bunds.

Kunu Siphon, on the Chambal scheme in India, is the largest prestressed concrete structure in Asia. It is 1758 m (5768 ft) long, 6 m (19·7 ft) in diameter and carries a flow of 90·6 m³/s (3200 ft³/s).

Zermatt inverted siphon in Switzerland, part of the system of catchwater aqueducts supplying Grande Dixence Reservoir (see page 209) has a length of 1893 m (6211 ft) and descends 266 m (872 ft). It consists of tunnels within the mountains on each side, and subterranean pipes across the valley.

water-supply and treatment

WATER-SUPPLY

Largest systems. Moscow claims the largest water-supply system in the world, 4·5 million m³ (158·9 million ft³) per day. One-third of the total is consumed by industry. Demand is rising rapidly. Half of the supply comes from the Volga River, some from the Moscow River.

A new western supply system is being developed for 0·8 million m³/day (28 million ft³/day) of potable water, and a new south-eastern supply system for industry. The population supplied is 7 million (350 l/day (77 Imp gal/day) *per capita*).

On 1 April 1974, the Thames Water Authority – the largest of ten new regional water authorities serving England and Wales – took responsibility for the water-supply and main drainage of about 12 million people living in the area shown on page 220, and including some 7 million people in Greater London. The volume involved is 4 million m³/day (141 million ft³/day or 1000 million gal/day). Previously this work had been done by about 200 water undertakings and sewage and river authorities. The new authority's area is 12950 km² (5000 miles²) in extent and is served by about 60 water-treatment plants and over 450 sewage-treatment plants.

Water-supply and main drainage of London. Water-supply works are coloured; they consist of the Thames and Thames valley reservoirs, the Lee and Lee valley reservoirs, the Thames–Lee tunnel and the New River; wells drawing from the chalk are not shown. Main drainage consists of interceptors running east–west to Beckton or Crossness, supplemented by high-level sewers, not all of which are shown

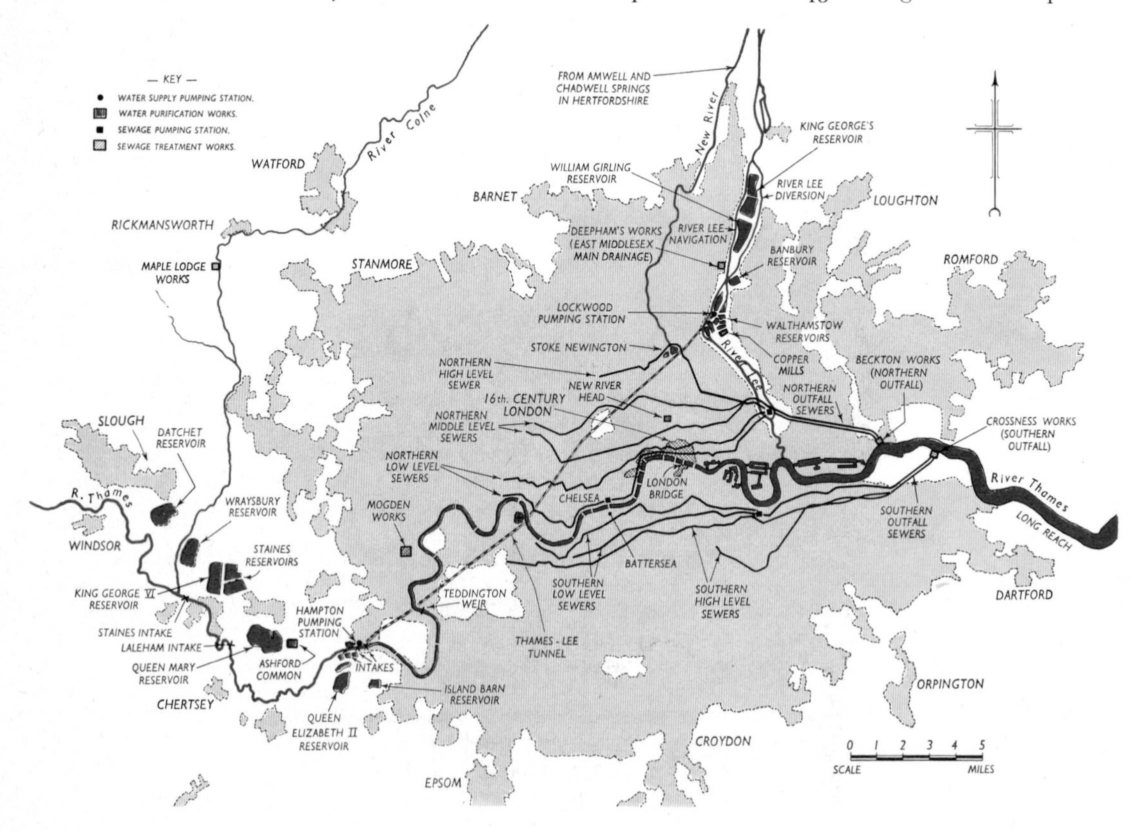

The figure of 4 million m³ of water supplied each day is an average that includes 1 million m³ of cooling water abstracted from, and returned to, the Thames at various power stations. The quantity supplied through the domestic mains is about 3 million m³ (average for 1974, 720 million gal, and peak on 7 August 1975, 934 million gal (3·27 and 4·25 million m³ respectively). This figure corresponds to a consumption per head of 250 l/day, and is also equal to the quantity supplied for domestic use quoted for Moscow. The Thames Water Authority's main source of supply is the Thames, but water is also taken from the Lee, and from wells in the chalk.

Within a radius of 100 to 150 km (60 to 95 miles) of Paris, 50 sources of water are tapped, and transported along about 600 km (375 miles) of aqueducts (including pressure pipes) with 14 pumping stations (about 11 MW total power) for treatment at three plants. The quantity supplied is 0·64 million m³/day (22·6 million ft³/day).

Aqueducts supplying Paris with water	**Year of Construction**	**m³/day**	**ft³/day × million**
Dhuis	1863–5	20 000	0·71
Avre	1890–3	160 000	5·65
Vanne	1867–74	150 000	5·30
Loing–Lunain	1898–1900	100 000	3·53
Voulzie	1923–4	100 000	3·53

The oldest aqueduct, Rungie (built 1613–24), is incorporated in the Vanne route, and where they are above ground, the two are carried one above the other on a common structure consisting of a series of arches.

Paris's distribution network consists of:

	Potable		**Non-potable**	
	km	**miles**	**km**	**miles**
Mains in conduit (mainly sewers that can be inspected)	1604	997	1515	942
Buried mains	88	55	83	52
Service reservoirs	9		8	
Distribution systems	10		10	

A population of 2·5 million, in addition to about 2 million that work in Paris, are supplied – 160 i/day (35 Imp gal/day) *per capita.*

Wells

The largest well in Canada yielded 15·3 million (US) gal/day (58 000 m³/day or just over 2 million ft³/day) on test. It consists of a reinforced concrete caisson 4·9 m (16 ft) in diameter and 30·5 (100 ft) deep, with spoke-like collectors at the base. The well is at Prince George, British Columbia, and penetrates the Nechako Aquifer.

The deepest well is in chapter 2 (page 153).

The largest well in Britain is at Bexley in Kent. It is able to produce 31 800 m³/day (7 million gal/day), but its normal output is about 4 million/gal/day (18 000 m³/day). The normal daily output of a well at Deptford is 4·18 million gal/day (19 000 m³/day).

Sewage treatment works at Acherès, Paris. The pretreatment stage is in the foreground, with sludge drying beds to the left of it. Secondary settlement tanks are to the right, with aeration basins, primary settlement tanks, power stations and sludge-digestion tanks progressively towards the background. This works serves virtually the whole of Paris and is one of the world's largest, as shown by table 50. *Photo: Préfecture de Paris, Service Technique de L'Assainissement*

WATER TREATMENT

The earliest method of attempting to purify water was the slow-sand filter-bed. The first successful one was built at Paisley, Scotland, in 1804. Permanent slow-sand filters were installed at Chelsea in 1829 to treat water from the Thames, and that event marked their general introduction. Chemical reagents appear to have been used for the first time on a practical as opposed to an experimental scale, at Bolton in 1881, when sulphate of alumina was used as a coagulant, although this use was preceded by the use of active carbon. Sterilisation by the use of chlorine began in 1897 at Maidstone, where mains were disinfected after an outbreak of enteric. The first permanent use was at Lincoln in 1905, but disinfection did not become general until after the First World War. The first water-softener in a public supply was used at Plumstead in 1854.

The world's largest water-treatment plant is at Chicago, Illinois, USA. It was completed in 1957, and it can treat 960 million US gal/day (3·638 million m³/day or 128·4 million ft³/day).

The largest in Britain is the Thames Water Authority's Coppermills Works at Walthamstow in the Lee Valley, completed in 1972. It has a reliable output of 108 million gal/day (0·49 million m³/day or 17·3 million ft³/day). Two other treatment works serving London are of comparable size, Ashford Common and Hampton.

main drainage and sewage treatment

MAIN DRAINAGE

Main drainage appears to be as ancient as water-supply. Mohenjo-Daro, a city on the Indus River, dating from 2500 BC had houses each of which had a bathroom and a privy flushed with water. Chutes carried the dirty water away to covered drains in the streets. In ancient Mesopotamia, houses were equally well served, and there were brick-lined sewers beneath the streets. The drainage of ancient Rome was elaborate.

In the modern era, the starting-point is the appearance of the world's most important invention, the water-closet. Sir John Harington, the Elizabethan poet, invented it in 1589. He supplied one to Queen Elizabeth, his godmother, apparently as a ploy for a literary exercise, for he described his device in a work called *The Metamorphosis of Ajax – a Cloacinian satire* (a 'jacks' was a privy). Later on, Bramah and others developed Harington's eccentricity into a reliable and ubiquitous necessity for dignified living.

That necessity, however, brought onerous problems. Early water-closets discharged into open ditches in the streets. It was illegal in England until 1815 to discharge sewage or other offensive substances into sewers, which were for surface water only. In the mid 19th century, the waterborne causation of cholera and typhoid was discovered. This was the starting-point of public health engineering, and of the responsibility and eventually

London's Northern Outfall Works at Beckton. Full treatment is now given to the entire flow, since the completion in mid 1975 of the latest secondary treatment plant. *Photo: Thames Water Authority*

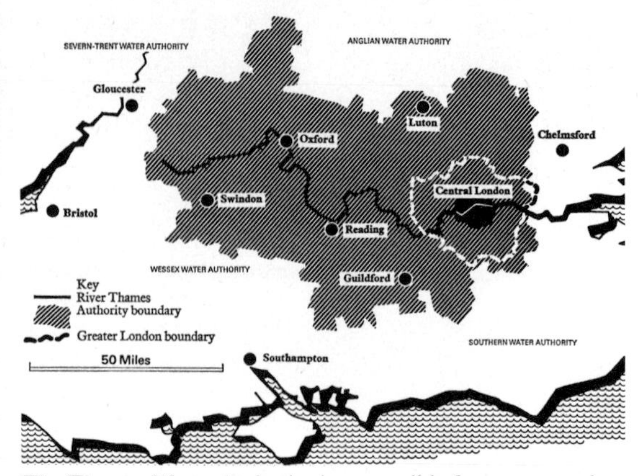

The Thames Water Authority is responsible for water supply, sewerage and sewage treatment, and for the conservation of river waters, in the area shown here

London's Northern Outfall at Beckton when completed in 1865, consisted simply of a large covered reservoir, in which sewage was stored until the ebb tide, and then released. It was believed that it would be swept out to sea on the ebb tide, and never seen again. That belief proved to be a major mistake. The sewage stayed in the estuary, and oscillated up and down with the tides

achievement, of the civil engineering profession, of ridding great cities of cruel epidemics of waterborne diseases. The water-carriage method of disposing of sewage in sewers, as it was called in the early days, was not by any means universally accepted. Although its supporters won the day, they left a legacy of foul rivers – as their opponents had forecast – because insufficient resources were given to the construction of sewage-treatment works. That problem persists. The main drainage of many major cities was established at that time.

In London, the master-plan (of Sir Joseph Bazalgette) envisaged an outfall sewer running towards the east, along each bank of the Thames to a treatment works beyond the city. In the central area, the Embankment was constructed, and the Northern Outfall Sewer incorporated in it. The Northern Outfall at Beckton consists of five oval sewers of 2·7 m (9 ft) major diameter.

By contrast the topography of Paris favoured a layout in which all the sewers discharge into a single treatment works.

The sewers of Paris were built so that they could be inspected. Lines of pipes are carried on brackets in the upper parts of these sewers. They consist of: potable water mains; non-potable water mains; telephone cables; compressed-air mains; cables for police and fire service warning systems and for coordinating the control of traffic signals. Gas or electricity mains are never put in the sewers

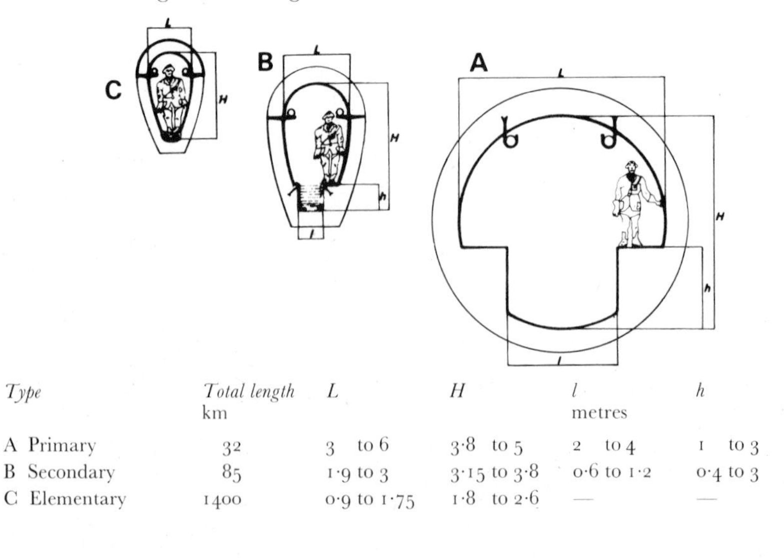

Type	*Total length* km	*L*	*H*	*l* metres	*h*
A Primary	32	3 to 6	3·8 to 5	2 to 4	1 to 3
B Secondary	85	1·9 to 3	3·15 to 3·8	0·6 to 1·2	0·4 to 3
C Elementary	1400	0·9 to 1·75	1·8 to 2·6	—	—

Even sewers have to be purged. A 'sluice boat' in use in a primary sewer in Paris, in an operation that removes sand and silt from the bottom of the sewer. *Photo: Préfecture de Paris, Service Technique de L'Assainissement*

The largest sewers in Paris have inside dimensions of 5 m (16·4 ft) high by 6 m (19·7 ft) wide (see sketch). Cloaca Maxima, the main sewer of ancient Rome is reputed to have had dimensions of about 10 m (33 ft) high by 4 m (14 ft) wide, apparently rather larger than the main sewers of Paris.

The smallest sewer was a sewage pumping main, only 31 mm ($1\frac{1}{4}$ in) in diameter. It was used as an experiment for rural sewerage at Loddon, Norfolk, from May to November 1962. Its use depended on a pump – the Mutrator – that was able to shred or reject solid material. The pump ran successfully for 406 hours during the trial.

The Chicago drainage canal, built between 1890 and 1900, is claimed as the first great excavating job in the United States. Special plant was designed for it. The canal is 45 km (28 miles) long and connects the Chicago River with the Des Plains River, a tributary of the Illinois River in the Mississippi catchment.

Chicago's problem was that it abstracted its water from Lake Michigan, but also discharged its sewage into the lake via the Chicago River which was virtually a sewer. By diverting the river to another catchment an immediate solution was found. When the canal was opened, the death-rate in Chicago from typhoid dropped sharply, but pollution of the Illinois River caused much protest, and the diversion of water upset the Niagara Falls Power Agreement. A Supreme Court ruling dictated the remedy.

Today, Chicago boasts the world's largest sewage-treatment works (details are in table 50). The Metropolitan Sanitary District of Chicago plans to construct underground storage where storm sewage (i.e. storm-water draining into sewers and so mixed with sewage) can be stored until the storm subsides and it can be treated.

The plan includes 200 km (120 miles) of tunnels 3–12 m (10–35 ft) in diameter, 75–90 m (250–300 ft) below the surface; 254 shafts from the surface to the tunnels; 3 reservoirs, with a combined working capacity of 157 million m^3 (about 5500 million ft^3). It would serve the central portion of Metropolitan Chicago, an area of about 1000 km^2 (375 miles^2). Three of the smaller tributary tunnels have been built, totalling 20 km (12 miles)

The world's largest sewage treatment plant at Chicago, USA. *Photos: Metropolitan Sanitary District of Chicago*

Mexico City's Central Outfall tunnel sewer compared with the statue of Carlos IV. It has an area in cross-section of 55 m^2 (592 ft^2) and an equivalent diameter of 6·5 m (21·3 ft). It is 50 km (31 miles) long. Its construction demanded the excavation of 2 700 000 m^3 (3 500 000 yd^3) of spoil. It carries a flow of 200 m^3/s (7063 ft^3/s) and it appears to have no rivals to the claim of being the world's largest sewer

in length. If money became immediately available, ten years would be required for construction, the estimated cost being approximately $1500 million.

SEWAGE TREATMENT (table 50)

The earliest form of sewage treatment was the sewage-farm (irrigation). Solids are first settled out – primary treatment – and the liquid is then spread over the land, through which it percolates to underdrains. It is purified by oxygenating bacteria as it percolates. Other widely used processes – percolating filters and activated sludge – virtually concentrate the same process into less space. The following résumé of the historical development of sewage treatment has been taken from the introduction of a recent paper by Mr J D Cargill (*A history of sewage works machinery*, Institution of Water Pollution Control, West Midlands Branch, February 1974).

'It was not until 1847 that the first Act was obtained, making it compulsory to drain houses into sewers. Works of sewerage were carried out during the mid-19th century to improve the sanitary condition of urban areas, but transferred the pollution to the rivers. In attempting to relieve the pollution, the Victorians were not merely concerned with "improving the environment" but endeavoured to utilise the manurial value of the sewage for profit. The principal methods employed at that time were irrigation, filtration and chemical precipitation.

'Irrigation had been carried out on the Craigentinny Meadows at Edinburgh since the 17th century. In 1844 part of the flow was used to operate a water-wheel for raising sewage to irrigate higher meadows. A number of farms were provided with steam-engines, pumps and a system of underground pipes, hydrants and flexible hoses, but it was not found to be profitable and was eventually given up.

'It was realised by a number of workers in the 1880s that complete purification of sewage was only possible by the action of bacteria. W J Dibden developed the contact-bed at Barking in the early 1890s and the first full-scale works of this type were constructed at Sutton in 1896. The septic-tank process was developed by Donald Cameron at Exeter, the first installation at the St Leonards Works being brought into use in 1896. The effluent from the septic-tank was treated on contact-beds.

'Many workers contributed to the development of the percolating filter which eventually proved far more successful than the contact-bed or the septic-tank process, and which has basically remained unchanged for the last 70 years.

'Aeration processes for purifying sewage had been investigated by a number of workers, the air often being introduced into contact-beds or other filters, but in 1912 Arden and Lockett at Manchester developed the activated-sludge process and by the 1920s it was in large-scale use.'

Table 50. The world's largest sewage-treatment works.

Name	Flow fully treated $m^3 \times 10^6$/day	Flow fully treated Imp Gal $\times 10^6$/day	Population served – millions	Year of completion	Country and location	Other details
West–Southwest Plant	3·16	835 (US gal – average actual flow in 1973)	2·7	1940	USA, Chicago, Illinois	Conventional activated sludge plant; 85 to 95 per cent of bod is removed. Two other smaller plants also serve the city. Claimed to provide a higher degree of secondary treatment to a greater flow of sewage than any other plant in the world
Moscow	2	440	7	–	USSR	Rated capacity. Actual flow is a little less
Achères	1·5 (present–Stage 3)	330	8	1940 (Stage 1 $0{\cdot}2 \times 10^6$ m^3/day)	France, Paris	Constructed in five stages, the last two of equal size. Stage 4 working in 1978. Conventional activated sludge plant
	2·7 (ultimate)	594	–			
Northern Outfall	1·14 (dry weather flow)	250	–	1975	England, Beckton (serving London)	Date given is for completion of latest extension. Works has been in use since 1865
	2·73 (storm capacity)	600	–	–		
Southern Outfall	0·98	216	2	1964	England, Crossness (serving London)	Bazalgette's works, completely reconstructed. The dry weather flow is 96 million gal (0·4 million m^3) per day, but the works has the capacity to treat fully the quantity stated, to cater for wet weather
Detroit	1·4	375		u.c. 1975	USA, Detroit	Present capacity is twice the stated figure for primary treatment only. World's largest pure oxygen aeration plant
	3·7	1000 (US gal)	3	1977		
Blue Plains	2	550 (US gal)	2	1978	USA, Washington, DC	Upgrading to stated capacity

chapter 4

ENGINEERS AND CONSTRUCTION

great civil engineers

Few admirers of Gothic architecture know the names of the greatest designers and builders of the Gothic cathedrals. A similar tradition of anonymity tends to have persisted, and persists today, in the appreciation of civil engineering feats. It is an attitude that contrasts with the personal attribution of architectural works, which not uncommonly has extended to accrediting the design of, say, a bridge, to an architect who was consulted, but played a minor role in its conception. Assessing personal contributions to great civil engineering works is made more difficult today by the professional ethics of engineers, and by the team effort that is necessary and characteristic of today's great schemes. In this last sense, the day of the individual is over; so one would think, until problems multiply, and only one member of the team has the audacity and perseverance to overcome them and to inspire others.

The subjective limitations of the critic are another factor that shape a choice and judgment of the greatest names. This selection, therefore, is a sample – not complete by any means, not fully objective. Nevertheless, it will at least indicate the sense of vocation and the hard toil that have gone into the conception and construction of every one of the structures listed in this book.

Re-emphasising the factor of personal judgment, the author would first give three names that seem to him to head the list. Great civil engineers leave their own monuments; they are remembered for what they build. These three have further claims on posterity. Each of them was responsible for great works that alone were sufficient to establish his reputation; but in addition, each changed, on a world-wide basis, the practice of his profession.

Thomas Telford (1757–1834), British, first President of the Institution of Civil Engineers, and father figure of the British civil engineering profession. He established the professional ethos of the civil engineer – the Engineer with a capital E in the contract documents, not himself a party to the contract between client and contractor, but administering it, giving advice in the interest of his client, yet in a position of independence, and arbitrator in matters of dispute.

British civil engineers have followed this tradition to the letter, and the strength of British consulting engineers over more than a century, and latterly of the world-wide consulting organisation FIDIC (Fédération Internationale des Ingénieurs Conseils) derive directly from it, and depend on it today.

Telford built bridges, roads, harbours and canals. His career was over before the railway age. His greatest works include the Menai Suspension Bridge and Pont y Cysyllte Aqueduct (the most spectacular structures on his London to Holyhead trunk road and Ellesmere Canal respectively), the Gotha Canal in Sweden, the Caledonian Canal, and many Scottish roads. He was the first and greatest master of the iron bridge.

Eugène Freyssinet

Eugène Freyssinet (1879–1962), French, is renowned for brilliance in engineering design, for example the Plougastel concrete arches and the slender Marne bridges such as Esbly and Lucanzy. His name is primarily linked, however, with prestressed concrete, which he invented, persisted with, and developed into a highly competitive and reliable form of construction. Prestressing is a philosophy; it conditions an engineer's understanding of how structures behave, how they can be handled, and the manner in which they can be designed. By demonstrating the success of this philosophy in practice, Freyssinet changed the thoughts of engineers, and so permanently benefited structural engineering.

Karl Terzaghi (1883–1963), Austrian and US. Analysis of the behaviour of soils and foundations was not entirely empirical before Terzaghi's work, but the modern science of soil mechanics is a discipline that he established and developed – again by successful application in practice. Modern virtuosity in handling soils and designing massive, heavily loaded foundations with confidence and safety developed directly from his basic contribution. The blend of theory and practice in his career is remarkable. He made a major contribution in an academic sense, yet that was linked with, and depended on, his success as a practical engineer.

Bjerrum records: '. . . His period of searching ended abruptly on a day in March, 1919. . . . On some sheets of paper he listed a number of possible ways of testing soils, made sketches of the equipment needed, and suggested how the results could be interpreted. These sheets of paper represent the birth of soil mechanics . . . he never showed them to anyone . . . they were found after his death at the bottom of a shoe-box filled with notebooks.'

THE PATRIARCHS

The earliest engineer known by name in the ancient world is **Eupalinus** of Megara, according to G E Sandström (in his book *The History of Tunnelling*, Barrie and Rockliff, 1963). He built a tunnel 1000 m (3000 ft) long for water-supply on the island of Samos in the 6th century BC. Herodotus regarded this tunnel as one of the three great engineering feats in the Greek world (the other two were a breakwater 2 km (1·25 miles) long and the temple, the largest one built by the Greeks, both also on Samos), but Sandström is much more critical. He says that the tunnel was driven from both ends, and the teams missed each other by 5 m (16 ft) in the middle. 'The surveying devices available to Eupalinus should have produced a much more accurate result' he adds, and mentions one or two other 'rather foolish' things that were done by Eupalinus.

Jan Leeghwater (1575–1650), Dutch, is the first great name to emerge from the tradition of reclamation of land from the sea in Holland. He was the leader in this during his lifetime, and drained numerous small lakes. He prepared a scheme for draining the Haarlem mere using 160 windmills.

Pierre Paul Riquet (1604–80), French, was the builder of the 240 km (150 mile) Languedoc Canal, which is generally regarded as the prototype of the modern era. The canal was started in 1666, opened in 1681, and finished in 1692. It started at the Garonne at Toulouse and rose 63 m (206 ft) to the summit, with 32 locks over a distance of 51 km (32 miles). The summit continued for 5 km (3 miles) and had to be supplied with water. The canal then descended to the Mediterranean over a distance of 185 km (115 miles), losing 189 m (620 ft) in height through 74 locks. The first canal tunnel was built at Malpas (157 m (515 ft) long) on this canal.

Sébastien le Prestre de Vauban (1633–1707), French, has been described as the greatest military engineer that ever existed. He built fortifications, and such is the tortuous evolution of language that, in the modern sense, his work was civil engineering, since it was one specialised aspect of designing and building structures. He made Dunkirk impregnable and fortified the town and citadel of Lille, to mention only two of many examples. Few of his works exist today.

Jean Rodolphe Perronet (1708–94), French, was perhaps the leading figure of, in his day, the leading country in civil engineering. The Ecole des Ponts et Chaussées – a unique and powerful influence on that leadership – had been formed in 1747, and was significantly reorganised by Perronet in 1760. However, he was prolific and talented as a practical engineer. Perronet was a specialist on piling. He built, for example, the Canal de Bourgogne, and some exceptionally fine bridges. His Pont de Neuilly in Paris, built in 1774, was in use until 1939. His Pont de la Concorde of 1791 has been reconstructed and enlarged, but preserved in its original form.

Romney's portrait of Smeaton is owned by the Smeatonian Society of Engineers and hangs in the library of the Institution of Civil Engineers in Great George Street, Westminster

John Smeaton (1724–92), British, was the first 'truly great' British engineer and the first man to call himself a civil engineer, at least in the English-speaking world. He was trained as a lawyer, but then took up mechanical engineering and instrument-making, at a time when educated people had nothing to do with such vulgarities. His reputation was established by his success in designing and building Eddystone Lighthouse. An active career then unfolded, in which for the first time, he established professional standards in investigations, reports and designs, mainly for harbours, canals and bridges.

Benjamin Wright (1770–1842), American, was the father of American civil engineering and the first US civil engineer recognised as an international authority. He was appointed chief engineer of the 364 mile (586 km) Erie Canal begun in 1817, and his contribution to this well-known project established his reputation. He was either chief engineer or principal consultant for several important US canal schemes of the early 19th century, and he helped to build railways in New York, Virginia, Illinois and Cuba.

Wright's legacy to America was not only his many projects or the high professional standards he established, but also his practice of sharing his engineering knowledge with others. He was the professional mentor of many of the leading US civil engineers of the 19th century.

Joining the rails of the Transcontinental Railroad at Promontory Point, Utah, USA. The photo was taken on 10 May 1869. A golden spike was driven to celebrate joining up 2841 km (1766 miles) of railway. Construction had started at Sacramento in 1863, the civil engineer in charge of the work being Theodore Judah. The western terminus at Sacramento (Central Pacific Railroad) is one of the American Society of Civil Engineers' national historic landmarks. So is Promontory Point – 'It signaled the opening of the West and the emergence of a unified nation' the ASCE says

GREAT VICTORIANS

The railway age brought with it need and opportunity, and thus an exuberant blossoming of genius among civil engineers; first in Britain, where the railways first became established, and where the original fund of experience was to be found, but later in the United States with the construction of the great trans-continental railways.

In the mid 19th century the two leaders of the profession in Britain were **Isambard Kingdom Brunel** (1806–59) and **Robert Stephenson** (1803–59). Their careers are well known and well documented. Brunel is a unique figure. The breadth of his invention and the works he accomplished are remarkable enough, but his uniqueness seems also to have lain in what today is fashionably called charisma. Recent presidents of the Institution of Civil Engineers have exhorted the membership to emulate the spirit of Brunel. Controversy about his work was as keen on the hundredth anniversary of his death as it was in his lifetime. If the Council of the Institution of Civil Engineers had powers to canonise, he would be their first saint, and perhaps their only one.

Robert Stephenson was more the expected professional *éminence grise* of railway construction, but his work, too, was not without controversy, as the remarks in chapter 1 show. He became a Member of Parliament, but did not shine as a parliamentarian – he is on record as opposing the Suez Canal, for example.

Both men were the sons of distinguished national engineering figures. **Brunel** senior (**Marc Isambard** (1769–1849)) being remembered for the Thames Tunnel and **Stephenson** senior (**George** (1781–1848)) as the most famous figure of the early days of railways.

The most prolific of the 19th-century engineers was, however, **Sir John Hawkshaw** (1811–91), a railway-builder of extraordinary fluency and competence – the bridges at Charing Cross and Cannon Street are his work. He is sometimes quoted – by consulting engineers of today with pure nostalgia – for his remark to his clients who queried what was due to the contractor: 'Gentlemen, what John Hawkshaw signs, you pay.'

BRIDGE-BUILDERS

The art of bridge-building has always seemed to draw out talent and adventurous initiative. A score of names could easily be mentioned:

The **Grubenmann** brothers, who built the long-span timber bridges in Switzerland in the 18th century mentioned in chapter 1, or **Wernwag**, the great American exponent of timber bridges. **Eads**, Engineer of the great bridge at St Louis; **Eiffel**, master of the iron arch, and, around the turn of the century, **Lindenthal** in the United States, followed in the 20th century by **Ammann**, builder of the George Washington Bridge, and today by the Italian engineer **Morandi**, builder of the Wadi Kuf and Maracaibo bridges that outspanned any of their contemporaries in concrete by a considerable margin. In detail, one example will suffice:

The Roeblings, American. John August Roebling (1806–69) and his son Washington Roebling (1837–1926) were outstanding among many talented and daring US bridge-designers of the second half of the 19th century. Evolution of the suspension bridge in its modern form and with spans of great length was an achievement that they cannot claim as their own, but they played the greatest part in it. Roebling senior was educated as a civil engineer in Germany (that in itself is somewhat remarkable at that date) and emigrated to the United States when a young man. He made wire rope in 1841, and first sold it to replace the hemp ropes, up to 0·23 m (9 in) in circumference that were used to haul the cars on the inclined planes of the portage railways which carried canal boats and passengers over the Allegheny Mountains. His first wire-cable suspension bridge was an aqueduct built in 1844–5 to carry the Pennsylvania Canal over the Allegheny River. From that time, he went from strength to strength with suspension bridges, until, with Brooklyn Bridge, one sees all the essential points of the modern suspension bridge, as explained in chapter 1. Roebling senior died from tetanus following an accident at Brooklyn Bridge, then at an early stage, and the main burden of building it – including the use of compressed-air caissons – fell on his son's shoulders.

INTO THE TWENTIETH CENTURY

Dr C Lely (1854–1929), conceived the master-plan of the Zuider Zee reclamation, as it has been carried out without fundamental changes (the work is still in progress) and shares with Telford the distinction of having had a town named after him. Lelystadt is the capital city of the polder province in the Ijsselmeer. Telford is a new town in Shropshire. A third member of this exclusive club* is Machado, the engineer of the railway from Lourenço Marques to Machadosdorp (Mozambique–South Africa). Lely was conspicuously unsuccessful as a young, practising engineer. He was on the point of emigrating to Brazil, when, at the age of 36, he was appointed Minister of the Rijkwaterstaat, because of the merit of his proposals for the Zuider Zee. However, the scheme was not started until 35 years later and the enclosing dam was closed about three years after his

*The custom in Brazil was to name a railway station after the engineer who built the railway. At least one of these stations so named – Ewbank de Camara – has grown into a town.

death. He thus exemplifies, not the designer nor the constructor but the mind that conceives a master-plan and doggedly promotes it until it wins acceptance.

Weetman Pearson (1857–1927), first Viscount Cowdray, English, exemplifies the successful civil engineering contractor in the tradition of the Victorian contracting empires of Brassey, Peto and Aird. Pearson's father had a small building business in Bradford; he turned it into the country's leading firm of civil engineering contractors. His firm achieved fame by completing in 1908, the East River tunnels in New York City in the face of extreme difficulties. He built Dover Harbour – three long breakwaters enclosing 247 ha (610 acres) of water, then the largest artificial harbour in the world, and he built the canals that drain Mexico City. His friendship with Diaz, the Mexican dictator, was the basis for an extensive series of public works in Mexico, including the development of oil-fields more in the style of 'economic development' in the modern sense than just contracting. He wound up his contracting company, still at a peak of success, about the time of his death, but the investment trust he established remains influential today. He became a figure in public life – a Member of Parliament and President of the Air Board, being the guiding hand in establishing the Air Ministry, and one of the creators of the Royal Air Force.

THE THEORETICIANS

The great names who have advanced the basic scientific theory on which civil engineers rely – that is theory of strength of materials, structures, hydraulics, and of soil machines – have (with the notable exception of Terzaghi) been renowned primarily in some other field of applied science or as mathematicians or physicists. The application of scientific knowledge to civil engineering was, historically, say 1750 to 1850, a French achievement based on appreciation of the work of scientists such as **Poisson** (1781–1840), **Navier** (1785–1836) and **Coulomb** (1736–1806). In France a civil engineer was a savant – in England a tradesman. The academic tradition spread to Germany, but only slowly to Britain.

The first attempt to analyse an actual structure was in 1742–3 by three investigators who wished to ascertain the cause of cracks and damage round the springing of the dome of St Peter's in Rome. They tried to calculate the horizontal thrust exerted by the dome, but according to Straub, in his *History of Civil Engineering* they were 'still wholly in the dark as to the nature of elasticity'. Mainstone disagrees, and says that the analysis was substantially correct and therefore epoch-making. The basis of elastic analysis is the law formulated by (and named after) **Robert Hooke** in 1678 (and independently by **Mariotte** in 1680).

Eaton Hodgkinson's work in the 19th century, some of it in collaboration with the engineer Fairbairn, made possible a more rational approach to structural design. This trend of development culminated in the design of the Forth Bridge by Baker and Fowler, the first major bridge which relied on analytical calculations as the basis and proof of its structural performance. Parallel developments in the United States included **Whipple**'s treatise on trusses.

Castigliano (1847–84) enunciated strain energy analysis and **Müller-Breslau** (1851–1925) the theorem of three moments. **Hardy Cross** invented his moment distribution analysis in 1930, an approach developed later with great generality in the physical world. Today's analytical tool that is its lineal descendant is finite element analysis.

The computer era has helped the structural analyst immensely. One of his problems was always to handle great numbers of simultaneous equations and this can now be done with relative ease, so permitting much greater freedom in choosing the physical form of one's structure. The cable-stayed girder with many stays radiating from each tower is an example of this freedom.

Daniel Bernoulli (1700–82) formulated the basic equation for energy of flow in open channels.

Osborn Reynolds (1842–1912) made a major contribution to the knowledge of fluid flow and practical application came with the work of **A H Gibson** who, at Manchester University pioneered the use of hydraulic models and many other practical applications of hydraulic theory.

More recently, the development of theoretical work on elasticity and fluid flow has been primarily linked with aeronautics, but with a measure of application to earth-bound structures. The foremost names are **Timoshenko** (1878–1972), **Prandtl** (1875–1953) and **von Karman** (1881–1972).

THE MODERNS

Sergei Yakovlevich Zuk

Sergei Yakovlevich Zuk (1892–1957), Russian, was head of the Hydroproject Institute of the USSR from 1932 to 1957, and the Institute has been named after him. During the latter part of his period as head of the Institute, the development of the Volga was initiated, and in particular the two enormous schemes now called V I Lenin and 22nd Congress power stations (originally known as Kuibyshev and Stalingrad) were designed and constructed. These achievements marked the emergence of the USSR as a world leader in the multiple-purpose development of large, continental rivers.

R G LeTourneau (1888–1969), American, built the first bulldozer, the first rooter, the first scraper, the first rubber-tyred earth-moving plant, the first US jack-up platform for offshore oil-drilling, and many other highly original machines. The 'triumph of logistics over intellect' referred to on page 68 was largely of his making – or at least that part of it that depends on the family of machines based on the bulldozer and the scraper. But the way he put it was: 'There are no big jobs; only small machines.' The ability to move mountains of muck quickly and cheaply, fundamentally affects the scope and concept of a 'big job' that a designer can consider to be practical; today it is taken for granted. His genius for devising totally original, and generally enormous, machines for special purposes was unique. He was a deeply religious man, and his own self-assessment was 'I am just a mechanic that the Lord has blessed.'

André Coyne (1891–1960), French, and **Carlo Semenza** (1893–1961), Italian, were the greatest designers of arch dams. The author can recall an open discussion at an international congress that was virtually a conversation between the two of them. Coyne's consulting firm, Coyne and Bellier, thrives today. Semenza was Chief Engineer of the Società Adriatica di Elettricità, now a part of the Italian nationalised electricity organisation. Coyne built arch dams in the three mountain regions of France when construction of hydro-electric schemes was at a peak in that country. His work spread overseas; for example, Kariba Dam is, in essence, a Coyne dam and the Daniel Johnson Dam bears his firm's imprint. Semenza built his dams in the region served by his company in north-eastern Italy, notably Lumiei,

Pieve di Cadore and Vajont dams. He died before the disaster at Vajont. When Malpasset Dam failed (Coyne had been consulted in its design, but had not supervised its construction) Coyne was mortally ill. These two disasters had an enormous effect on engineering opinion. The arch dam, as its exponents, and notably Coyne, affirmed, was fundamentally the safest type, and one had never failed. After those two events, virtually every arch dam was looked at again. Confidence has re-emerged in recent years, as the latest arch dams in chapter 1 clearly show.

Sir Ove Arup, born 1895 at Newcastle upon Tyne, son of the Danish consulate (and Norwegian mother). Sir Ove founded and built up the world's largest consulting firm (page 269). If one fact could account for the firm's growth, maybe it is that he served architects with understanding and sympathy – backed up, needless to say, by engineering of a quality that made the firm's offices, in its early days, a magnet for talented young engineers from overseas, who were keen to study under the master for a pittance. This approach was exemplified in extreme form in his treatment of Sydney Opera House, which would still be just a set of drawings if he had not been its engineer. His firm is, of course, catholic in its expertise nowadays, but one notable point of emphasis in its policies is its liking for 'total design' in which the divisions between the different professional disciplines that contribute to the creation of a building are, for practical purposes, made to disappear.

Ulrich Finsterwalder, German. Chief designer of the German contracting firm of Dyckerhoff und Widmann has spent the whole of his professional career with the one firm and has been with them for 52 years. He was co-designer with Dischinger and Bauersfeld of the first shell roof (see chapter 1) and has a world-wide reputation as a designer of daring and original structures, mainly of prestressed concrete. One of his best-known structures is Bendorf Bridge; he is the inventor of the stressed ribbon in its modern form. Inventiveness is not only expressed in the design of a structure. Finsterwalder's career exemplifies this fact in the development of a novel form of steel bar, adaptable to many uses on site and equally suitable for reinforced and prestressed concrete. He has been one of the foremost exponents of 'design and construct' – his firm negotiates, designs and builds, for as much of its work as it can.

Fazlur R Khan (born 1929), designer of the Sears Tower, John Hancock Center, Chicago, and One Shell Plaza, is a partner in the firm of Skidmore, Owings and Merrill; at one time he was the only engineer partner, the other nineteen partners being architects. He was born at Dacca, Bangladesh, and studied engineering in Calcutta. He was the first designer to replace the traditional steel skeleton of beams and columns with the tubular concept in which the exterior framing resists wind loads. His designs yielded dramatic savings in the structural costs of very tall buildings and were instrumental in bringing in today's vogue for buildings 40, 50 or more storeys in height.

Pier Luigi Nervi (born 1891), Italian, is perhaps the only designer of structures today, who is claimed equally by architects as an architect and by engineers as an engineer. His work has a strong personal style, characterised by the intimate interdependence of form and structure. He developed his own version of reinforced concrete – ferrocement – and applied it to create elegant and adventurous buildings. Equally, he applied other techniques when the design problem demanded them. The Burgo Paper-mill (see table 21) is one of his works. He has reached the peaks of architecture with the beauty of buildings such as the Hall of Audience at Vatican City.

P P Jansen, Dutch, was the first chief engineer of the Delta Service in Holland, and was thus responsible for initiating actual design and construction of the Delta scheme. Apparently he would have been the first to suggest that this outstanding scheme was a team effort; nevertheless, he pressed through the construction of the Haringvliet Barrage, despite its great originality and arduous duty, and so the success of the whole scheme owes much to him. He was also responsible for closing the breach in the Walcheren dikes, an epic post-war task. Distinguished contemporaries of Jansen are the engineers **Maris** and **Thijsse** and the present chief engineer of the Delta Service, **Ferguson**. The Delta scheme was first suggested by **van Veen**, whose book has been quoted in chapters 2 and 4.

Fritz Leonhardt (born 1909), German, has a world-wide reputation for original and brilliant structural design. The German television towers shown in the end-paper diagrams are mainly of his design; he has more than 150 bridges to his credit. His consulting firm, Leonhardt und Andra, has designed the first cable cooling tower, and the Pasco-Kennewick Bridge in Washington State, USA. This bridge is, in the author's view, the most advanced design the world has seen since the Severn Bridge.

Leonhardt's specialities have been design in prestressed concrete, and cable roofs; he was Professor of Concrete Structures at Stuttgart University from 1957 to 1974, his academic work involving research as well as teaching.

great schemes

CIVIL ENGINEERING PROTOTYPES IN CHINA

Irrigation schemes, canals and bridges were built in China on a greater scale and in advance of Western civilisation. This résumé is based on the information given in *Science and Civilisation in China*, by Joseph Needham* who says 'No ancient country in the world did more in civil engineering both as to scale and skill, than China, yet very little has been done towards making known the history of it.'

The Great Wall. The Great Wall is the best known among these works. The Chinese use the same word for 'city' as for 'city wall'; every city was surrounded by a wall, and city walls are known as early as the 15th century BC. A wall 20 m (65 ft) wide at the base and enclosing about 1750 m^2 (2100 yd^2) is of that period. Of 163 city walls built before 722 BC, ten were in use in 1928.

The Great Wall does not have a character similar to the typical cross-section shown in the sketch in its western parts, where it has no stone facing, and is almost buried by desert sand in some places. Dr Needham quotes estimates that there are 20000 wall towers and 10000 isolated towers, close to the wall, still standing, and that at maximum strength there were at least 5000 more of each of these two types of structure.

As the wall now stands, its main length dates from the mid 14th to mid 17th centuries, but both the history of its construction and maintenance

*Part 3 of vol. 4, with the collaboration of Wang Ling and Lu Gwei-Djen, *Science and Civilisation in China* is published by the Cambridge University Press.

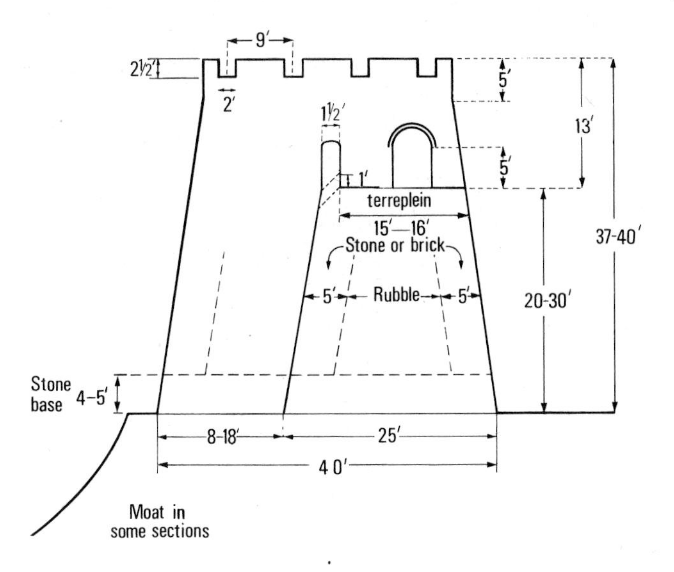

Typical dimensions of the Great Wall and wall tower in the Eastern region along the Hopei and Shanshi borders. Granite blocks up to about 4 m (14 ft) long by 1 m or so (3 or 4 ft) in width were used for the masonry. Brickwork is seven or eight courses in thickness. There are eight to twelve towers per mile

(if known) and its route (with loops and spurs) are complex. In summary, it had a period of importance from the 3rd century BC to the 3rd century AD, then little maintenance until major reconstruction in the 6th and 7th centuries, then seven centuries of neglect, followed by the main period of construction. Its purpose was primarily to keep out troops of nomadic horsemen, for which it was very effective.
The best estimate of the length of the wall is:

main line only	3460 km (2150 miles)
including all branch and spur walls	6320 km (3930 miles).

This may be compared with 550 km (350 miles), the length of the longest Roman wall, between the Rhine and the Danube across south Germany, and 117 km (73 miles) for Hadrian's Wall in the north of England. Hadrian's Wall was faced with stone, but the German wall was of earthwork with timber forts.

Roads. In the 3rd century BC, there was a network of Imperial roads in China consisting of six routes with a total length of some 6840 km (4250 miles). This length compares with the continuous length of 6020 km (3740 miles) of first-rate Roman road from the Antonine Wall in Scotland to Jerusalem. The most spectacular road-building of this period was across the mountains (the road reaching 2150 m (7000 ft) above sea-level) between Shenshi Province where the capital city was at that time, and Szechuan. Timber trestling was used extensively and the road was cut, or cantilevered on timber, out of sheer faces.

By about the 2nd century AD, the length of main roads in use in China was between 30000 and 40000 km (20000 and 25000 miles) compared with a similar total of 78000 km (48500 miles) in the Roman Empire (of which 14500 km (9000 miles) was in Italy and 3900 km (2400 miles) in Britain).

Bridges. The use of a single stone as a beam was carried to the extraordinary length of span of 21 m (70 ft) in a series of at least a dozen bridges – some of them more than 1220 m (4000 ft) long and up to 6 m (20 ft) in width – in Fukien Province, built in the 11th and 13th centuries.

An Chi Bridge built by Li Chhun in AD 610. Its span is 37·5 m (123 ft) and the rise of the arch is 7·2 m (23–7 ft). The spandrels are lightened by subsidiary arches. Typically, the masonry of Chinese arched bridges consists of relatively thin skins of dressed stone, with stiffening walls and rubble infilling. An Chi Bridge has recently been restored and the carved balustrade reconstructed

The maximum weight of one of these stone beams is about 200 t, the corresponding cross-section being about 1·8 m (6 ft) wide and 1·5 m (5 ft) deep. They are of granite. The art of handling them was lost, and although some bridges survive and are in use, in some cases they have settled, and in others, where the great beam has collapsed, it has been replaced by shorter spans.

There are, however, more dramatic and sophisticated advances than these megalithic monsters. Dr Needham stresses the advance made in building arched bridges when the semicircular shape is no longer considered obligatory. The segmental arches of Ponte Vecchio (Florence), Pavia, Castelvecchio (Verona) and Trezzo (table 2), built from 1345 to 1375, established a new step forward. However, a bridge comparable in structural form and in architectural accomplishment was built by the Chinese engineer Li Chhun in about AD 610 across the Chiao Shui River near Chao-hsien in Hopei Province, at the edge of the North China Plain. The illustration shows more clearly than words the graceful line of the bridge, the fine proportioning of the detailing, and why the Chinese regard it as one of the greatest achievements of their ancestors. There are nearly twenty comparable segmental arches in China, but mostly of lesser span and mostly dating from about the 12th century. One of them is the bridge that Marco Polo thought (in AD 1280) was the finest in the world. Ten mounted men could ride abreast on it, and its marble parapet boasted 283 stone lions. Li Chhun's bridge emerges from this account as one of the world's great structures.

The invention of wrought-iron chains was made in south-west China perhaps in the 1st century AD, but not later than the 6th century. Dr Needham lists 'some interesting' iron bridges (24 of them) which include one of 69 m (225 ft) span in Yunnan Province that had iron chains by AD 1470, and one of 110 m (361 ft) span built in 1701–06. These are samples; no complete list has ever been assembled although about 50 iron-chain bridges have been listed. The dates clearly make the spans of every bridge prior to Telford's Menai suspension bridge rather feeble. The longest span cited in China (but see below) is 131 m (430 ft) being one span of a two-span bridge.

Krasnoyarsk Dam and Power Station, at present the world's largest power station (table 45). In contrast to home-based criticism of hydro-electric construction in the USSR, this station exceeded the output expected from it, and had paid for itself by 1972, four years after its completion. The power station is on the left in the upper view, two penstocks serving each of the turbine-generator sets; beyond the power station is the spillway. The lower view shows the alternators in the machine room of the power station

Traditional iron-chain suspension bridge over the Yangtze River in Yunnan Province. The span is 100 m (328 ft) and the road is carried by eighteen chains. The masonry at the anchorage is designed for a variation in the level of the river of 20 m (say 70 ft) or more

Chi-Hung iron-chain suspension bridge over the Mekong River in Yunnan Province. Its span is 69 m (225 ft) and it has had iron chains since AD 1470. The chains are twelve in number (plus two serving also as hand rails) and consist of links 0·3 in (1 ft) long of 19 mm (0·75 in) diameter

The wrought-iron chains of these bridges consisted of hand-forged links made from bar 2 to 3 in (50 to 80 mm) in diameter. I-bar links are not unknown; one bridge 91 m (300 ft) long had bars 5·5 m (18 ft) long and 57 mm (2·25 in) in diameter, pinned together. In general the chains installed in the bridge could not be tightened, and were fixed in massive stone anchorages. Historical accounts refer to a bridge with iron chains built in AD 65. Whether or not that is acceptable, Dr Needham insists on 'The indubitable existence of long-established iron-chain suspension bridges on the route to India by the beginning of the 6th century.' He also cites a cluster of bridges in north-west Yunnan, built in the 6th century, where one route was the responsibility of an Iron Bridge Commissioner!

These bridges were basically similar to bamboo cable bridges, but at least one detailed account is available of a bridge – in Tibet, not China – that used the chains in the manner of the Western suspension bridge, that is in catenary, with a deck slung below. It dates from AD 1420, and its span was 137 m (450 ft). Its deck was wide enough for pedestrians, and it was in use in 1878, but had been removed in 1903 (table 2). There seems to have been even less study of these bridges in Tibet than in China.

Canals and irrigation. Irrigation, and specifically wet rice agriculture, lie at the traditional heart of life in China. The power of the Chhin State in the 3rd century BC was largely built on extensive irrigation works. There is reference to irrigation as early as the 8th century BC. By the 6th century BC, impressive irrigation works were well under way.

In northern Anhui, south of the city of Shouhsien, there exists a reservoir measuring 100 km (62 miles) round its circumference and known today as Anfang Thang, but in ancient times as Shao Pei (Peony Dam) that was built between 606 and 586 BC, and is the earliest known irrigation reservoir in China. The Hung Kou or Pien Canal, linking the Huai and Yellow Rivers, about 800 km (500 miles) long, was built between 361 and 353 BC according to some scholars, but 200 years earlier 'might be a reasonable estimate', Dr Needham says.

The Chang River irrigation system, involving a diversion of that river north-eastwards so that it did not debouch wastefully into the Yellow River but irrigated a region of Honei, was built in the 4th century BC.

The Han Kou Canal connecting the Huai and Yangtze rivers (to the north-east of the canal just mentioned) initiated in 486 BC as a military measure, was completed in the 4th century BC; this is an early part of the Grand Canal.

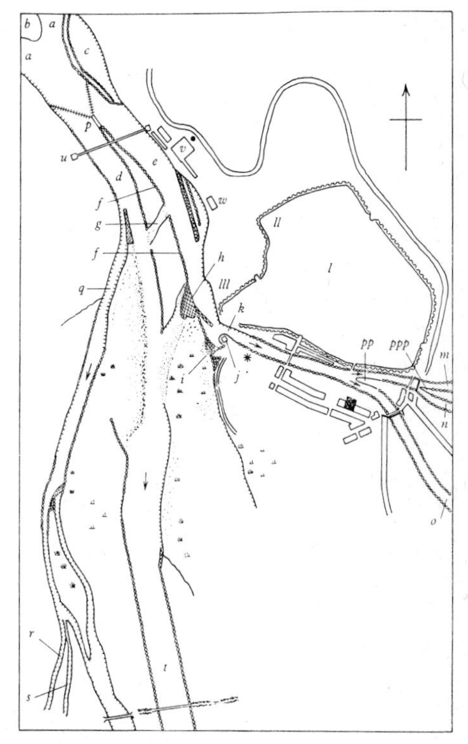

Plan of the Kuanhsien Irrigation System Headworks (Tu-Chiang Yen) as given by Needham

- *a,a* Min Chiang (Min River)
- *b* Han-Chia Chü (Han-family island)
- *c* Pai-Chang Thi (Thousand-foot dyke)
- *d* Wai Chiang (Outer feeder canal; old course of river)
- *e* Nei Chiang (Inner feeder canal)
- *f,f* Chin-Kang Thi (Diamond Dyke)
- *g* Phing-Shui Tshao (Water-level by-pass or adjusting-flume)
- *h* Fei-Sha Yen (Flying Sands Spillway)
- *i* Jen-Tzu Thi (V-shaped spillway)
- *j* Li-Tui (Separated hill) and Fu Lung Kuan (Tamed-Dragon Temple; the votive temple of Li Ping)
- *k* Pao Phing Khou (Cornucopia Channel; the rock cut)
- *l* Kuanhsien City
- *ll* Yü-lei Shan (Jade Rampart Mountain)
- *lll* Fêng-Lou Wo (Phoenix Nest Cliff)
- *m* Phu-Yang Ho (derivatory canal)
- *n* Po-Thiao Ho (derivatory canal)
- *o* Tsou-Ma Ho (derivatory canal). New sluice-gates were installed across this, as shown, in 1952.
- *p* Tu-Chiang Yü Tsui (Fish Snout; primary division-head of piled stones)
- *pp* Thai-Phing Yü Tsui (left secondary division-head)
- *ppp* Ting Kung Yü Tsui (left tertiary division-head)
- *q* Sha-Hei Tsung Ho (right main derivatory canal)
- *r* Sha-Kou Ho (derivatory canal)
- *s* Hei-Shih Ho (derivatory canal). The feed into both of these is assisted by a spillway higher up, overflow rejoining the Chêng-Nan Chiang
- *t* Chêng-Nan Chiang (old course of river, flood course, etc.)
- *u* An-Lan So Chhiao (suspension bridge)
- *v* Erh Wang Miao (Temple of the Second Prince; the votive temple of Li Erh-Lang)
- *w* Yü Wang Kung (Temple of Yü the Great)

The two lines, drawn in the convention usual for railways, which connect the Thousand-foot dyke (*c*) and the right bank of the Min R. respectively with the Fish-Snout primary division-head (*p*), represent the positions of the temporary *ma chha* dams set up when the water is low for clearing the beds of the Nei Chiang (*e*) and the Wai Chiang (*d*), the former in January, the latter in November. A steel cable, anchored at the point marked with an asterisk, guides rafts, floating tree-trunks, etc., into the Phu-Yang Ho (*m*), avoiding the Tsou-Ma Ho (*o*). On the hill above the Temple of the Second Prince (*v*) there is a Taoist temple.

Table 51. Statistics for the Kuanhsien irrigation scheme.

Total area irrigated:	1940	200 000 ha	500 000 acres
	1958	375 000 ha	930 000 acres
	ultimate extent	1 780 000 ha	4 400 000 acres
Total length of subsidiary feeders in 1940		1175 km	730 miles
Flow of River Min			
Average, December/March		200 m³/s	7063 ft³/s
June and July		7500 m³/s	265 000 ft³/s
Minimum river flow for irrigation (starting in April)		585 m³/s	20 660 ft³/s
This flow divided:			
inner canal		305 m³/s	10 770 ft³/s
outer canal		280 m³/s	9888 ft³/s
Flow in inner canal in July		1000 m³/s	35 300 ft³/s

Examples of irrigation works include the Chengkuo irrigation canal completed in 246 BC with a length of nearly 160 km (100 miles) and commanding the irrigation of 160 000 ha (400 000 acres). Today, about 140 km (87 miles) of main canal command about 34 000 ha (85 000 acres). The intake of the canal is on a tributary of the Yellow River and the canal was named after its engineer.

Kuanhsien scheme. The Kuanhsien irrigation scheme in Szechuan, utilising the Min River is 2200 years old, and is described by Dr Needham as one of the greatest of Chinese engineering operations which can be compared only with the ancient works of the Nile. An area 65 × 80 km (40 × 50 miles) has been made to support 5 million people, without danger of drought or flood. There can be few irrigation schemes in the world that have functioned for so long, and as far as can be judged, with an ecologically benign régime.

In modern irrigation engineering, the typical river scheme starts with a barrage across the river that heads up the level, so permitting diversion of water into the main feeder canals. The intakes to these canals are controlled by sluice-gates. When the river floods, the gates of the barrage are opened, so effectively lowering the river-level, and flow into the feeders can be restricted by their gates.

The Chinese system achieves similar control without benefit of gates. Typically it consists of a 'snout' – that is a V-shaped groyne that divides the river, so diverting some of its water into the feeder. Feeder and river form the two branches downstream of the snout, the feeder, in general, at a flatter slope than the river. The feeder then has one or a series of spillways – typically low dikes on the river-side – so that when the river floods, excessive flow diverted at the snout into the feeder spills back again into the river.

This arrangement serves to divert the correct flow required for irrigation over the whole range of river flows; the irrigation flow can be varied by adjusting the levels of the spillways. The main difference between this system and the modern one is that only a relatively modest proportion of the river's flow can be utilised in the feeder. The old method also entails maintenance – not in the least surprising, nor a fact that has changed. These features are exemplified by the Kuanhsien scheme, but in recent years it has apparently been modified and enlarged, and the benefits of gated control of flow applied to it in greater measure.

The plan of the scheme plus the legend have been borrowed unaltered from Dr Needham's text. The feeder is formed by the channel 'e' at the Fish Snout and the wall 'f' is higher than the flood-level in the river. There are three spillways that can return water to the river. The principal control of flow is at 'h' (Flying Sands spillway) the crest of which is adjusted for optimum flow in the feeder. As the flow rises, the V-shaped spillway at 'i' comes into operation, and finally, for the highest floods, the spillway at 'g' which appears to have plenty of capacity.

At 'k' – the Cornucopia channel – is the main work of diversion to the lands to be irrigated. The feeder canal has been excavated through a steeply sloping hillside of rock, the depth of cut reaching a maximum of over 60 m (200 ft) from the high side down to formation level, and about 30 m (100 ft) on the low side of the canal, the canal being about 15 m (50 ft) wide at the low water level.

Maintenance is carried out rigorously each year from October, when the river is low and the growing season is over. The region round the snout is thoroughly refurbished, being dewatered in stages by cofferdams. The beds of the channels are relevelled and the stone banks realigned, accurate geometry and levels evidently being the key to proper functioning of the diversion. Levelling devices include iron bars set in the bed of the channel.

The Grand Canal. Like the Great Wall, the Grand Canal has gone through periods of importance and decay and has been enlarged, reconstructed, and rerouted, for political or economic reasons, or because silting or other defects have occurred. In recent years it has again been made to serve useful purposes, and for example 2000 t vessels now ply on it. Its original purpose was to collect tax in the forms of rice grains, tax-collecting vessels being the principal traffic (for example 165 000 t of grains in AD 735). One early section (4th century BC) has already been mentioned. The sections were linked together about AD 600. More than 5 million people were mobilised at that time to excavate the main work, the new Pien Canal.

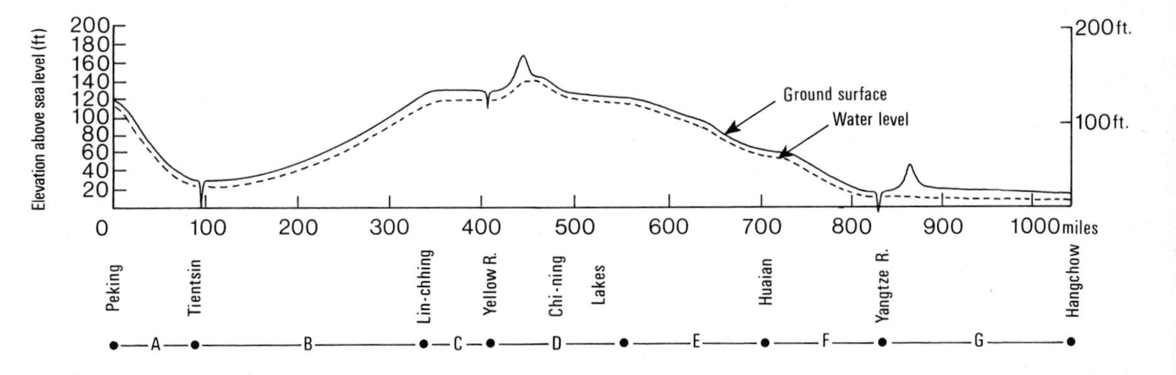

Longitudinal section of the Grand Canal

A Canalised River Pai (AD 1293); B Canalised River Wei (*c* 1290); C Hui Thung (1289); D Chi Chou (1283); E Huan Kung (1280); F Shan-yang (350 and 587); G Chiang Nan (610)

Bonnet Carré spillway on 14 April 1973, spilling for the seventh day of the flood. For data on this structure see table 44. The river is to the left of the photo. The spillway discharges into Lake Pontchartrain, about six miles away. It was operated in the floods of 1937, 1945, 1950 and 1973, and was opened on 14 April 1975 for the 1975 flood

Morganza spillway flooding in May 1973–see also table 44. The floodway is 6·5 km (4 miles) wide and flows into the Atchafalaya Basin floodway. It is much larger than the Bonnet Carré spillway, and was operated for the first time in 1973. Opening the spillway takes 15 hours
Photos: Lower Mississippi Valley Division, US Corps of Engineers

Including the Yellow River itself up to the city of Sian, a Y-shaped route with a total length of some 2500 km (1560 miles) was then in use.

Remodelling to provide a route from Hangchow to Peking and the northern end of the North China Plain – i.e. a broad canal for nearly 1750 km (1100 miles) – say New York to Florida – and with a summit 42 m (138 ft) above sea-level, took place in the 13th century. North of the Huai River, a more easterly route was selected, and the canal was virtually reconstructed to provide a north–south artery across a vast region where the rivers run predominantly west to east.

The new construction was not merely excavation. One section of about 80 km (50 miles) required twenty lock-gates. The summit section had problems of water-supply, so a reservoir was built with a mile-long dam in 200 days with a labour force of 165000. Four subsidiary reservoirs, called 'water boxes', were built beside the canal. In a fall of 35 m (116 ft) there were 21 gates. By AD 1327, the final form was delineated – 1700 km (1060 miles); but somewhat longer, 1782 km (1107 miles) according to another source. The summit canal, completed in 1283, is the oldest successful, fully artificial canal in any civilisation.

PREVENTING FLOODS IN THE MISSISSIPPI VALLEY

Fertility and destructiveness are the dual characteristics of the alluvial plain of a big river. When New Orleans was founded in 1717, the site chosen was opposed by the engineer La Tour, because the Mississippi flooded it from time to time. He was overruled (it is surprising how well defined such flood plains are, and how unerringly they are chosen for homes, cities, etc.) and so started to build the first levees (i.e. embankments) along the Mississippi. By 1735, there were levees on both sides of the river from 30 miles (50 km) above, to 12 miles (20 km) below New Orleans, but they were insufficient, and were breached by the floods of that year.

Throughout the 19th century, navigation and flood protection were given considerable attention, but a co-ordinated system of levees was not started until 1882. The most disastrous floods on record occurred in 1927; 67000 km² (26000 miles²) of land were flooded (i.e. three-quarters of the lower valley). The present scheme started in 1927 and is now in operation. It consists of the elements listed overleaf:

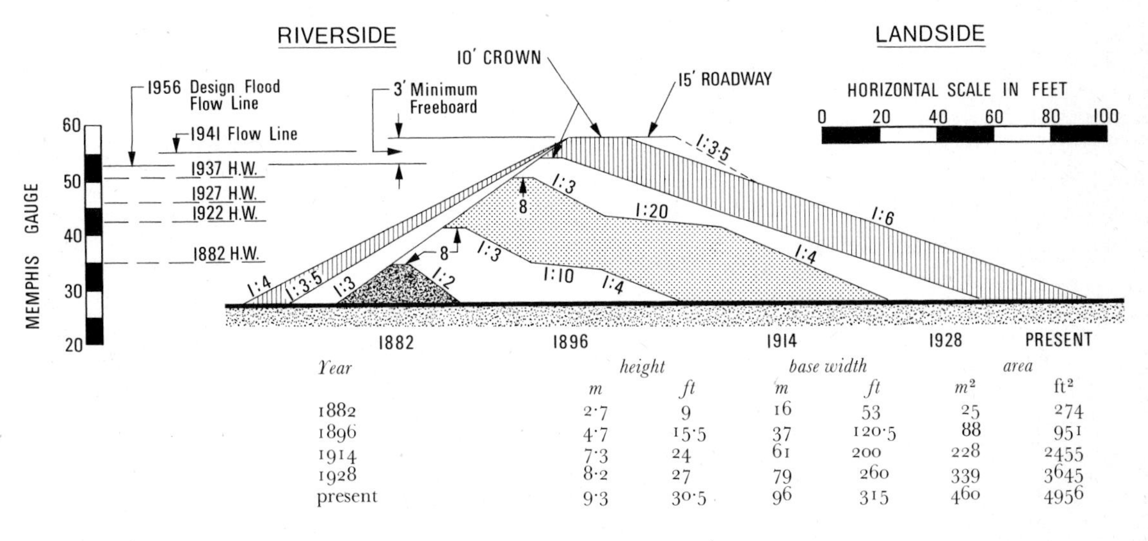

Year	*height* *m*	*ft*	*base width* *m*	*ft*	*area* *m²*	*ft²*
1882	2·7	9	16	53	25	274
1896	4·7	15·5	37	120·5	88	951
1914	7·3	24	61	200	228	2455
1928	8·2	27	79	260	339	3645
present	9·3	30·5	96	315	460	4956

Embankment cross sections

● **Levees**, to contain the main river, from about Cape Girardeau, Missouri, down to the vicinity of the mouth of the river at Head of Passes.

The longest continuous length of levee in the world (pierced in four places by locks) begins at Pine Bluff, Arkansas, on the western side of the main stream, follows the south bank of the Arkansas River and continues along the Mississippi River to near its mouth, a distance of over 1045 km (650 miles).

On the east bank, the longest continuous length of levees is 434 km (270 miles) from just below Memphis to just above Vicksburg.

In certain places there are so-called fuse plugs – levees at a lower crest level that can be overtopped as the flood builds up, so drawing off some flood-water into storage in a flood reservoir.

● **Floodways:** Channels to carry water past critical reaches. The principal floodways are two channels leading into the Atchafalaya Basin floodway, and the Bonnet Carré spillway, leading to Lake Pontchartrain.

● **Channel improvement:** Straightening and stabilising the alignment of the main river to increase its flood capacity. By cutting through the necks of meanders, the Mississippi has been shortened by 274 km (170 miles). Flexible concrete mattresses are used extensively to resist scour, and so stabilise the straightened channel; training works in the channel and dredging also contribute to stabilisation.

● **Tributary works:** Works similar in character to those already listed, but on tributary streams, and also including one other important element – reservoirs for flood control. Drainage works, pumping stations, etc., are also included. Extensive works have been carried out on tributaries that join the Mississippi within the area of the lower valley, such as the St Francis River

Table 52. Statistics for the Mississippi River.

Catchment: total area	3 200 000 km²	1 245 000 miles²
as proportion of continental land area of US	41%	41%
administrative spread	31 states plus 2 Canadian provinces (New York to Montana)	
Alluvial plain of lower river:		
length	965 km	600 miles
width	50–200 km	30–125 miles
area	90 600 km²	35 000 miles²
extent	Cape Girardeau southward – parts of seven States	
Rivers with catchments larger than Mississippi	Amazon, Congo, Nile	
Flood control: 1928 Act		
'project' flood: at Cairo	66 830 m³/s	2 360 000 ft³/s
at Old River	85 800 m³/s	3 030 000 ft³/s
'project' flood compared with actual flood, 1927	123–129%	
'project' flood without reduction by reservoirs in the upper catchments (at Cairo)	80 700 m³/s	2 850 000 ft³/s

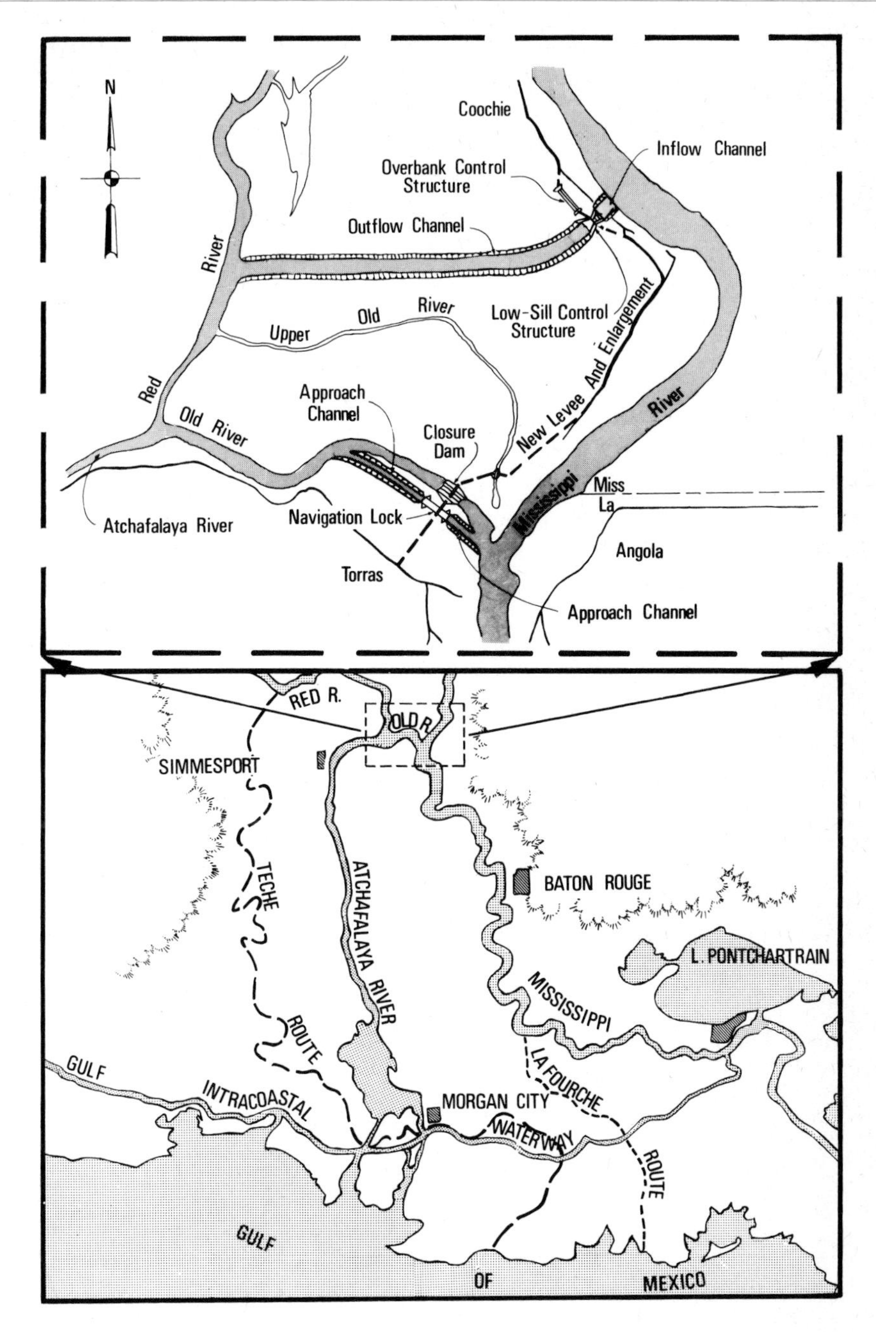

Sketch showing works to prevent capture of the main stream by the Atchafalaya River

(in Missouri and Arkansas) to name only one of about a dozen streams, some of which have required works that in other contexts would be major flood-control schemes in themselves.

Dams in the upper regions of the catchment – along the main river, the Missouri, and in the Tennessee catchment for example – contribute to flood protection, but these works in the upper catchment are not included in this description.

One problem encountered on the lower river – that of preventing a river capture that would have virtually ruined the whole scheme – is unique, at least to river engineering in the United States.* Old River was formed in 1831, by cutting off one of the many loops of the Mississippi to shorten navigation. The upper portion of the loop eventually silted up, leaving the lower portion as the connecting link between the Red, Atchafalaya and Mississippi rivers as shown in the sketch. When the Red River was high and the Mississippi low, flow in the Old River was eastwards, and vice versa. Reversals became less frequent as the Atchafalaya enlarged itself by capturing the Mississippi's waters, and no easterly flow occurred after 1945. If left alone, the Atchafalaya would have become the main channel of the Mississippi below Old River; the costly flood-protection works would have become useless, the former main stream would have become a salt-water estuary, massive disruptions would have occurred in the Atchafalaya Basin, and many towns would have been virtually destroyed. The uncontrolled link – Old River – was dammed and replaced with a controlled connection as shown in the sketch. The main flow now passes down the Mississippi, low floods are spilled into the Atchafalaya at the low-sill control structure, and high floods by this structure and at the overbank structure beside it. Old River is closed by a dam (the closure was made in 1963) with a lock to cater for navigation.

The total estimated Federal cost of the scheme is $5100 million (1974 estimate) and the amount allocated to June 1975, was $2200 million. The benefit to cost ratio of the scheme is 11·9 to 1. Indirect benefits are vast – to navigation, and in establishing confidence and hence the growth of an industrial community.

'WATER IS OUR EVERLASTING ENEMY'

There is no parallel in human history with the reclamation of Holland, the economic strength that grew from it, and of its reaction with the country's habits and politics. The first attempts – the mounds in north-eastern Holland – were mentioned in chapter 2. Dr van Veen† distinguishes four main stages:

● **About 400 BC.** The mounds, which were probably dependent on the use of iron spades. Iron was introduced at about that date.

● **Between AD 800 and 1000.** The first attempts at building sea-walls or dikes, and the first use of sluices.

● **About AD 1600.** General use of windmills for land drainage, i.e. for draining polders lying below sea-level. The first use of a windmill for land drainage was in 1405, but for grinding corn (in Holland) more than 400 years earlier.

● **About AD 1850.** General use of the steam-engine, to replace windmill drainage, to provide mechanical power for excavation, and specifically for dredging. This is the start of the modern era.

From the start, the sea was gaining – a fact not realised by the pioneers, who thus had to struggle harder to retain what they had already won. History thus shows repeated flooding and disaster, sometimes with great

*See chapter 3 The Farakka Barrage and Hooghly River scheme which is similar.

†*Dredge drain reclaim. The art of a nation*, by Dr Joh van Veen Martinus Nijhoff, The Hague, 1952.

The sea pouring through a breached dike in the 1953 flood

loss of life. The mound-dwellers had to start dike-building about AD 1000 or they would have had to evacuate. In about AD 1300, the Zuider Zee was formed, largely in a single incident it is generally supposed, when a storm surge, like the one in 1953, occurred. It was not recaptured until 1932.

Until the reclamation of the Zuider Zee, losses exceeded gains. Dr van Veen sums up thus:

	Acres	**Hectares**
Gains: along the sea-shore	940 000	380 000
by pumping out lakes	345 000	140 000
reclamation in the Zuider Zee	550 000	223 000
Total	1 835 000	743 000

Reclamation is still in progress in the Zuider Zee and part of this acreage has still to be won (see below). Against this, he puts losses since AD 1200 as 1 400 000 acres (567 000 ha). In the first quarter of the 17th century, 80 000 acres (32 000 ha) were gained and held – the best progress ever. The technique is to plant spartina grass along the tidal foreshore; accretion then gradually takes place. Andries Vierlingh, a dikemaster of the 16th century, was a pioneer and exponent of this method. Dr van Veen points out that although the present generation is the greatest reclaimer of land, it should be with modern knowledge and techniques. Even so, it has neglected Vierlingh's low-cost techniques for a century, and they should be revived.

In the modern era, the Dutch tradition can be crystallised by describing two great civil engineering feats – the reclamation of the Zuider Zee and the Delta scheme.

Draining the Haarlem Mere

By way of introducing the modern era as just defined, it is worth mentioning the largest of the lake reclamations, draining the Haarlem Mere, an area that today includes, for example, Schiphol Airport. This scheme was the greatest land-drainage scheme of its day, and of special interest because it depended on the use of steam-engines.

At that time there were said to be 18 000 windmills in Holland, mainly used for land drainage. Steam-engines were used occasionally, but were unpopular because of the heavy cost of their fuel – 20 lb (9 kg) of coal per horsepower-hour of useful work. Large pumping engines, used to drain the Cornish tin-mines, were much more economical; so the Dutch Commissioners of the scheme visited Cornwall to find out at first hand and to measure the economy of the Cornish engines. Eventually three engines of unprecedented size were ordered and installed. They pumped for 84 years and required practically no repairs, consuming 2½ lb (1·1 kg) of coal per horsepower-hour of useful work. One of the three engine-houses has been preserved. Each engine had a high-pressure cylinder 84 in (2·13 m) in diameter working inside an annular low-pressure cylinder 144 in (3·66 m) in diameter, worked four double strokes per minute and developed between 400 and 500 horsepower. Pumping started in 1848; by the middle of 1852 the whole area had been drained – 19 000 ha (47 000 acres) with an average depth of water initially of 4 m (13 ft).

The Zuider Zee

The enclosing dam was built in 1927–33. Its volume was unprecedented, but it was not volume alone that made its construction so epoch-making. The scale of the tidal closure was a much more daunting problem than that of simply placing such a great volume of material in the dam. The tidal range at the dam is 1·6 m (5·33 ft).

Two important contributions to its success offer a comment on the interaction of theory and practice. Lorentz, the mathematician, devised methods of calculating the heights of the tides, before and after construction of the dam, that proved to be precisely accurate. Boulder-clay was discovered, and used in vast quantities as the ideal local material to resist scour.

In two deep regions sill dams were built first – clay banks covered with fascine mattresses on which a layer of stones of high density had been laid. The sill dams extended up to 3·5 m (11·5 ft) below mean sea-level.

At the same time, the dam was built in the shallowest regions, where it affected the tides the least. Side walls of boulder-clay were dumped on the sea-bed and raised above the water, then sand was pumped between them to form the body of the dam. The sloping faces of the dam were faced with fascine mattresses and stone.

The dam did not affect the tides strongly in the early stages of construction. As the work progressed, the cubature – the volume of water moving in or out on each tide – became less and was much reduced when the time had come to close the dam. The tidal race through the final gap in the dam flowed at a velocity of up to 4·5–6 m/s (15–20 ft/s) and the mattresses, stone and clay, could withstand it for a short period with acceptable losses of material by scouring. The closure was successfully made on 28 May 1932. The volume of material in the dam is about 38 million m³ (48 million yd³). The first major reclamation – Wieringermeer Polder – was dry in August 1930, after six and a half months of continuous pumping, but was not finished until 1940 – in good time for the Germans to flood it during the war! A dam linking the island of Wieringen with the mainland was built in 1920–4; it is 2·4 km (1·5 miles) long. An experimental polder of 100 acres (40 ha) had been reclaimed in 1927.

When the polder land dries out for the first time, the job of reclamation is less than half done. The technique today is to sow reed seed from the air as the water recedes, so that reeds, rather than heterogeneous weeds, grow initially and can be ploughed in later. Land drainage – open ditches and canals, and miles of land drains – follows, along with roads. Then come services of all kinds and houses. Today the South-eastern Polder is growing crops; it has larger fields and more spacious planning than earlier polders.

The capital of the new polder lands, Lelystadt, is occupied, but still has a frontier aura about it, and has not yet had time to grow into a mature town. The South-west Polder has been drained, but is not yet productive land. The last polder has still to be reclaimed. So the scheme has already taken 55 years since the first dike-building. There is no comparable land reclamation, anywhere in the world. The scheme has always had two other important aims in addition to the reclamation of land. The Ijsselmeer – the freshwater lake formed behind the enclosing dam – is of major importance to Holland. The freshwater reservoir alone justified the cost of the enclosing dam. It is a source of water; it forms part of a system of hydrological management linked with the Rhine and other water sources of Holland, and it has excluded salt from a vast region. One of the worst troubles of reclaimed land is poisoning by salt. The other important aim is, of course, sea defence; this again, would alone justify the enclosing dam.

The Delta Scheme

Sea defence is the principal aim of the Delta scheme which was started after the disaster of the 1953 storm surge. There is virtually no land reclamation involved, but the exclusion of salt and the scheme's contribution to the freshwater economy are important. Nevertheless, its main purpose is to shorten the line of defence against the sea, and to heighten the dikes so that they are able to resist extreme conditions that occur once in 10 000 years.

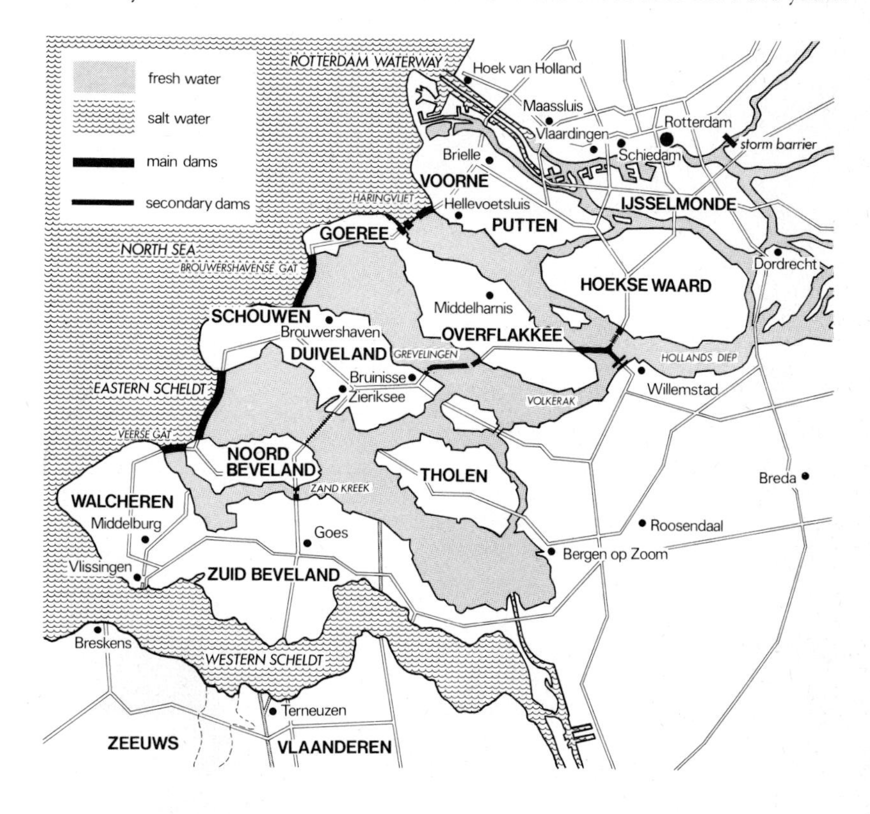

Plan of the Delta scheme

The port of Rotterdam is left open. River discharge (additional to that which discharges through the port of Rotterdam) is controlled at a barrage in the Haringvliet. The other sea openings of the Rhine Delta are permanently closed by dams. The scheme has been completed, except for the largest dam of all – the Eastern Scheldt Dam – which is under construction. Dutch civil engineers have given of their best in the Delta scheme, and the whole story of its concept and realisation have raised the traditions of Dutch engineering to a new peak. The author rates the scheme number one in today's world for engineering excellence.

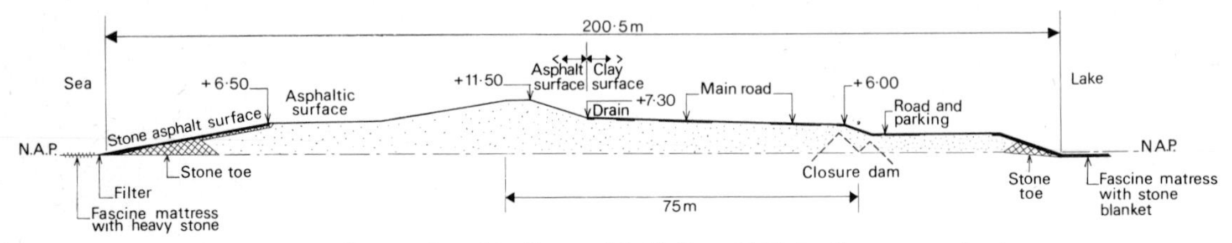

Cross-section of the Eastern Scheldt Dam. NAP signifies mean sea-level

Veerse Gat Dam; placing the last of the seven gated caissons at slack water

Veerse Gat Dam completed

The method of gradual closure. A cableway car dumps its load of stone

The stone weir emerging above the water, and sand placed behind it – a stage in the construction of Grevelingen Dam

Haringvliet Dam

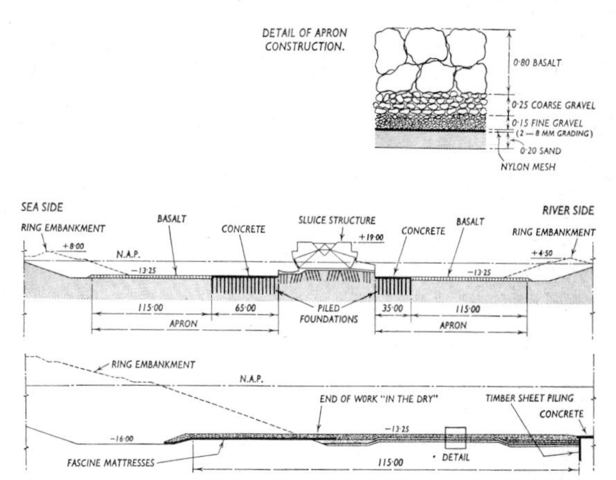

Sketch showing the protective aprons at Haringvliet Dam. The cofferdam in which the sluice section of the dam was built is illustrated in chapter 2

A fascine mattress of willow twigs and polypropylene ready for sinking

Special-purpose vessel – the *Jan Heijmans* – for laying a mattress on the sea-bed

The scheme was considered essential for the safety of Holland; but it was not known how it could be built; it was decided to go ahead and to gain experience as the work progressed, then apply that experience to the major problem – that of closing the largest tidal openings – when that stage was reached. Two possible methods were known in principle, and were tried out full-scale on smaller dams early in the programme. These two methods – gradual closure and the use of gated caissons – are shown in the sketches and captions.

A different set of problems was faced with the Haringvliet Barrage. The sea-gates have to resist the forces of waves from the open sea, spanning sluiceways that are large enough to pass ice during a winter flood. The gates are carried by prestressed concrete beams that also carry a motor road, and are a world record for the loads that they can carry, the wave loads from the gates being the greatest factor (see chapter 1). The barrage was built within the shelter of an enormous cofferdam. Scour protection alone was an epic problem. (Details are given with the pictures.)

The last and greatest task is the Eastern Scheldt Dam (see tables 12 and 15). Closure was, until 1974, planned for 1978. It was to have been by the method of gradual closure, with extensive protection of the sea-bed against scour by new methods. This bottom protection was already being laid in 1974.

But in that year the scheme met a new hazard, imposed by Dutchmen, not the forces of nature. The Dutch Government decided that to meet environmental objections, the dam would be modified, so that the tides would still flow through it. It is to be redesigned with gated caissons that will stay permanently open and will be closed only when an abnormal tide is expected or has occurred. It is not yet known whether the new idea is practicable or possible. Its main novelty is that it retains permanently the most dangerous stage of the closure operation. All this has been described

Brouwershavense Gat Dam

Photos: Acknowledgment is made to the Delta Service of the Rijkswaterstaat (the Dutch Ministry for hydraulic works) for the various photos of the Delta scheme and for the 1953 photo from their archives

in some detail in *The Consulting Engineer* (August 1974) and the author's conclusion is that the sooner the decision is reversed and the dam built as originally conceived, the better. The reasons for the change are relatively trivial, the cost of it enormous, and the hazards probably insurmountable. The irony of it is that it is solely the unqualified success of the whole of the Delta scheme that has made it possible for such irresponsible tinkering to be considered.

THE SUPREME DEMAND

The most complex problem for the structural designer is not necessarily the longest span, nor the heaviest loads, nor even building on the poorest subsoils. The two examples now described were each, in their turn, claimed to be the ultimate in difficulty that had ever been faced. Each was 'the supreme example of architectural inspiration making an absolute demand on the capability of the engineering collaborators'. Sydney Opera House was 'an adventure in building . . . not really of this age and in concept more appropriate to the product of autocratic rule of a former era'. The Barbican Arts Centre in London 'more than supersedes Sydney Opera House both in foundation complexity and the technical response which it demands from its architectural and engineering designers'. The quotations are from Sir Ove Arup or his partners.

In 1957, an international competition – assessed by four architects – for Sydney Opera House was won by Jørn Utzon. The drawings he submitted were simple to the point of being diagrammatic.

His submission was unaided by structural engineering advice, and strictly, his intuitive assessment of the structure was erroneous. He conceived the roof shapes as thin shells; but a shell in structural terms is a shape that is under direct stress, generally compression, and bending is substantially

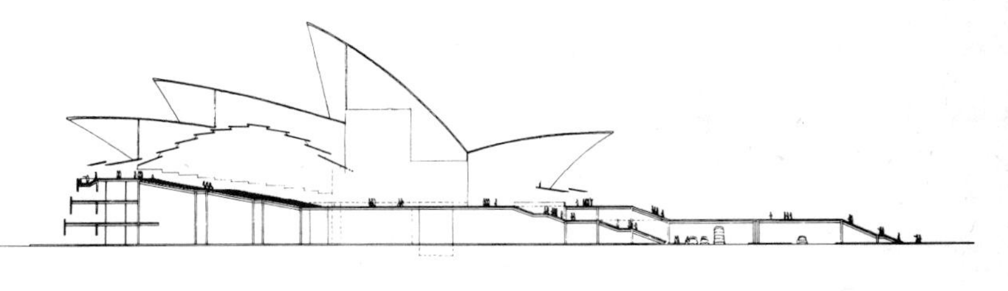

Utzon's sketch of the proposed opera house at Sydney, submitted to the competition which he won

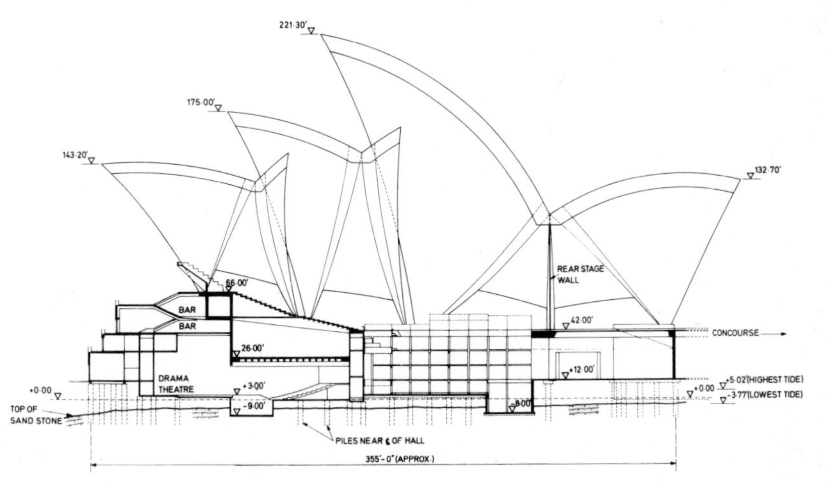

The opera house as built, with a structure of fan-like ribs of precast prestressed concrete.
Sketches: Ove Arup and Partners

eliminated from it by its geometry. It turned out that Utzon's shapes introduced large bending moments. Several structural solutions were evolved; models were tested; but no answer resulted. Several years of work were thrown away, and a reappraisal commenced. Two possible structural solutions then became apparent; the one chosen by Utzon, in which the main shells were formed from a series of fan-like ribs of precast prestressed concrete, was the one constructed.

Utzon resigned from the job in 1966. Other architects took over his work, but the engineers, Arups, and the contractors, Hornibrooks, stuck to their last. The building was inaugurated in 1973 after nearly fifteen years of construction, and at a cost of about 40 times the original estimate. The assessors of the competition were convinced that their choice could become one of the world's great buildings. Perhaps the critics, and maybe the world at large, are inclined to agree with them, except for those purists who maintain that architectural form should be rationally based on structural necessity or function, and that to devise a complex structure to support an imaginative sculptural shape is architecturally dishonest.

The Barbican Arts Centre is still under construction; and it is not yet possible to judge it architecturally. The problem was rather more specific and practical in this case – it was to get a quart into a pint pot. A great deal of accommodation was needed; a high building was ruled out, so it had to go into the ground.

The engineers thus had to evolve a foundation design that would be stable at all times during construction as well as after completion of the building, and to substantiate their scheme by rigorous analysis. In their view, this is where the challenge lay, and where they penetrated beyond the boundaries of established practice.

The fundamental structural problem was to resist safely the large forces by the ground on the excavation, i.e. to create a stable hole in the ground, 137 m (449 ft) long, 76 m (249 ft) wide and 15–18 m (50–60 ft) deep, but 23 m (75 ft) deep in one region. This was done by a novel use of diaphragm walls in T, L and U shapes, prestressed in various ways to resist or transfer the loads from the ground, such that large-scale temporary bracing and shoring were unnecessary. Unlike more normal deep-basement construction, there were no intermediate floors that would serve as struts, and loads had to be carried under and round the theatre and concert hall that, in the completed building, sit in the big hole.

SOME RIVER-VALLEY SCHEMES

The purposes behind some of the great schemes described in this chapter are ironic in relation to that famous definition of a civil engineer's profession: 'The art of directing the great sources of power in nature for the use and convenience of man.' The Great Wall was to keep out plunderers; the Grand Canal was to assist legal plunder (taxation); the primeval Dutch settled on their clay heaps to avoid paying taxes, and that started their fight against an encroaching sea; the Mississippi flood protection started because New Orleans was put in the wrong place. Perhaps it will not be a surprise to learn that what is today the largest civil engineering undertaking ever constructed is, in the exact form in which it has been built, largely unnecessary in engineering terms.

When Britain left India and the sub-continent was divided into two countries, there existed in the plain of the Punjab the largest irrigated area in the world. Fifty million people depended on the irrigated lands, which were watered by the five rivers of the Indus catchment. These works reached their peak over 40 years ago with the completion of the Lloyd Barrage at Sukkur on the Indus, the main statistics of which are in table 44.

At partition, the problem was that the rivers used for irrigation in Pakistan flowed most of their courses in India; India needed their water. The resources of rivers farther north had not been exploited. So, in the Indus Waters Treaty it was agreed to build two large storage reservoirs in Pakistan and link canals across the country, controlled at five barrages, to feed irrigated lands in the Beas, Sutlej and Ravi catchments in Pakistan. The waters of these three rivers could then be tapped in India. Thus what was indubitably one single system in engineering terms has been split into two separate systems for political reasons.

There seems no doubt whatever that the Treaty proposed a correct solution. It has worked well. Nevertheless, the link canals entirely and the two great reservoirs, at Mangla and Tarbela, to a limited extent – for storage is essential in the full utilisation of any river catchment – are political works. All three have been featured in earlier chapters since they are oozing with superlatives. So has the complementary development on the Indian side, the great Bhakra–Nangal scheme.

Construction is, in 1975, complete except that there are some troubles still to clear up at Tarbela. Some of the world's best engineering talent has been deployed to build these works. International co-operation was at the root of it, and there exists a group of donor countries led by the World Bank that co-ordinate the aid on which the project, in Pakistan, has depended. The Bhakra–Nangal scheme in India is logistically a separate story since it exploits *de novo* a rich source of hydro-electric power and commands vast new acreages of irrigation. In Pakistan they are largely 'replacement' works. The author was rash enough to grade the Delta scheme as number one in the world, but it is by a short head only. Some of the problems of the Indus Basin have been immensely difficult and have involved adventurous engineering of great merit.

If the largest physically, and politically unique, the Indus Valley development is in other respects similar to many river-valley developments that collectively are characteristic of the mid 20th century. The formula is that storage is created by dams as high in the catchment as possible. The flow of the river can then be regulated and floods stored until the water is needed. The regulated flow can be made to serve as many uses as are thought necessary – generation of power, irrigation, navigation and recreation for human benefit, and conservation of flow, swamp or wilderness for the benefit of riverine creatures and their habitats.

Generally the primary incentive for such development, and the most profitable one financially, is the exploitation of hydro-electric power. Many examples can be cited; among them the development of the Tennessee Valley stands out as the notable prototype. Development of the river was taken as the key to a rebirth of the whole area in economic and human terms. The engineering works were massive and crucial, but nevertheless quite secondary to these wider aims. The political animosities of the TVA (Tennessee Valley Authority) are all in the past now, and its success is complete and enduring. It supplies an effective answer to the ironies listed above. The work of the TVA is well documented, so there is no need to go into detail.

Grande Dixence Dam, the world's highest dam. Details have been given in chapter 1, and the aqueduct system that supplies it is shown on page 209. However, it needs a picture like this to bring alive the massive scale and heroic architectural quality of such a structure.

The Snowy Scheme
Completed in 1974 after 25 years of construction and the largest scheme ever carried out in Australia, its capital cost amounting to A$800 million. Water from the Snowy River that would otherwise flow south to the Pacific Ocean is diverted at Island Bend either to the west, to Geehi Reservoir, or to the north to Lake Eucumbene. Both of these reservoirs are in the headwaters of the Murray catchment and there is extensive demand for irrigation in the dry regions to the west, lower down the catchment. A major reservoir at Jindabyne on the Snowy impounds more Snowy water, and Jindabyne pumping station, which feeds into Island Bend Reservoir, increases the flow diverted. The westward leg of the scheme from Geehi Reservoir consists of two stages of power generation. The northern leg consists of: another diversion into Lake Eucumbene from Tantagra Reservoir on the Murrumbidgee (the purpose of this diversion is to increase the volume of water flowing through the power stations of the northern leg); a tunnel from Lake Eucumbene to Tumut Pond Reservoir; a similar tunnel from Tooma Reservoir to Tumut Pond; from Tumut Pond, four stages of power generation. The power stations generate at low load factor (i.e. for peak load only) and the generous provision of reservoirs gives flexibility in managing river flows so that demands for irrigation are also met. The scheme has 16 dams, 12 tunnels, 7 power stations, and 2 pumping stations. Its installed capacity is 3·74 GW and its annual output 5 TWh. The additional water it makes available for irrigation is sufficient to serve 2600 square kilometres (1000 square miles) of land, that land having a primary production valued at A$60 million per year

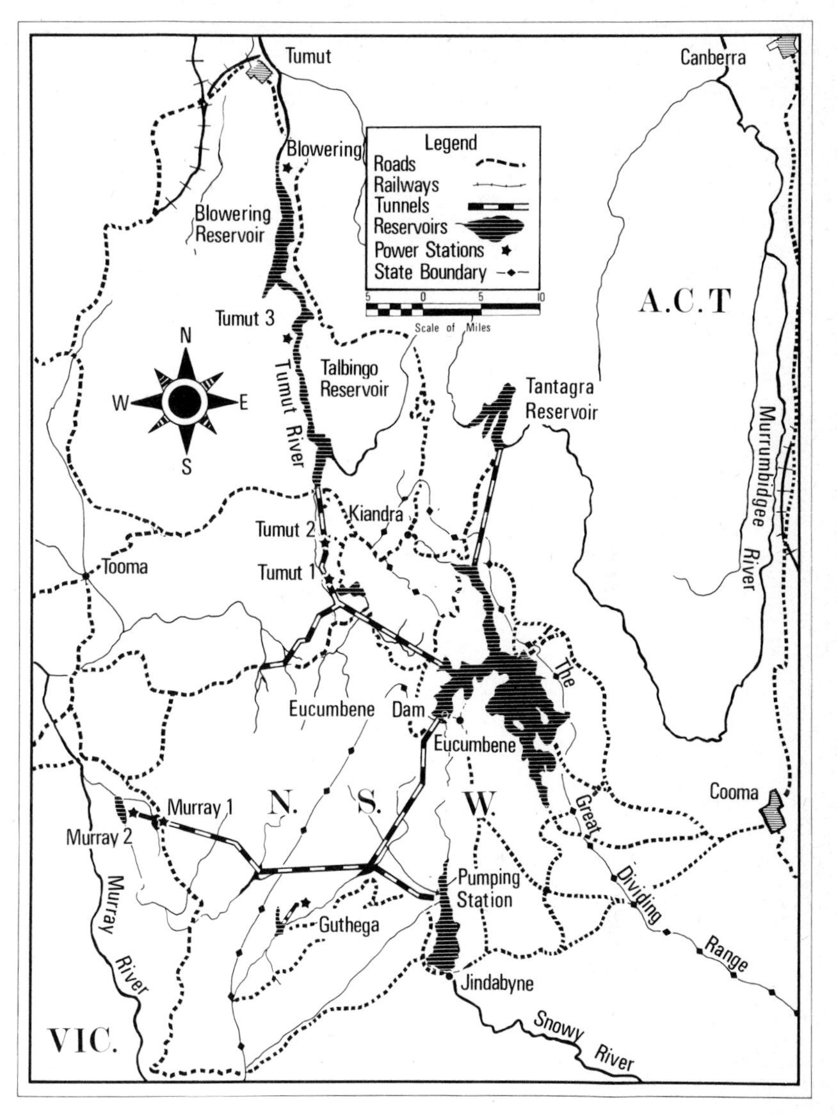

Other examples could be given that have proved to be immense sources of wealth and benefit. The Rhône Valley in France is claimed by French engineers to pre-date the TVA; the Damodar Valley in India is another example. The great dams on the Missouri could be high on this list but they served in chapter 2 as examples of constructional merit. In Australia the story takes a different twist with the work of the Snowy Mountains Authority. Hydro-electric power was again the key, but water was diverted northward (see map) and power generation arranged so that arid areas could be supplied with water.

Looking ahead, the greatest schemes that seem likely include the Ganga–Cauvery project in India, explained in chapter 3. and diversion of the rivers of Siberia in the USSR. The basic idea is to irrigate the arid plains of great areas of southern Siberia and around the Caspian by diverting the flow of major rivers south, so that the rivers do not run to waste on the northern, arctic coast of Russia. Such a proposal is bigger than anything yet, but within the competence, technically, organisationally and financially, of the USSR. Yet it poses an interesting problem.

Diverting the rivers could have climatological effects. The efflux of fresh water into the Arctic Ocean (Kara Sea) would cease. Saline water would replace it on the surface of the sea, and so the ice cap would cease to exist and the whole area would become warmer. Atmospheric circulation would be diverted by this change, the warmer sea attracting moist, warm air to it, it is argued. This would increase the rainfall over the sea, where the rain would fall instead of where it does now, in Europe, but also in the gathering grounds of the catchments of the rivers that would have been diverted. If that happened the flow of the rivers would be reduced, and the effectiveness of the scheme diminished. This argument seems to be significant, but is not proven, nor has the magnitude of the effect, if it will occur, been estimated. Diversion of rivers to the arid lands in the south of the USSR is under study but has not yet reached a stage of decision.

SUEZ AND PANAMA

De Lesseps' Suez Canal was the great prototype of the ship canal of the modern era. Basically, it was a feat of excavation, but one attended by many setbacks; for example an outbreak of cholera in 1865. Picks and baskets were used in the early stages of construction. Couvrex used excavators of his own invention. Borel and Lavally operated up to 60 dredgers at a time, some 'long-arm' dredgers. In 1869 the Bitter Lakes were filled, which took five months. Their total capacity is about 1·5 km³ (1962 × 10⁶ yd³), but a much greater volume of water was needed to make good losses from evaporation and seepage.

The canal has been continuously improved throughout its life. The figures quoted here are from the Suez Canal Company's records.

Excavation for the Suez Canal up to 1956:

	Total quantity $m^3 \times 10^6$	$yd^3 \times 10^6$
During construction 1859–75	74	97
Improvement 1876–1939	153	201
Maintenance* 1874–1951	158	208
Improvements 1950s	65	85
Total	450	591

*About half of this in the channel and harbour at Port Said.

In 1956 the canal was closed for a time by the Suez War (when Egypt nationalised it) and again in 1967 by the Six-Day War in the Middle East. The first ships passed through it again towards the end of 1974 and it was opened again in June 1975.

Improvement was again the policy under Egyptian ownership and major rehabilitation was carried out from 1959. Between the two closures an additional volume of excavation of 100 million m³ (130 million yd³) was completed, and the canal was deepened to 15·5 m (51 ft) so that ships of about 200000 dwt could use it. Clearance of the canal at the end of 1974 was part of yet another major programme costing the equivalent of US$288 million, which will restore navigation to the level prevailing before the closure, provide improved communications, buildings, etc., and assess future needs. It is expected to be completed in 1978. Extensive urban developments at each end of the canal are envisaged.

The Panama Canal. 'The canal is a project crystallised from a great multitude of enterprises and is indisputably the greatest of them all.'* Unlike Suez, the canal was similar to many modern projects in its complexity, if larger in scale than most of them. Its cost to the United States including purchase from the French and all charges, was $375 million. The canal was started under the direction of M. de Lesseps who, following his success at Suez, at first intended to build a sea-level canal, without locks. Two years of preparatory work were carried out in 1881–2; then the contract was annulled. Work continued, but by mid 1885 less than one-tenth of the excavation had been completed. Nominal amounts of work continued under a new company after that time. In 1902, the Spooner Law became effective in the US, enabling the US to purchase all rights, take control and build the canal. The US took possession in 1904 and the lock canal – substantially the project as built – was approved by Congress in 1906.

The canal rises to approximately 26 m (85 ft) above mean sea-level; Gatun Lake extends for 38 km (23·5 miles) at this level. It is formed by an earth dam, and served:

- to minimise excavation over its length,
- to carry, via the Gatun spillway, floods from the Chagres River,
- to provide a store of water to operate the locks.

There are twelve locks on the canal, in pairs – that is, a vessel making the transit has to pass through six locks. From the Atlantic side, the canal leads to a ladder of three pairs of locks which rise to Gatun Lake. From Gatun Lake the canal passes to Culebra Cut, where the greatest feats of excavation were necessary, to a maximum depth below ground level of 69 km (227 ft) to the bed of the canal, on its centre-line. Then, at Pedro Miguel, descent through one lock (i.e. there is one pair of locks) gives access to Miraflores Lake, 17 m (55 ft) above mean sea-level. This lake also is impounded by dams. Two pairs of locks lead from it down to tidewater-level, and thence, past a long breakwater, to the open Pacific.

Excavation was carried out by steam-shovels and dredgers. At Culebra Cut, 46 shovels, half of them with 5 yd³ (3·8 m³) buckets, were in use in 1912,

*The quotation is from Reuben E Bakenhus, US Navy civil engineer in Bakenhus, Knapp and Johnson, *The Panama Canal*, John Wiley and Sons Inc., 1915.

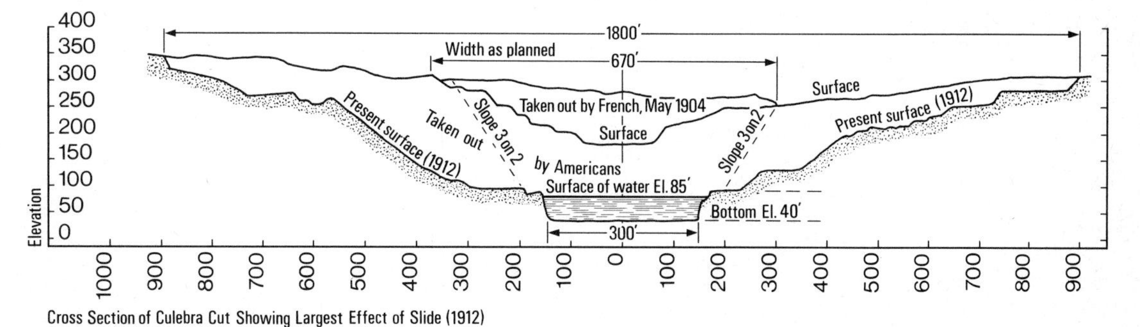

Cross Section of Culebra Cut Showing Largest Effect of Slide (1912)

Table 53. Suez and Panama Canals.

Name	Over-all length of canal route km	miles	Length of canal proper km	miles	Other ruling dimensions	Years of con-struction	Country and location	Other details
Suez	174	108	146	91	Initial maximum draught: 7·5 m (24·6 ft). Up to end of 1951, authorised maximum draught 10·36 m (34 ft); minimum width at depth of 10 m (33 ft) 60 m (197 ft). 1952 draught increased to 11 m (36 ft)	1859–69 (to opening) 1875 (completion)	Egypt, Port Said to Suez (Mediterranean/ Red Sea)	See text; the canal is without locks, the only obstruction (before 1956) being a swing bridge. For 1951 draught, depth was never less than 11·7 m (38·4 ft) at lowest tides. A large vessel steaming at 14 km/h (7·5 knots) settles 0·9 m (3 ft) so giving 0·45 m (1·5 ft) margin
	Length consists of: Port Said to Suez Roads		161	100				
	Dredged channel in Port Said		9	5·5				
	Dredged channel in Suez Roads		4	2·5				
	Total		174	108				
	Including Bitter Lakes for about		15	9·3				
Panama	81	50·4	67 within shore lines	41·5	Locks 305 × 30·5 m (1000 × 100 ft). Minimum depth 12 m (40 ft) to 10·7 m (35 ft) at low tide, on Pacific side	1882–1914 (opening)	Panama (US Canal Zone) (Atlantic/ Pacific)	See text. Rises to 26 m (85 ft) above sea-level
	Length consists of: At sea-level		23	14·5				
	Gatun Lake		38	23·5				
	Locks and approaches		5	3				
	Miraflores Lake		2	1·5				
	Culebra Cut		13	8				
			81	50·5				

with elaborate muck-handling railways; 4896 cars were handled in one day on these railways in 1912. The total quantity of explosive used on the entire canal was 25 000 t (56×10^6 lb).

There were dramatic slides of material into Culebra Cut necessitating much extra excavation. A slide in October 1914 blocked the canal when it had been in use for several weeks, but it was quickly cleared by dredgers. The West Culebra slide covered 27·5 ha (68 acres) and involved $8{\cdot}4 \times 10^6$ m^3 (11×10^6 yd^3) of excavation; others were nearly as big (see cross-sectional sketch). Except for the years 1889–95, excavation was in progress in Culebra Cut for over 32 years.

The ladder of three pairs of locks plus approaches at Gatun needed $1{\cdot}5 \times 10^6$ m^3 (2×10^6 yd^3) of concrete – an unprecedented quantity at that time, and unsurpassed until the construction of Hoover Dam.

The canal can take ships with gross weights up to about 80 000 t and is expected to remain in service until about the end of this century.

OFFSHORE

Lighthouses, caissons for breakwaters, sea-forts and the mystery towers are examples of structures that have been floated out to sea and sunk in position on the sea-bed. Another type is the jacket – a steel tower widely used to drill for oil beneath shallow water and consisting simply of a braced tower with four tubes forming the uprights at the corners. When the tower has been positioned, a pile is driven down each of the tubes to pin it to the sea-bed – hence the name jacket. A revolution was forced on this family of structures with the discovery of oil beneath the North Sea in depths of about 100–150 m (300–400 ft) of water. Production of oil and gas is also to

Kish Bank lighthouse, as positioned off the Irish coast in 1965, with the tower raised. This type of construction, which originated in Sweden, made it possible to build the lighthouse inshore by making it like a telescope. It was floated out, sunk on the sea-bed, and the inner sections raised to extend the 'telescope' and so form the lighthouse tower

Two 'mystery towers' constructed at Shoreham Harbour in 1918. The bases consisted of hexagonal cells of reinforced concrete. The idea was to construct an anti-submarine barrier across the English Channel. One of these towers was installed as the Nab light tower, and is in use today

Chesapeake Bay Bridge-Tunnel see page 56. *Left:* Portal of Chesapeake Tunnel. *Right:* View from Cape Charles on the eastern shore of Virginia looking towards the southern terminus of the Chesapeake-Norfolk area, with Cape Henry on the horizon and the Atlantic Ocean to the left. The two bridges and the four islands each of which accommodates a tunnel portal, a causeway and many miles of trestle, can all be seen. *Photos: Sverdrup and Parcel and Associates Inc.*

The Chesapeake Bay Bridge-Tunnel crosses 28 km (17½ miles) of open sea. It is the only crossing of any kind, bar one, over such a great distance of open water, its only rival until the completion of the Seikan Tunnel being the Zuider Zee Dam which is of similar length and carries a road, but is not sited in such open or deep water. Originally a suspension bridge was proposed, but was ruled out by the US Navy. The crossing is two-lane roadway consisting of: 20 km (12½ miles) of low-level trestle; two tunnels; four artifical islands, one accommodating each tunnel portal; a causeway 2·4 km (1½ miles) long; a high-level bridge 1158 m (3800 ft) long; and a medium-level bridge. There are also several miles of approach roads. The scheme was completed in 1964. It is a very considerable engineering feat, clearly designed and built with economy in mind. The islands are each 457 × 70 m (1500 × 230 ft) in size, and the tunnels are immersed tubes that were prefabricated in 88 m (290 ft) elements and are each over one mile in length

Ekofisk 1 under construction. The outer, perforated wall attenuates the forces imposed by waves. The inner nine-celled cylinder can store one million barrels of oil. The nine cells contain either sea water or oil, and the structure would refloat if they were emptied. It is of prestressed concrete, 92 m (302 ft) in diameter, and 102 m (335 ft) in height; it displaces 215000 t when floating

Left: Highland 1 being floated out on its raft, from the dry-dock at Nigg Bay; in the foreground is the world's largest dock gate (see page 188). *Right:* Upending Highland 1. The two spheres prevent overturning or yawing by virtue of the buoyancy they give as they enter the water; their function is solely to control the initial stage of upending. The clusters of ring shapes at each corner are pile guides

be started in other parts of the world in similar or greater depths. The production platform has to stand in this depth of water, exposed to all that the elements can do, yet provide plant to drill, pump and operate, and quarters in which the crew can live. Events are moving swiftly in this subject; this review dates from the spring of 1975.

Production Platforms of Steel

These are broadly scaled-up versions of the jacket, retaining its name but precious little else. The largest one ordered is to be installed in the Santa Barbara Channel about 175 km (110 miles) north-west of Los Angeles, USA. It will be 288 m (945 ft) high and will stand in 260 m (850 ft) of water. It will be pinned to the sea-bed with piles 1·2 m (4 ft) in diameter and extending 104 m (340 ft) into the sea-bed. Two platforms over 300 m (1000 ft) in height are expected in the same area.

A large steel platform – Maui A – has been built for gas production off the New Zealand coast, in 110 m (360 ft) of water. Other work is mainly for the North Sea where, in the northern basin, nearly 30 steel platforms have been ordered. About a dozen of them are being built in the United Kingdom. The largest so far installed are the two platforms for the Forties Field, one built at Nigg (Cromarty Firth, Scotland) and the other at Graythorp on Teesside. These platforms are now standing in 128 m (420 ft) of water. The jacket alone (i.e. without the decks and prefabricated 'modules' of equipment of all kinds that are lifted on to it when it is in position) weighs some 20 000 t, and was floated out on a raft weighing 10 000 t. At site, the raft and platform were upended, and the raft recovered for a second use.

The largest steel platform under construction for the North Sea so far is Graythorp 3 for the Thistle Field. It will stand in 162 m (530 ft) of water, and will be self-buoyant with vertical legs on one side that will have a large diameter and give buoyancy during the float-out, and then be used to store oil. This platform will be 50 per cent larger than the ones already built.

Production Platforms of Concrete

A dozen concrete platforms are under construction for the northern North Sea Basin. One – strictly not a production platform, since it is used only for storage – is in position, and has behaved well, namely the Ekofisk 1 structure in the Norwegian sector. The same basic system of construction is being employed for the production platform for Ninian Field, now under construction at Loch Kishorn, which will, when floated out to site in 1977, have a displacement of over 400 000 t. It will store 1 million barrels (16 000 m³) of oil and will have an over-all height of 250 m (820 ft).

Other Developments

One submerged device, with pumps remotely controlled, was placed in position experimently in a depth of water of 52 m (170 ft) in the Gulf of Mexico in 1974. If successful, it could make the platforms obsolete. Proposals not so far adopted include floating platforms, prestressed by stays anchored to the sea-bed, using the principle described for floating bridges.

constructional miscellany

HEAVY LIFTING (table 54)

Largest Cranes

The Japanese twin-boomed floating crane *Musashi* has a capacity of 2722 t (3000 short tons). It was used to lift the Arakawa span (table 54) and the cantilever sections of Nanko Bridge.

Crane ships of 2000 t capacity include the *Thor*, *Ocean Builder I*, and the pipe-layer *Blue Whale*, all at work in the North Sea.

The largest power-station crane is of 1723 t (1900 short tons) capacity at Grand Coulee No. 3 Power Station, USA. Two travelling revolver cranes of 726 t (800 short tons) capacity are in use at Graythorp Building

Floating crane Musashi of 3000 short tons lifting capacity, shown lifting a side span of the Oshima Ohasi Bridge, in a 10 knot current. The side span weighs 2100 short tons and spans 212·5 m (697 ft). The main span of the bridge of 325 m (1066 ft) was due to be placed in position in October 1975. The bridge is not entered in chapter 1; it is in the Yamaguchi Prefecture and is scheduled to be completed in 1976. *Photo: Nippon Kokan (NKK)*

Graythorp dock, showing two revolver cranes and the first portion of BP's platform being erected on top of the raft. The cranes each stand on towers 48·8 m (160 ft) high. They have a walking mechanism for travelling round the dock. The main boom of each crane is 59·4 m (195 ft) long but can be increased to 68·6 m (225 ft) by using the fly jib. There are three lifting points, for 60, 135 and 800 short-ton hooks respectively. The nominal capacity of the crane is 800 short tons; that weight can be lifted at a radius of 20–21 m (65–70 ft). At a radius of 24·4 m (80 ft), 660 short tons can be lifted; at a radius of 61 m (200 ft), 125 short tons

In the Jackblock system, a building is 'extruded' upwards, rather like squeezing toothpaste out of a tube. The plan shape of the walls is defined by an installation of jacks at ground level. The topmost coping of the wall of the building is built on them. The jacks are operated, and it is pushed up; then building blocks are inserted and the jacks retracted. This cycle is repeated many times until the building has been jacked up to its full height. Construction continues for a few storeys just above the jacks, but at about the third or fourth storey up, everything has been done and a complete building ascends slowly to its ultimate height. The figure given in the table for the weight lifted is the maximum weight lifted by the jacks, just before completion of the building. This photo shows the head office of an insurance company in The Hague completed by this technique, and another similar construction in progress on the left. *Photo: Hollandsche Beton Groep NV*

At Methil, Fife, erection of the steel structure for the Brent A North Sea oil production platform involved a 'lift' of 2820 t in March 1975. This weight was in fact rotated through about 90 degrees, and not physically lifted, four 40 t winches and two 110 m (360 ft) towers being used. At a later stage a frame weighing about 3450 t was moved horizontally about 100 m (330 ft)

Dock, on Teesside, England. Used in tandem, they have lifted 1100 t to a height of about 90 m (300 ft). The largest mobile crane is the Rosenkrantz of 1000 t nominal capacity. Its height is 202 m (663 ft) and it can lift 30 t at a radius of 160 m (525 ft).

Greatest Lift in Victorian Times

Each of the main girders of the Britannia Bridge was floated into position, then lifted up the towers of the bridge by jacking. The weight lifted was 1250 tons. This was done for each of the four large girders. Saltash Bridge involved two lifts each of about 1000 tons.

Heaviest Weight Ever Dropped

The suspended span (195 m (640 ft) long) of the Quebec Bridge fell when being lifted into position in 1916. Its collapse was caused by a faulty casting at one corner of the span. Thirteen men were killed in the fall. Within a year, a new span had been erected.

Ancient Feats of Lifting

The monoliths of the Chinese bridges described on page 233 weighed up to 200 t each. The largest monoliths at Stonehenge weigh about 45 t. The largest single stone quarried and moved is one of 818 t (805 long tons) at Baalbeck, Lebanon.

Table 54. The world's heaviest lifts (over 3000 t).

Name of scheme	Weight lifted metric tonnes	Weight lifted other unit	height lifted m	height lifted ft	Year of lift	Country and location	Other details
Velodrome	37 190	41 000 short tons	0·1	0·3	1975	Canada, Montreal	Lifted to strike centering. See chapter 1.
Jackblock system	32 000	–	0·7 or 0·5	2·3 or 1·6		Holland	See caption. Height given is stroke of the jack
Barras Heath	8800	–	0·7	2·3	1963	England, Coventry	Only British example of jackblock
Condeep (concrete production platform for North Sea oil)	22 000	–	±0·8 at each tower	±2·6	1975	North Sea	The platform has three towers on which the deck is supported; 96 jacks of 200 to 360 t each are installed to level the deck. The capacity of the jacks is 24 000 t
BMW headquarters building	12 000	–	3·82	12·5	1972	West Germany, Munich	Four-leaf-clover-shaped tower, built storey by storey at ground level and hoisted from a central core
Lüneburg Shiplift	5700	–	38	124·6	1975	West Germany	See chapter 3 total moving weight – caissons plus counterbalance weights – is about 11 400 t
Fremont Bridge	5443	6000 short tons	52	170	1973	USA, Oregon	Mid-span bowstring section 275 m (902 ft) long lifted in 40 hours.
Ponte Presidente Costa e Silva	5275	–	52	170	1974	Brazil, Niteroi	Two 292 m (958 ft) long box girders plus two ring beams were lifted by twelve 450 t jacks; the process was repeated on the other side; finally the central, 3200 t, section was lifted from the newly erected cantilevers
Quebec Bridge	5080	5000 tons	40	130	1916 1917	Canada, St Lawrence River	Two attempts. The first span collapsed (see text)
Expo Building	4800	–	30	98	1968	Japan, Osaka	Steel space-frame roof, 292 × 108 m (958 × 354 ft)
Narita No. 1 Hangar	4536	5000 short tons	20	66	1972	Japan, Tokyo, International Airport	Roof 190 × 90 m (623 × 295 ft) lifted from nine temporary towers, using jacks, strand and anchorages adapted from prestressed concrete
Nanko Bridge	4536	5000 short tons	52	170	1974	Japan, Osaka	Trussed cantilever bridge, centre span 513 m (1683 ft) and two 236 m (776 ft) side spans. Suspended span 187 m (614 ft) long lifted in three and a half hours, using eight winches
Frodsham bridges	4500	–	–	–	1969	England	The largest bridge rolled sideways into position
Arakawa	4400	4850 short tons	9	30	1975	Japan, Tokyo	196 m (644 ft) span floated to site on a barge lifted by 3000 short ton floating crane in tandem with two 1500 short ton barge mounted cranes moved forward 100 m (330 ft). Heaviest of seven spans erected in this way

Table 54. cont.

Heinrichenburg Shiplift	4370	–	14·25	47	1962 initially	Germany, Dortmund–Ems Canal	Weight lifted in normal operation. The Ronquières Shiplift (see chapter 3) lifts approximately the same weight
Gunpowder Tower	4000	–	2 (lift) 300 (moved)	6·6 985	late 1975	Denmark,	Ancient tower moved, using hydraulic jacks to make way for new dry-dock.
Hangar for engineering base	3200	–	20·4	67	1970	England, London Airport	Roof supported by four columns spanning 91·5 m (300 ft) in each direction. It was raised by 64 jacks in four groups automatically synchronised

Note: 1 metric ton (tonne – t) = 1000 kg = 2204·62 lb = 0·9842 long tons = 1·1023 short tons
This table excludes – with the exception of Heinrichenburg Shiplift – lifting by flotation. The weights lifted by floating docks are greater than any listed here (chapter 3).

House moving in Canada. Two house-moving machines were used to move 525 houses when the St Lawrence power station was built in the late 1950s. To move a house, beams were placed under it after driving channel holes through the foundations. Jacks were then used to lift the house off its foundations and the machines' U-shaped frame was backed in under the beams. A hydraulically operated system in the frame corrected the level of the house as the machine travelled up or down hill. Travelling speed was six miles per hour. The larger machine could carry 150 short tons up to widths of 12 m (40 ft). The smaller one had a capacity of 100 short tons for houses measuring up to 10 m (34 ft) wide. *Photo: Ontario Hydro*

Eight jacks, each of 200 short tons capacity, were used at each of the four corners of the cantilevers of Fremont Bridge, to lift the central section of the span into position. The jacks raised suspension rods in 0·6 m (2 ft) cycles. When the lift had been completed, one side span was jacked horizontally at the springing to redistribute loads in the permanent structure. The completed bridge is shown in the picture on page 38, from which it can be seen that the centre section of the arch was the piece lifted into position

The roof of this hangar at London Airport was jacked up from ground level (table 54)

According to the records, the ancient Egyptians set up obelisks weighing more than 800 t. Ones over 500 t are known. The largest known obelisk weighs 1170 t, but it is still on its side, in the quarry at Aswan, having cracked during the quarrying. An obelisk weighing 330 t, 22 m (71·5 ft) high, was moved from Heliopolis to Rome. Its re-erection was the subject of a famous engraving dated 1588.

The Inca cities of Tiahuanaco and Sacsayhuaman in Peru have dressed blocks of masonry of 100 t in weight, but von Däniken (see chapter 2) also refers to a block 'as big as a four storey house', estimated to weigh 20000 t; 'impeccably dressed it has steps and ramps', yet it is not the right way up nor does he suggest that it was ever moved by humans.

Returning to strictly engineering aspects of lifting, it is known that the timber bridge at Schaffhausen, Switzerland, listed in table 2, was repaired in 1783, the method of repair being to lift and support the bridge by jacks – no mean feat for a bridge 122 m (400 ft) long.

CASTING SUPERLATIVES IN CONCRETE

The largest quantity of concrete placed in one continuous operation is 20185 m³ (26400 yd³). That quantity was placed in seven and a half days, under water by tremie, for the construction of the east anchorage of the Delaware Memorial Bridge in the USA (table 1). The anchorage was an open cofferdam, 30 × 68·6 m (99 × 225 ft) in area, and concrete was placed in it to a depth of 9·75 m (32 ft).

The cofferdam of Pier W2 of the West Bay Bridge, San Francisco, was sealed with a continuous pour of 14374 m³ (18800 yd³) of concrete lasting nine days, and ending on 21 September 1933. The concrete seal was 16 × 37 m (52 × 121 ft) in plan with an average thickness of 24 m (79 ft). When the pier had been completed to the level of +40 ft, it contained 19592 m³ (25624 yd³) of concrete.

The largest pour in one day was achieved when concreting an airfield runway at Dallas/Fort Worth Airport, Texas in 1973. A quantity of 9657 m³ (12630 yd³) was placed in fourteen hours.

Roads and runways are concreted continuously with specialised plant. The record rate for paving is 4025 m (13205 ft) of road paved in seventeen hours. This record was achieved at Osseo, Minnesota, on 29 August 1969, using a slip-form paver. The average rate of concreting was 4 m (13 ft) of road per minute. A concrete slab 0·23 m (9 in) thick and 7·3 m (24 ft) wide was constructed, reinforced with mesh and with dowelled joints at intervals of 11·9 m (39 ft). The total quantity of concrete placed was 6728 m³ (8800 yd³), and the rate of placing was 396 m³/h (518 yd³/h).

Largest Pour in a Building

Concreting a building is a more complicated task than the great mass of a bridge foundation or the relatively straightforward strip of a runway or road. Heavily reinforced, complicated shapes are needed. Also, the concrete is relatively rich in cement, and so generates more heat when it sets. Nevertheless, the difficulties have been overcome in current British practice, the basis of the technique being to insulate the newly placed concrete, so that it is all at about the same temperature. If this is done, it will not crack.

The British record is 2910 m³ (3806 yd³) in one operation lasting 23 hours on 16 November 1974, to build the basement raft of an office building in Cannon Street, in the City of London. Thirty ready-mixed-concrete trucks and six concrete pumps (two stand-by, four working) were used. One feature of the work was the use of blast-furnace slag to replace 70 per cent of the cement, with the intention of reducing the heat of hydration.

Fastest Rates of Concreting

Although dams contain the largest quantities of concrete, there is no compelling reason to arrange concreting in large, continuous operations. Hence no record pours, but record rates of progress measured over the job as a whole or over a period of time; for example, at Grand Coulee Dam, the rate of concreting reached 306 000 m³ (400 000 yd³) per month, which is an average of about 9940 m³ (13 000 yd³) per day over a 31 day month. The daily output clearly exceeded the Dallas record, but it was probably placed in a number of different places in the dam.

Concreting of Grande Dixence Dam reached a peak in 1959. The daily rate was then 9300 m³ (just over 12 000 yd³).

At Tarbela Dam 380 000 m³ (almost 500 000 yd³) of rollcrete – a type of lean concrete mixed by dropping the materials down a tower – were placed in 44 days. The maximum rate of output was 870 m³ (1 137 yd³) per hour.

Highest Lift

Newly placed concrete behaves like a liquid and exerts pressure on the formwork in which it is placed. The depth of wet concrete that can be placed in one operation – called a lift – is thus an important limitation. Apart from slip-forming, the highest lift of concrete placed in one operation appears to be 23·5 m (77 ft) at Barrett Chute Dam on the Madawaska River in eastern Ontario in 1945. Up to 9250 m³ (12 000 yd³) of concrete was not uncommonly placed on one pour. This high-lift technique was developed by Ontario Hydro and used on many of their jobs, for example 11 m (36 ft) lift when building the St Lawrence Power Station at Cornwall.

23·5 m (77 ft) lift of concrete at Barrett Chute Dam, Ontario

Slip-forming Records

Slip-forming (that is, continuously sliding shuttering) overcomes the limitation just mentioned. The shutters are hung on vertical rods embedded in the concrete and extending above it, and are jacked up the rods continuously. The depth of the shutters and the rate of climb are adjusted, so that the concrete at the bottom of the shutter has set and has enough strength to support the work above. The technique is used to build structures such as grain silos or chimneys, which then sprout up like mushrooms in an incredibly short time.

The greatest height concreted in one continuous operation (with the qualification mentioned below) by this technique is 446 m (1464 ft), the height of the concrete section of the CN Tower in Toronto (see page 87). The tower rose at an average rate of 6 m (20 ft) per day and was 25 mm (1 in) out of plumb at the top of the concrete structure. It is the most daring and accomplished example of slip-forming. Although 'continuous' in its character, the slip-forming at CN Tower actually continued day and night for five days at a time, and then closed down for the week-end. It has been found that it is the human element that is the fallible link in these operations when they go on too long, so it pays to accept the complication of starting the slide again, after the men have had a rest. A recent application of slip-forming is to the concrete production platforms for North Sea oil, and slip-forming of the McAlpine-Sea-Tank platforms at Ardyne Point in Scotland is claimed as the largest slide yet, in terms of the complexity of the structure that is being continuously cast.

Concrete cast in place, or an arch of masonry or a structure of precast pieces, have to be supported by temporary structures, called falsework, until they can carry their own weight. The steel centering for Gladesville concrete arch (see page 33) is shown here. The bridge (span 305 m – 1000 ft) consists of four arched ribs, composed of precast elements, and the centering was moved sideways to support each of the ribs in turn, as it was erected

Wire-winding machine for Hartlepool nuclear power station

Another of these structures – the Ninian Platform being built at Loch Kishorn involves slip-form shutters of 102·8 m (337 ft) in diameter without radial or central support.

Unique Twist

There are innumerable reasons why an engineer can build a structure that is inaccurate. The most unusual – and one that was anticipated – was at the CN Tower. The tower rose sufficiently fast for the plastic concrete (i.e. concrete in the process of setting) at the top to be affected by the rotation of the earth. This effect caused the slip-form assembly to twist, and a continuous correction had to be applied by adjusting cables attached to the slip-form at one end, and at the other, farther down the tower. The total correction applied in this way was equivalent to about half a dozen degrees of twist.

Biggest Wire-Winding

The largest prestressed concrete pressure vessels are for the AGR nuclear power stations of the British electricity system. At Hartlepool there are two pressure vessels, each a cylinder 25·9 m (85 ft) in diameter and 29·3 m (96 ft) high with walls 6·4 m (21 ft) thick. The design pressure they withstand is 4·4 N/mm^2 (644 lb/in^2). These vessels are prestressed by winding tensioned wire round them with a special machine – 6436 km (4000 miles) of 5 mm (0·2 in) wire disposed in 20 channels, 120 and 119 turns of wire alternately per layer, and up to 50 layers per channel.

One 33-layer channel has a force of 9650 t in the tendon of wound wire, and the tendon has an ultimate strength of 13 700 t.

Strongest Concrete

Seven beams, one of which, when tested to destruction, had concrete with a strength of 127·45 N/mm^2 (18486 lb/in^2) were made by Pierhead Limited in 1966. They were made with Alag synthetic aggregate and high alumina cement, with a water/cement ratio of 0·33.

Concrete with a compressive strength of 200 N/mm² (29000 lb/in²) has been made experimentally in France (the work was reported in 1970) using the sand-lime process, but with clay instead of sand and modified heat treatment. The strongest material in this category is, however, neat Portland cement, subjected to very high pressure and then cured under water. Strengths of 372 N/mm² (54000 lb/in²) have been measured.

ORGANISATIONS: CONTRACTORS

The world's largest firm of civil engineering contractors is United Engineers and Constructors Incorporated of Philadelphia, USA (a subsidiary of Raytheon Company). This claim is based on the firm's contracts awarded during 1974 (as reported in *Engineering News Record*) which were valued at $6247·5 million. Sixteen US contractors were awarded contracts worth more than $1000 million during 1974.

The largest contracting firm based in Europe is the British company, George Wimpey and Company Limited. In 1974 work carried out by the firm was valued at £358 million. The year's net profit before taxation was about £33·5 million and the value of work carried out overseas £69 million.

The largest single contract ever placed for civil engineering construction was for Tarbela Dam in Pakistan. The contract was awarded in May 1968 to a consortium of international contractors known as TJV (Tarbela Joint Venture) led by the Italian firm Impregilo. The value of this contract, in 1968, was US$ equivalent 623 million (£260 million). This record was overtaken in December 1975, with the award of a contract valued at £310 million to a joint venture of HAM BV of Holland and Dredging International of Belgium by the Saudi Arabian Ministry of Communications for a new harbour and entrance channel for the port of Dammam. Announcement of an even larger contract in Saudi Arabia is expected as this book is passed for press.

ORGANISATIONS: BUREAUCRATS

The largest organisation in the world that designs and constructs is the Property Service Agency (PSA), a British civil service organisation that functions within the Department of the Environment and is responsible for all Government property management, construction, maintenance, and appropriate supplies. It has a staff of about 60000 (a quarter overseas) including 2300 professional staff. Its annual expenditure on new works (1973–4) is approximately £295 million. Total value of new work in progress, 1973–4 was £1250 million.

One of the largest professional engineering organisations in the USSR is the Zuk Hydroproject Institute, the State organisation that designs the country's river-control schemes. The Institute has offices in ten Soviet cities, and numerous site and exploratory offices. It has teams available for about 50 complex or specialised investigations and surveys of various kinds. One hundred and seventy hydro-electric power stations have been built from the Institute's designs (35 GW total installed capacity) including the largest in the Soviet Union. In addition, the Institute has been responsible for the design of a score or so of schemes overseas, the best known among which is the Aswan High Dam.

ORGANISATIONS: PROFESSIONALS

The world's largest firm of consulting engineers – i.e. a professional firm independent of any links with contracting or business – is the Ove Arup Partnership. Its present strength on a world basis is a professional staff numbering 2500.

The US firm Dames and Moore and the British firm W S Atkins and Partners, each have a professional staff numbering about 1500.

Consulting engineers in the UK earn in foreign exchange (so-called invisible exports) more than all the other professions combined – the net amount was £113 million in 1975. At the end of 1975 they were responsible for work overseas with a total capital value of £12 000 million.

The largest single commission ever awarded to an independent firm of consulting engineers was in 1975 to W S Atkins and Partners for the design and supervision of construction of the El Hadjar iron and steelworks in Algeria for SNS (Société Nationale Sidérurgie). The commission is worth some £25 million in fees for construction that will cost about £300 million.

VOLUME OF NEW CONSTRUCTION

In the UK, the annual output of the construction industry is £6500–7000 million. (1973 estimate) divided thus:

	%
civil engineering	11
new housing	24
general building work	31
repair, maintenance, direct labour	34

All except one-tenth of civil engineering is in the public sector.

A total sum equivalent to US$5000 million is transferred annually from the industrialised countries to the less-developed countries to finance economic development of one kind or another. This sum is matched by a similar amount contributed by the less-developed countries themselves. Economic development projects generally depend on a substantial amount of construction. About one-tenth of the total annual expenditure equivalent to some US$10000 million is spent on fees to professional advisers and managers, among whom consulting engineers play the largest part.

The 400 largest US contractors signed contracts worth $75 600 million in 1974, a 37 per cent increase over 1973, and more than double the 1971 figure. Of this amount, $11 700 million was for work outside the United States. These figures are published by *Engineering News Record.*

MOST COSTLY SCHEMES

The US Interstate highway system is claimed to be the largest public works project ever. Its cost is $76 000 million, and it is 85 per cent complete (spring–1975).

The St Lawrence Seaway (including the power scheme) cost $1055 million and took four years to build. The rate of expenditure – $264 million per year (£110 million per year) – is the highest. The scheme was completed in 1959.

Spencer Works, the steelworks near Newport in South Wales, completed in 1962, was claimed at the time to be Europe's largest unit of investment. The cost of the civil and structural engineering at Spencer Works was about £48 million; it was built in two and a half years. It has since been exceeded in size by the British Steel Corporation's (BSC) Anchor project at Scunthorpe, completed in 1973. The corresponding cost of Anchor was £76 million. Redcar, BSC's latest venture, is twice the size of Anchor. The cost of the Thames Barrier is estimated to be about £171 million at 1974 prices.

Thomas Telford, F.R.S. This head and shoulders is part of a larger picture by Samuel Lane which hangs in the Lecture Theatre at the Institution of Civil Engineers

These figures need to be adjusted – say by reference to the cost index below, to compare costs at different times.

THE COST OF CONSTRUCTION

An index of the cost of construction has been used for half a century by the US journal *Engineering News Record*. The index is an average of the cost in twenty US cities, of a mixture of the costs of certain measures of unskilled labour, steel, cement, and timber, valued at $100 in 1913. The history of the index is:

Year	Index
1903	94
1907	101
1913	100 (base year)
1919	198
1930	203
1935	196
1940	242
1945	308
1950	510
1955	660
1960	824
1965	971
1970	1386
1974	2020
1975	2120 (average for first three months)

HUMAN COSTS

Feats of construction tend to be cited as the positive achievements of despotic rulers. The achievement is often genuine, but the cost to those involved may be overlooked. The Roman tunnels described in chapter 2 were built with slave labour – human life was expendable. This use of slave labour has persisted and it is only mechanisation with plant of high capacity that makes it less attractive to despots.

Examples in the modern world are the Todt organisation that carried out construction in Nazi Germany and the occupied countries, and the work done by political prisoners in the USSR, as described in *The Gulag Archipelago.* Solzhenitsyn's theme is a human one; not surprisingly he does not enthuse about great works of construction in the USSR, but some of those works in the 1930s, although he barely mentions them, are the background of his book.

Without the compulsion of imprisonment or slavery, there remains a tradition of roughness. The navvys – i.e. navigators, meaning canal-builders – were a class apart, feared by and segregated from the communities where they were working. They were paid well, but lived in primitive camps, were milked in tommy-shops (the traditional perks of the old-time contractor's agent was the sale of beer in the tommy-shop) and were expendable in terms of accidents. The last British job of the tommy-shop era was the Grampian hydro-electric scheme.

Today, the industry has a bad record for accidents when compared with other industries, but it is a paradise if compared with, say, 75 years ago. Even so, it is a sad fact that 102 lives were lost in the building of the Orange–Fish Tunnel (see table 25).

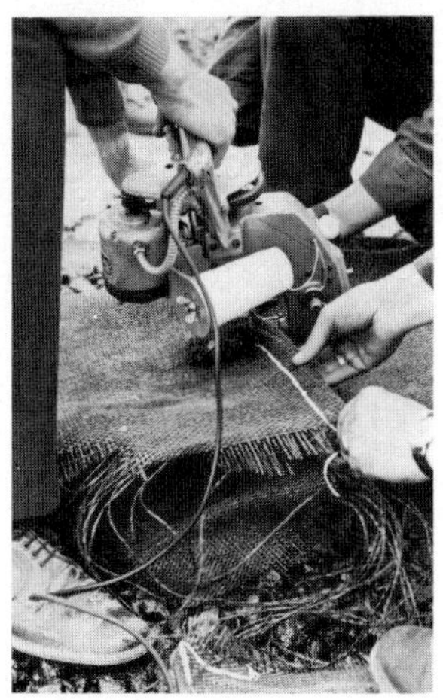

The ultimate oddity in constructional plant. Among the usual powerful collection of heavy plant that was at work building the Haringvliet Barrage, was one unusual item, shown here. It is a sewing machine. Its purpose was to join together the sheets of nylon mesh that form a barrier permeable to water yet impermeable to sand, at the base of the apron; this construction is shown on page 249

An account of the appalling conditions in ancient Egyptian gold mines – the miners were condemned criminals – and of similar conditions in Swedish copper and silver mines is given by Sandström (see page 225 for reference). However, medieval Saxon mining, he says, was a cultural achievement of the highest order, and was the golden age of mining with conditions that have never been achieved before or since.

A SENSE OF APPRECIATION

'Lie heavy on him Earth! For he laid many heavy loads on thee' (epitaph to Sir John Vanbrugh).

'The stability of a building is inversely proportional to the science of the builder' (Tredgold).

First code of practice.

The laws of Hammurabi, King of Babylonia, in 2200 BC included the following:

'If a builder builds a house that is not firmly constructed and it collapses and causes the death of the owner, that builder shall be put to death.

'If it causes the death of the son of the owner, a son of the builder shall be put to death.

'If it causes the death of a slave, he shall give the owner a slave of equal value.

'Whatever he did not make firm and collapsed he will restore at his own expense.

'If a man opens his irrigation sluice to let in water, but is careless and the water floods the field of his neighbour, he shall give grain to his neighbour equal to the yield of his neighbour's field.'

Paddy's wigwam. *Photo: Taylor Woodrow*

The three ugly sisters of Marsham Street. *Photo: Crown copyright*

nicknames

The ordinary vulgar human man does not always seem to treat great constructions with respect. Here are some examples of nicknames, applied with more or less popularity locally, or coined by journalists or engineers.

The Mersey funnel, or *Paddy's Wigwam* – Roman Catholic Cathedral at Liverpool, built of reinforced concrete and completed in 1967.
The three ugly sisters of Marsham Street – Government offices in Westminster completed 1970.
Westminster Gas Works – round the corner from the three ugly sisters.
The corncobs (or *les choux*) – fourteen-storey blocks of flats at Creteil, France.
The wisdom tooth – building 140 m (460 ft) high at Karl Marx University, Leipzig.
Galloping Gertie – the first Tacoma Narrows suspension bridge
The dropsy and the megrims – William Hazlitt's description of the Royal Pavilion at Brighton. Another contemporary first impression: 'The dome of St Paul's has gone down to Brighton and pupped.'
The ginormous squircle – the Louisiana Superdrome (its shape in plan is partly square and partly circular).
The stranded whale – Sydney Opera House, also known as the *peeled orange*.
The gallows – John Hancock Building, Chicago (chapter 1).
The Gog and Magog of water-power – surge tanks at Fort Peck Dam disguised by *soignée* architectural treatment.
The pregnant oyster – Bell System Pavilion at New York World Fair, 1964–5.

Westminster gas works. *Photo: Crown copyright reserved*

Gog and Magog

The corncobs

The pawnbroker with nine balls – the Atomium built for the Brussels exhibition, 1958, to symbolise the mysteries of the atom.
The environmental valve or *Buckminster's bubble* – Buckminster Fuller describes his geodesic dome as 'a prototype environmental valve . . . a benign physical microcosm'.
The time machine – a name given to Niteroi Bridge by the *cariocas* (Rio's equivalent of cockneys) because crossing the bridge, they say, is like going back 100 years.
The world's most successful foundation failure – the leaning tower of Pisa (see chapter 1. (Six other towers in Italy are reputed to lean as much or more.)
The giant's crutch – War memorial in Parque Flamingo, Rio de Janeiro, Brazil.
High-rise heaven – Singapore, where nearly half the population of 2 million, live in buildings 14 to 24 storeys in height.
Minhocão (the big worm) – winding, elevated steel expressway in São Paolo, Brazil.

The wisdom tooth

The pregnant oyster (officially the floating wing pavilion)

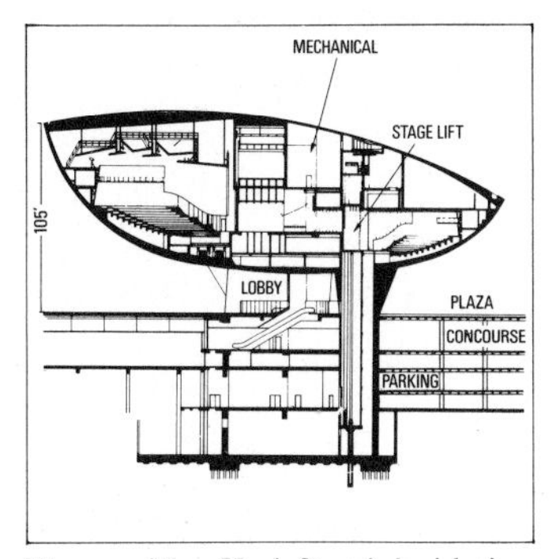

The egg, New York State's legislative and cultural building at Albany, NY, USA. It is a concrete shell 85 m (280 ft) by 61 m (200 ft) in plan, carried on a single column.
Acknowledgement: Engineering News Record

The Atomium

INDEX

(Entries in italics refer to illustrations)

Abu Simbel temples, Egypt 137
Acherès sewage works, France *218*
Acqua Felice, Rome 211
Admiralty Floating Dock, Genoa 183
L'Aigle Dam, France *196*
Akashi–Kaikyo Bridge, Japan 27, 29, 51
Akosombo Dam, Ghana *66*, 67, 165, 166, 167
Alaskan pipeline *213*
Albany, USA 156; legislative and cultural building *275*
Albemarle Sound Bridge 57
Alcantara Bridge 31
Aleppo, Syria 137
Alexander Bridge *52*, 60
Alexandrina Aqueduct, Rome 211
Allt na Lairige Dam *64*, 73, 132
Almendra Dam, Spain *64*, 66; Tunnel 115
Almonacid Dam, Spain 83, 84
Alsietina Aqueduct, Rome 211
aluminium bridge, 37, roof largest *91*, span 97
Alva B. Adams Tunnel, USA 111
Amazon Canal, Brazil 189
Amazon River 200
Ambassador Bridge, USA–Canada, 29
Ambursen Dam 69
Amistad spillway, Mexico–USA 193
Amizade Bridgè, Brazil 33
Amlwch, Anglesey 178
Ammann 228
Amu Darya catchment, USSR 204
Amsterdam: North Sea Canal 178, 190, 192; port 177, 178; ship canal (first) 189, 192; ship canal locks 186; Town Hall 164
An Chi Bridge 103, *234*
Anchor project (BSC), Scunthorpe 271
Anfang Thang Reservoir (Shao Pei Dam) 237
Angostura Bridge, Venezuela 28
Anhembi Park roof, *91*
Anio Novus Aqueduct, Rome 211
Anio Vetus Aqueduct, Rome 210
Antifer, France *175*, 180
Antonine Wall, Scotland 233
Antwerp: land reclamation 154; locks 186, Zandvliet Lock *188*; port 177, 178
Appalachia Tunnel, USA 112
Appia Aqueduct, Rome 210
aqueducts 209–11; British, first 211, longest 210; historical 210; Roman 210–11
Arakawa span 262
arch bridges 11, 12, 13, 18, 22, 24, 26, 32–4; longest built-in span 32
arch dams: earliest multi-arch 76; first historically proven 75; greatest radius 68; only failure 79; slenderest 68, 73
arches: earliest 103, 104; shallowest 60
Archibald Mine, Germany 172
Arden 222
Arlberg Tunnel, Austria 113
Arpa Sevan Tunnel, USSR 113
Arrabida Bridge, Oporto 33
Arrowrock Dam, USA 83, *84*
Arthur Kill Bridge, New York 54
Artificial island salt-loading installation, Brazil *183*
Arup, Sir Ove 231, 251
Arvida Bridge, Canada *37*
Ashford Common water-treatment works 219
Aspdin, Joseph 99
Astoria Bridge, USA *23*, 39, 60
Astrodome, Houston 96, *110*
Aswan High Dam, Egypt 67, 70, 171–2, 269; quarry 266; Tunnel 116
Ataturk Bridge, Istanbul 27
Atchafalaya Basin floodway 242, *243*, 244
Atkins and Partners, W S 270
Atomium, Brussels exhibition *275*
Auburn Dam, USA *63*, 68, 80
Augusta Bridge, USA 34
Australian irrigation 255
Autolite-Ford Parts Redistribution Centre, Brownston, USA 99
Avre Aqueduct, France 217
Awe Tunnel 133
Ayub arch, Pakistan *36*
Azotea Tunnel, USA 133
Baalbeck, Lebanon 263
Babylon Tunnel 119
Baebolo Tunnel, Spain 119
Baglan Bay pipeline 212
Baidayevskiye Colliery, USSR 131
Bailey bridges 25
Baipaza Dam, USSR 73, 152
Baker 229
Baltimore Bridge 39
Baltimore Patapsco Tunnel 120
Bangladesh coastal embankment project 197–8
Bankhead Tunnel, USA 120
Bantry Bay 174, *175*
Barbican Arts Centre 251
Barking contact bed 222
Barlow 124
barrages and spillways 193–7; world's largest barrages 194; world's largest high-head spillways 193, 195
Barrett Chute Dam, Canada *267*
Barton Swing Aqueduct 54
Bascule bridges: Cadiz *17*; longest spans 53; most famous 53
Baton Rouge Bridge, USA 39
Bauersfeld 231
Bay Area Rapid Transit (BART) USA, *108*, 117, 119, 121
Bayonne Bridge, USA 32, *35*, 60
Bazalgette, Sir Joseph 220
Beach 124
beam bridges 12, 16–17
Beauharnois Canal, Canada 190, 191; power station 191, 194, *200*
Beaumont 129
Beckton Northern Outfall Sewer *220*
Bee Ness 180
Belfast: East Twin dry-dock 185; ship-building dock 184
Bell System Pavilion, New York World Fair 274, *275*
Belledonne Tunnel, France 114
Belmont steel towers 90
below ground constructions 105–73; excavation and dredging 141–55; foundations 155–65; ground engineering 165–73; tunnels and underground workings 105–40
Bendorf Bridge, Germany *14*, 45, 59, 231

Bernoulli, Daniel 230
Bexley water well 217
Bhakra Dam, India 64, 80
Bhakra–Nangal scheme, India 253
Big Muskie dragline *145*
Bimini, Straits of 200
Bingham Canyon Mine, USA 133, *141*, 142, 152, 154
Birchenough Bridge, Rhodesia 32
Bircherweid Bridge, Switzerland 46
Birkenhead port 178
Bitter Lakes 256
Bjerrum 225
Black River Bridge, USA 53
Blackbrook Dam 73
Blackwall Tunnel 108
Blanco River Bridge, Vera Cruz 25
Blanco Tunnel, USA 112, 133
blanket, Tarbela Dam, 171
Blenheim Palace Bridge 41, 42
Blue Whale 262
Boeing Assembly Hall, Everett, USA 99
Bollman, Wendel: truss bridge *21*
Bolton 219
Bolzano elevated road, Italy 58
Bonnet Carré spillway 194; *240*, 242
Borel 256
Borrodale Bridge 14
Bosporus Bridge 51
Bou-Hanifia Dam, Algeria 171
Boulby Mine, Yorkshire 172
Bouzey Dam, France 78
Bowland Forest Tunnel 112
Brabazon Assembly Hall, Filton 98, *101*
Brahmaputra River 200
Bramah 219
Brassey, Peto and Aird 229
Bratsk Dam, USSR 67, 80; power station 201
Braunfels, Germany 214
Brazo Largo Bridge, Argentine 44
Breadalbane Tunnel 133
breakwaters: Bombardon for Mulberry harbours 179; Brasher air 179; Harris floating 179, *183*; longest world, Britain 180
Brent A oil platform *263*
Brent pipeline 213, 214; production platform *263*
brick structures: load-bearing brickwork 94; tallest modern reinforced, non-reinforced 94, in Europe 94
bridges, 11–60; beauty of line 25; 'bridge factory' *13*; builders 228; cantilevering *13*; Chinese wrought-iron chain 234–6; compressed air, first use of 125; costliest 40–2; earliest 11; failures and disasters 52; heaviest loads 26, 35; longest arch spans 32–4; longest cantilevers 39; longest continuous suspended length 27; longest spans 27–31; longest as opposed to spans 56–8; most ambitious group of projects 51–2; oldest surviving 11; slenderest 6, 14, 59; structural continuity 60; structural types 12–13; tallest 33; ugliest 35–6; unusual duties 25
Brighton Chain Pier 43, 49
Brindley, James 122
Bristol West Dock lock 186; lock-gates 189
Britannia Bridge 2, *20*, 263
British Gas 213
British Steel Corporation 270
Bronx–Whitestone Bridge, New York 28
Brooklyn Battery Tunnel *128*
Brooklyn Bridge 24, 29, *34*, 49, 158–9, 228
Brooklyn Queens Expressway elevated road 58
Brouwershavense Gat Dam, Holland 82, *250*
Brown, Capability 42
Brown, Sir Samuel 43
Brunel, Isambard Kingdom *2*, 227
Brunel, Marc Isambard 21, 99, 119, 123, 227; his shield *123*
Brunings, Christiaan 152–3
Brunkeberg Tunnel, Stockholm 172
bucket-wheel excavator *142*, 143
Buffalo Bill Dam, USA 83
Buffelsfontein Eastern upcast shaft, South Africa 140
Building Research Establishment 165
buildings and towers: floor areas, greatest 99; largest in timber *102*; largest in world 99; leaning 104; tallest in world 85–7, 89, in Britain 86–7, 90, in Europe 88, 89; tallest free-standing tower in Britain 90; tallest by materials 92
Bukhtarma Dam, USSR 67
bulldozers *149*; first 149–50
buoy, largest for single mooring 178
Burgo Paper Mill, Mantua 88, 96, 231
Burnsville, USA 135

cable structures 101–2
Cabora Bassa Dam, Mozambique 67; Power Station 138
Cadiz dry-dock 184
caissons: deepest 157; disasters 159–60; early United States 159; largest in Britain 158, in world *161*; use in under-the-roof tunnelling *156*
Caledonian Canal 225
California Aqueduct *207*
California Central Valley irrigation project 207–8, 215
California State Water project 191, *202*, 207–8, 210; Aqueduct *207*, 210; Canal 191; pumping plant *204*
Callender–Hamilton bridges 25
Cameron, Donald 222
Camarasa Dam, Spain 170
Canadian River project 210
Canal de Bourgogne, France 226
canals: first ship, 189, 192; irrigation and 237–40; longest 189, 191; longest routes 189; oldest 189; Suez and Panama 256–9
Cannes sewage outfall 212
Cannon Street Bridge, London 228; office building 161, *162*, 267
cantilever bridges 13; heaviest 39; longest spans 39, 53–4; longest concrete spans 45; longest continuous three-span truss claim 39
Canvey Island gas tanks 173
Cape Cod Canal Bridge 54
Cape Girardeau, USA 241
Cargill, J D 22
Carlsbad, USA, largest natural cavern 138
Carquinez Strait Bridge, USA 157
Carthage Aqueduct, Tunisia 211
Cascade Tunnel, USA 112
Castelbello Tunnel, Italy 115
Castelvecchio Bridge, Verona, 234
Castigliano 229
Catchwater Aqueduct *209*
Catskill Aqueduct, USA 106
cavern, largest natural 138
Central outfall tunnel, Mexico City 107, 109
Centre Nationale des Industries et Technologies (CNIT), Paris, 96, *110*
Chagres River 257
chain bridges 18, 22, 233-36
Chak-sam-chö-ri Lamasery Bridge, Tibet 30
Chambal scheme, India 215
Chambon Dam, France 83
Champlain Bridge 166
Chang River irrigation system 237
Channel Tunnel 105, 109, 129
Charing Cross Bridge 228

Chasma barrage, Pakistan 194
Chatelleraut Bridge 15
Cheeseman Dam, USA *68*, 83
Chelsea filter-bed 219
Chengkuo irrigation canal 238
Chepstow Bridge 125
Chesapeake Bay bridge/tunnel 56, 58, *260*
Chicago *85*, 219; Drainage Canal 147, 221; River Bridge 54; Ship Canal Bridge 54; sewage works *221*
Chicoasen Dam, Mexico 64
chi-Hung Bridge, China *236*
chimneys, concrete: greatest capacity 88; tallest: Europe 88, world 87–8
Chinese civil engineering prototypes 232–40; bridges 233–6; canals and irrigation 206, 237–9; Great Wall 232–*3*; roads 233
Chirkey Dam, USSR *62*, 63
Chivor Dam, Colombia 63
Chocolat Menier Factory, Noisiel-sur-Marne 92–3
Chryses of Alexandria 75
Chrysler Building, New York *86*, 89, 94
Chrysler, Walter P 86
Church Street Post Office, New York 162
churches and cathedrals, tallest 89
Churchill Falls Power Station, Labrador *138*–9, 201
Chute-des-passes Tunnel, Canada 7, 114, 131
Cincinnati Bridge 22, 29
City and South London Railway 124
civil engineering terminology 9–10
clapper bridges, Dartmoor and Exmoor 11
Clark, Bernard L 150
Claudio Aqueduct, Rome 211
Clear Creek Tunnel, USA 111
Clifton Bridge 51
CN Tower, Toronto 87, 92, *155*, 156, 267
Coblenz Bridge 52
Cochiti Dam, USA 70
Cochrane, Sir Thomas 125
code of practice, first 273
Coen Tunnel, Holland 121
cofferdams 165–7; deepest 167
Coignet, M 99
Colechurch, Peter de 41
Cologne Cathedral 89, 92
Cologne–Deutz Bridge 59
Colorado River, USA 209; Aqueduct 132, 209, 210
Colossus Bridge, Philadelphia 30
Commodore John J Barry Bridge, Pennsylvania 39
compressed air in construction 125–6
concrete: bridges 13–15, first in France, Britain 14, longest arch spans 33 and spans 45–6, shallowest arches 60, slenderest 59; fastest rates 267; greatest height in continuous operation 267; highest lift 267; largest pour in day 266; largest pour in a building 162, 266–7; largest prestressed pressure vessels 268; largest prestressed structure in Asia 215; largest quantity placed 266; paving rate record 266; slip-forming records 267–8; strongest 268–9; reinforced structures 99–100; tallest buildings and towers 86, 90; tallest concrete building of load-bearing blocks 94; tallest structure of light-weight 90; unique twist 268; wire-winding, biggest 268
Condit 35
Conneaut coal and ore installations, USA *181*
Consulting Engineer, The, 251
consulting engineers: largest firm 269; largest single commission 271
consulting firm, world's largest 231
Contra Dam, Switzerland 64, 66
contractors, civil engineering: largest 269; largest in Europe 269; largest single contract 269
Conway Bridge 21
Coppermills Works, Walthamstow 219
Copperthwaite 123
Corby ironstone mines 148
Corinth Canal, Greece 190
Corn Market, Paris 104
Cornalbo Dam, Spain 84
corncobs (building) *274*
Cort 22
Cortez 76
costs: cost of construction 272; costliest schemes 270; human costs 272
Couesque Dam, France 68
Coulomb 229
Couvrex 256
Coyne, André 230, 231
Coyne and Bellier 230
Craigentinny Meadows, Edinburgh 222
cranes, largest 262–3
Creteil flats, France *274*
Cripple Creek Tunnel, USA 132
Cross, Hardy 229
Croton Aqueduct, USA 106
Cruachan Power Station 139
Crystal Dam, USA *65*, 73
Culebra Cut, Panama 257, 259
Cumberland Basin, Bristol 134, *135*

Daimonike Dam, Japan 83
Dames and Moore 270
Dammam Pier, Saudi Arabia 180
Damodar Valley, India 255
dams: beneath glaciers 69; buttress, highest 69; earliest major tidal closure 82; extent of building 72; failures and disasters 77–9, 81, 84; first large-scale 78; first true 75; highest enbankment dam *74*; historical facts 75–7; largest concrete 80; largest of earth or rock embankments 70; longest 81; notable 73; only British one damaged by earthquake 73; river barrages 69; sea dams 69, major 82; worst disaster 84; *see also* arch dams
Daniel Johnson Dam, Canada *66*, 67, 169, 230
Däniken, Erich von 136–7, 266
Danube 206
Danube Canal Bridge, Vienna 23
Dartford–Purfleet Tunnel *124*
Deception Bay storage building, Canada *102*
Delair Bridge, USA 54
Delaware Aqueduct, USA *43*, 106–7, 109, 117, 132, 136
Delaware Memorial Bridge, USA 28, 266
Delta scheme 247–51, 273
Den Helder harbour, Holland 153
Denver, England 198
Deptford water well 217
Derwent Dam, Yorkshire 169
Detroit River Tunnel, USA 120
Dewey, US floating dock 183
Dhuis Aqueduct, France 217
Diablo Dam, USA 83
diaphragm walls: 168–9, deepest 169; greatest quantity 169; largest single contract 169
Diaz 229
Dibden, W J 222
Dickey Dam, USA 70
Dillenburgh Castle, Germany 214
Dinorwic Power Station 139
Dischinger 231
diving bells 125
Djatiluhur Dam, Indonesia 73
Dneprodzerzhinsk Dam, USSR 81
Dokan Dam, Iraq 169–*70*
Dolgarrog Dam 78
Dollan Baths *94*
domes 101; earliest example 103; first iron 104
Dortmund–Mengede Tunnel *173*
Dover Harbour 164, 229

draglines 147–8, *149*; Big Geordie 148; Bucyrus–Erie 147; Ransomes and Rapier 148
Drake, Sir Francis 211
Drake's Leat Aqueduct 211
Drax Power Station 88, 162
Dredge, drain, reclaim 151
drill-holes, deepest 153; on land 153; at sea 153
dry-docks: for North Sea platforms 184; world's largest 184–5
Dudley 202
Duisburg–Neuenkamp Bridge, Germany 44, 59
Dunkirk, France: fortifications 226; lock 186
Duplessis Bridge, Canada 52
Durban storm-water culvert, South Africa 134, *135*
Dworshak Dam, USA 65, 80
Dyckerhoff und Widmann 231

Eads 24, 228
Eads Bridge, USA *23*, 157
Earls Court Bridge 14
East Bay Bridge, California 39
East Omaha Bridge 53
East River, New York: bridges *34*; tunnels 108, 122, 229
Eastern Scheldt Dam, Holland 70, 82, 160, *248*, 250; Bridge 58
Ecuador tunnel system 136
Eddystone Lighthouse 226
Eder Dam, Germany 84
Edmonston pumping station 203, *204*
Edmonton–Buffalo crude oil pipeline 212
egg, the (building) *275*
Eider Dam, Germany 82
Eiffel, Gustave 228
Eiffel Tower, Paris 88, 89, 92
Ekofisk 1 production platform, North Sea *260*, 262
Ekofisk–Emden gas pipeline 213
Ekofisk–Teesside oil pipeline 171, 213
El Hadjar iron and steel works 270
El Segundo, USA 179
El Teniente Underground Copper Mine, Chile 140
Elbe Bridge, Hamburg 54
Elizabeth I 219
Elizabethtown Bridge, USA 47
Ely Ouse Tunnel 111, 134
Emley Moor TV Tower 90
Emmerich Bridge, Germany 51
Empire State Building, New York 85, 89, 93, 94
Enchanet Dam, France *65*, 68
Engineer, The 36, 40
Engineering News Record 269, 270, 272
engineers and construction: great engineers *2*, 224–32; great schemes 232–62
Erie Canal, USA 191, 226
Ernesto Nord Well, Italy 153
Esbly Bridge *13*, 59
Escraveos Bar, Nigeria 180; long mole 153
Etruria Pottery Works, near Stoke-on-Trent 122
Eucumbene–Tumut Tunnel, Australia 114, 133
Eucumbene Tunnel, Australia 114 133
Eupalinus of Megara 225
Europe Canal 189
excavation and dredging: bucket wheel excavator *142*, *143*, *144*; bulldozers 149–50; daily rate record 144; draglines *149*; dredging records 142–3, 144, 154–5; earliest 150; greatest depth 150, potential 150; drill-holes and wells 153; dumper trucks *148*; explosive 152; fastest marine 145; fastest rates 133, 143–5; greatest single project 143; greatest single structure 143; largest 141–2; largest land machine *145*; leaching *139*; manual prodigies 151–2; most productive excavator 142, *143*; nuclear power station *172*; river and tide use 152–3; scrapers *147*; steam shovel, first 145; stripping shovel *144*; tractor shovel *147*
Exeter sewage works 222
explosion, largest chemical 152

factories, underground, largest 139
Fairbairn 21, 229
Fairfax Falls Power Station, USA 139
Fairmont Bridge, Philadelphia 30, 31; Water Works 202
Farahnaz Pahlavi Dam, Iran 69
Farakka Barrage, India 168, 194, *195*
Federal Reserve Bank, Minneapolis *93*
Fédération Internationale des Ingénieurs Conseils (FIDIC) 225
Fehmarn Sound Bridge, Germany 25
Ferguson 232
Fiastrone Dam *65*
Fife, Colonel 78
Finnish Gulf Dam, USSR 81
Finsterwalder, Ulrich 231
First Canadian Place, Toronto 86
First National Bank Building, Chicago 86
First National City Corporation, New York 86
Firth of Forth BP tanker terminal 160, 175
Flathead Tunnel, USA 113
floating bridges, longest world 43
floating docks, 181–3
flood control and sea defence 197–200, 241–4; greatest protection schemes, Britain 197, 198, world 197
Florence Lake Tunnel, USA 111, 132
Florianopolis Bridge, Santa Catherina, Brazil 51
Florida Everglades flood-protection scheme 197
Foley, M 126
Folgefonna hydro-electric scheme, Norway 131
Folson Dam, USA 80
Foraky Limited 173
Ford, Henry 189
Fort Madison Bridge, USA 53
Fort Peck Dam, USA 70, *274*
Fort St Vrain Power Station, Platteville, USA 173
Fort Worth Airport, USA 266
Forte Buso Dam *65*
Forth Bridges: rail 24, 29, 36, 39, *48*, 299; road 26, 27, *48*, 49, 51
Forties Field platforms, North Sea 160, 161, 261; pipeline 213
Fortuna-Gardsdorf mine, Germany 141–2, 144, 154
40 Wall Tower, New York 86
Fos, Marseille, France 175, *176*, 177, 184, 185
foundations 155–65; caissons 156–60; horizontal loads 164; greatest settlements 165; most accurate measurement of settlement 165; piling 160–4
Fowler, John 14, 229
Foyers Power Station 139
Frankfurt Airport hangar *101*
freezing 172–3
Freiburg Bridge, Germany 46
Fréjus 79
Fréjus Tunnel, France–Italy 112
Fremont Bridge, USA 32, *38*, 264, *265*
Freyssinet, Eugène 15, 55, *225*
Fribourg Bridge, Switzerland 30
Frigg pipeline, North Sea 213
Frodsham bridges *25*
Frying Pan Aqueduct, USA 210
Fuller, Buckminster 275
Furens Dam 76
Furnas Dam, Brazil 84

Gage Dam, France 68
Galveston port, USA 180

Ganga River scheme, India 211
Ganga–Cauvery canal project, India 211–12, 255
Garabit Bridge, France 33
Gardiner Dam, Canada 70
Garrett Hostel Bridge, Cambridge 25
Garrison Dam, USA 70
Garry Tunnel 132
gates for locks and dry-docks: largest in Britain 188, 189; world's largest 188
Gateway Arch, St Louis 88
Gatun Lake, Panama 257
Gaunless Bridge, York Railway Museum 22
Gazelle Bridge, Yugoslavia 44, 59
Geislingen Bridge, Germany 44
Geneva, Lake 67
Genf-Lignon Bridge, France 46, 59
Genoa 180; floating dock 181
George P Coleman Memorial Bridge 54
George Washington Bridge, New York 27, 29, *35*, 49, 166, 228
Ghulam Mohammed barrage, Pakistan 194
Gibson, A H 230
girder bridges: box 13; longest box spans 44; longest stayed spans 45; slenderest stayed 59; slenderest unstayed 59; ugliest stayed 35
Gladesville Bridge, Sydney 33, 60, *268*
Glen Canyon Bridge, USA 32; Dam 65
Glomar Challenger 153
Glorywitz Bridge, Germany 18
Godavari Bridge, India 57
Gold of the Gods 136
Golden Gate Bridge, San Francisco 27, 42, 166
Goodyear Airship Hangar, Akron, USA *88*, 99
Gorky Dam, USSR 70
Gotha Canal, Sweden 225
Gothenburg floating dock 181; oil reservoirs 139
Grain Power Station, Kent 164
Grampian hydro-electric scheme 272
Grand Anicut weir, Cauvery River, India 206
Grand Canal, China 143, 151, 152, 189, 191, 237, *239*–41, 252
Grand Central Station train shed, Chicago 96
Grand Coulee Dam, USA *64*, *74*, 76, 80, 267; power station *200*, 201, *214*, 262
Grande Dixence Dam, Switzerland *61*, 62, 80, 83, *171*, 267; Reservoir 215
Grand Maitre Aqueduct, Fontainebleau *12*, 14
Graythorp dry-dock 184, 261, 262–*3*
Great Apennine Tunnel, Italy 111
Great Arch of Ctesiphon, Iran 104
Great Bridge, China 57
Great Eastern 41
Great Ouse flood-protection schemes 154, 197, 198
Great Wall of China 143, 232–*3*, 252
Greathead J H 123–4, 125
Green Cove Springs floating dock, Florida 183
Grevelingen Dam *248*
Grosvenor Bridge, Chester 34
ground engineering 165–73; coffer-dams 165–7; control of ground water 167–8; diaphragm walls 169; grouting and chemical treatment 169–73
ground-water control 167–8
grouting and chemical treatment 169–73; alluvial 171; largest and deepest cut-off 171; largest curtain 170; physical treatment 171
Grubenmann brothers 228
Grunenthal Bridge, Kiel 33
Gukow River Dam, China 84
Gulag Archipelago, The 272
Gulf Stream 200
Gulf of Mexico 262
Guri spillway, Venezuela 191

Haarlem Mere drainage 245–6
Habbaniyah flood-protection scheme 198
Hadrian's Wall 233
Hague: insurance company office *263*
Hall of Audience, Vatican City 231
HAM BV of Holland 269
Hamana Bridge, Japan 45
Hambach mine, Germany 143
Hamburg: dry-dock 184; floating dock 181; port 177; sewer tunnel 172, *173*
Hammurabi, King of Babylonia 273
Hampton Roads Tunnel, USA 121
Hampton water-treatment works, England 219
Han Kou Canal, China 237
hangar, world's largest *88*, 99; largest diagonal lattice grid *101*, 102–3; jacking *265*
harbours: deepest water 174; Japanese construction 178
Harecastle Tunnel, near Stoke-on-Trent 118, 122
Harington, Sir John 219
Haringvliet Barrage, Holland *26*, 35, 82, 162, 166, *167*, 168, 188, 194, 232, 248, *249*, 250, *273*
Harlem River Tunnel, USA 120
Harrison, Sir C S C 207
Harrsele Tunnel, Sweden 115
Harspranget Tunnel, Sweden 115
Hartlepool Power Station *268*
Haskin 125, 126
Haweswater Aqueduct 210
Hawkshaw, Sir John 227–8
Hazlitt, William 274
Heianza, Okinawa 174
Heinrichenburg shiplift, Germany 185
Hell Gate Bridge, New York 26, 32, 35
Henderson Tunnel, USA 112
Hernandarias Tunnel, Argentina 121
Herodotus 225
Hersent 126
Hetch Hetchy Aqueduct, USA 210
Highland 1, oil platform *261*
Hikoshima Bridge, Japan 45
Hirakud Dam, India 81; spillway 191
History of Civil Engineering 229
Hodgkinson, Eaton 21, 229
Hokkuriku Tunnel, Japan 112
Holderbank–Wildegg Bridge, Switzerland *6*, 46, 59
hole, world's deepest 153
Holland, land reclamation in 244–52; Delta Scheme 232, *247*–51; draining the Haarlem Mere 245–6; mounds 244; sea-walls 244; steam engines 244; windmills 244; Zuider Zee 246–7
Holland sea defence scheme 197
Holland Tunnel *136*
Hollandse Ijssel sluice gates *197*
Holyhead 180
Homersfield Bridge, Suffolk 14
Homs Dam, Syria 75
Hong Kong 101
Hood Canal Bridge, Point Garnbee 43
Hooghly River Bridge, India 44
Hooke, Robert 229
Hoosac Tunnel, USA 132
Hoover Dam, USA *63*, 64, *65*, 76, 83, 209, 259
Hornibrook's 252
Hornsea gas reservoirs 139
Housatonic River Bridge, USA 59
house-moving machine *265*
Howrah, Calcutta 39
Huai River, China 151; Bridge 25
Hudson Tunnel, USA 125–6
Huey P Long Bridge, USA 57, 58, 157

Humber Bridge 27, 29, 51, 59, 164
Humber gas main crossing 212
Humber–Tetney oil pipeline 212
Humberside–Aldermaston oil pipeline 213
Hungtze, Lake, China 81
Hunterston, Scotland 174
Hwang Ho Bridge, China 57
hydraulic works: aqueducts 209–11; barrages and spillways 193–7; flood control and sea defence 197–200; harbours and canals 174–92; hydro-electric power 200–1; irrigation 204; main drainage and sewage treatment 219; pipelines 212–15; pumping stations 202–3; water supply and treatment 216–19
hydro-electric power 128, 131, 133, 200–1; largest power stations 201; largest sites 200
Hydroproject Institute, USSR 230
Hyperion Tunnel, USA 119, 121

Ijsselmeer Lake, Holland 247
Impregilo 269
Indian flood-protection works 197; irrigation 252–4; water conservation 211
Indus irrigation scheme 154, 191, 207, 215, 253; Waters Treaty 253
Industrial Revolution 18, 43, 105
Infernielo spillway, Mexico 191, 195
Inga, Congo 200
Ingalls (Transit) Building, Cincinnati *90*
Inguri Dam, USSR *61*, 62
Innertkirchen Tunnel, Switzerland 128
Institution of Civil Engineers 224, 227
International Commission on Large Dams (ICOLD) 72, 77, 78
International Nickel Company chimney, Sudbury, Canada 87
Iranian irrigation 206
Irkutsk Dam, USSR 67
iron bridges: cast-iron, first 18, earliest in Europe and Britain 18, longest 33, shallowest arch 60; longest arch spans 33, longest span 22, 33; puddling process 22; railway bridge, first 22; wrought-iron, first 22; suspension 234–6
Iron Gates lock, Lower Danube 186
Ironbridge, Coalbrookdale *18*
irrigation: areas of land 205; Australian 255; California 207–8; earliest 204; historical 204; hydraulic 174; India 208; largest unified scheme 207; modern 206; Pakistan 154, 191, 194, 206, 207, 214, 253; pumping station, largest for irrigation *204*; world's largest scheme 206–7
Irtysh–Karaganda Canal, USSR 191
Isère–Arc Tunnel, France 129
Ishakari Wan, Japan 180
Itaipu Dam, Brazil–Paraguay 80; power station 201
Italy–Sicily pipeline 213
Iwakuni Bridge, Japan 31
Izmir Bridge, Turkey 11

Jackblock *263*, 264
jacket (steel tower) to drill oil 259
Jacques Cartier Bridge, Montreal *26*
James Bay power station, Canada 201
Jan Heijmaus 249
Jansen, P P 232
Japanese bridge building 51–2; irrigation 206
Jari Dam, Pakistan 70
Jarlan breakwater, Holland 178
Jena planetarium, Germany 104
Jerusalem 233
jetties, longest world, British 180
John Day Dam lock, USA 185
John F Kennedy Tunnel, Belguim, *107*, *108*, 121
John Hancock Center, Chicago *86*, 94, 274
Jose Leon de Carranza Bridge, Cadiz 53
Judah, Theodore 227
Juktan Tunnel, Sweden 114
Julia Aqueduct, Rome 210
Junnar caves, India 137
Justinian I, 75

Kadana spillway, India 193
Kaerumataike Dam, Japan 75
Kaiser, Henry 150
Kala (Balula), Sri Lanka 84
kanats 137
Kanev Dam, USSR 81
Kanheri caves, India 137
Kanmon Strait Bridge, Japan 28; Tunnel 113, 119
Kansas City underground storage 139
Kara-Kum Canal, USSR 191, 210
Kariba Dam, Zambezi 67, 230
Karl Marx University, Leipzig 274, *275*
Karlstad, Sweden 171
Karman, von 230
Kawasaki, Japan 178, 180
Keban Dam, Turkey 66, *127*, 169, 170
Kebar Dam, Iran 83, 84
Kemano Power Station, Canada 138–9
Kemano Tunnel, Canada 115, 132
Kennecott Copper Corporation 88
Kentucky Dam, USA 80
Khadakwasla Dam, India *64*, 78–9
Khan, Fazlur R 231
Kharg Island Terminal, Persian Gulf 194
Khosr River dams 75
Kiel Canal, 148, 190; dry-dock 184, lock 186
Kielder Tunnel 109
Kiev Dam, USSR 70, 81
Kilforsen Tunnel, Sweden 115
Kimberley 'Big Hole', 151
King County Stadium, Seattle 96
King George V dry-dock, Southampton 185
King Street Bridge, Melbourne 52
Kingsgate Footbridge, Durham 25
Kirklees Park Bridge, Leeds 18
Kish Bank lighthouse *259*
Kloof Gold Mine *140*
Knee Bridge, Düsseldorf 44
Kobe, Japan 178, 180
Kohlbrand Bridge, Hamburg 44
Koror Babelthaup Bridge, Pacific 45
Kosi Barrage, India 194
Krasnoyarsk Dam, USSR 67, 80, 197, *235*; power station 195, 201
Kuanhsien irrigation scheme *237*, 238
Kubiki Tunnel, Japan 113
Kuibyshev (now V I Lenin) power station, USSR 230
Kunu siphon, India 215

La Tour 241
Labelye 164
Lake Alban Tunnel, Italy 118
Lake Mathews, USA 209
Lake Michigan, USA 221
Lake Pontchartrain Bridges, USA: No. 1, No. 2 and trestle 56, 58; twin bridges 57; 241
Lake Washington Bridge, USA 43
lakes, world's largest artificial, world's largest 67
Lambot, M 99
Lan Chin Bridge, China 30
Langelinie Bridge *16*, 59
Languedoc Canal, France 226
Lansdowne Bridge, Pakistan *36*
Lar Kalan Tunnel, Iran 114
Lavally 256
Laviolette Bridge, Canada 32
Le Havre port 178; lock 186
leaching, excavation by *139*

Leaning Tower of Pisa 104, 275
Lee River 217
Leeghwater, Jan 226
Lely, Dr C 228–9
Lelystadt 228, 247
Leningrad floating dock 183
Leonardo da Vinci 16
Leonhardt, Fritz 232
Leonhardt und Andra 232
Lessep, Ferdinand de 256, 257
LeTourneau, R G 150, 230
Levensau Bridge, Kiel 33
Lewiston–Queenston Bridge, Niagara *19*, 29, 32
Li Chhun's Bridge, China 234
lifting, heavy: ancient feats 263, 266; bridge *265*; greatest Victorian 263; hanger *265*; heaviest weight dropped 263; highest lift of concrete 267; house *265*; largest cranes 262–3
lighthouse *177*, *259*; tallest 89
Lille fortifications 226
Lillebaelt Bridge, Denmark 28
Lincoln 219; Cathedral 89, 92
Lincoln Tunnel, New York *137*
Lindenthal 228
Lisnave dry-dock, Lisbon 184, 185
Littlebrook D Power Station, Dartford 162–3
Liverpool: dock 169; lock 186; lock-gates 188; port 178; RC Cathedral *273*, 274
living in high buildings 101
Lloyd Canal, Pakistan 191, 206–7; barrage 194, 253
Loch Kishorn 262, 268
Lochaber Tunnel 114, 128, 132
Lockett 222
locks: first 178; highest-head shiplift 187; invention 185; Welland Canal *187*, 192; working against highest heads 185; world's largest 186
Logensalya, Germany 214
Loing–Lunain Aqueduct, France 217
London, Greater 216–17; main drainage *216*, *220*
London, Port of 177
London Airport 102; hangar *265*
London Bridge 18, *40*, *41*, 202
London Bridge–Deptford Creek Railway 58
London Bridge Waterworks 202
Long Beach, USA 180
Longannet Tunnel 111
Longarone 79
Longview Bridge, Washington 39
Lorentz 246
Los Angeles 86, 209
Lotschberg Tunnel, Switzerland 112
Louis Hippolyte Lafontaine Tunnel, Canada 121
Louisiana Superdrome, New Orleans, *95*, 96, 274
Lower Maas River, Holland 82
Lower Zambezi Bridge 57
Lucin Cut-Off (Great Salt Lake) Bridge 56
Ludlow's Ferry Bridge, USA 161
Luiz I Bridge, Oporto 22, 33
Luling Bridge, USA 44
Lumiei Dam, Italy *65*, *68*, 230
Luneburg, Scharnebeck Shiplift, Germany *187*
Luxembourg Bridge, Germany 34
Luzzone Dam, Switzerland 66

Maas Tunnel, Holland 120
McAlpine-Sea-Tank platforms, Ardyne Point 267
McCall's Ferry Bridge, Pennsylvania 31
Machado 228
Machadosdorp, South Africa 228
Mackinac Bridge, USA 27, 166
Madrid 101
Maesgwyn Mine 155
Magna chimney, USA 88
Mahoning River Dam, USA 168
Maidenhead Bridge 60
Maidstone 219
Maillart 15
Mailsi Siphon, Pakistan *152*, *215*,
main drainage and sewage treatment 219–22; first septic tank process 222; largest sewage works *221*, 223; main drainage 219–20; Paris sewers *220*, *221*; sewage treatment 222–3
Maine-Montparnasse Tower, Paris 88
Mainstone 229
Malgovert Tunnel, France 128, 129
Malpas canal tunnel 226
Malpasset Dam, France 79, 231
Manchester 222; Ship Canal 147, 154, 190; University 230
Mandali Canal, Iraq 189
Mangfall Bridge, Germany 46
Mangla Dam, Pakistan 70, 133, 134; spillway *146*, 193; reservoir 253
Manhattan Life Building, New York 159
Manicouagan Dam, Canada 169
Manuel Belgrano Bridge, Argentina 45
Maracaibo Bridge 228
Marcia Aqueduct, Rome, 210, 211
Marco Polo 234
Mariakerke Bridge, Belgium 46
Marib Dam, Yemen 75
Marina City Towers, Chicago 90
Marine Parkway Bridge, New York *35*, 54
Mariotte 229
Maris 232
Marlette Lake water supply scheme, USA 214
Marne bridges, France 15, 55, 225
Marsham Street offices, Westminster *273*, 274
masonry bridges 34; longest brick structure in the world 58; longest span of stone 11; longest stone bridge *38*, 58; longest world, Britain 34; shallowest brick arch 60
masts: tallest 87, 89, 90; tallest stayed mast in Britain 90
Mauvoisin Dam, Switzerland *62*, 63, 65, 83, 170
May Dam, Turkey 81
Medway Bridge 59
Meer Alum Dam, India 76
Mekong River bridge 236
Melville pipeline bridge, USA 28
Menai Suspension Bridge *20*, 21, 30, 43, 225, 234
Menes, King 204
Merchandising Mart, Chicago 99
Mersey Tunnels 108, 119, 129, *134*, 162
Mesopotamia drainage system 219
Metropolis Bridge, USA 47
Metropolitan Milanese 169
Mexico City main drainage 107, *222*, 229
Mica Dam, Canada 62, *74*
Middle East Six-Day War 256
Milford Haven Bridge 52
Milford Haven–Seisden oil pipeline 213
mines: deepest mine 140; deepest British shaft 140; fastest shaft-sinking 140; largest gold, copper 140; largest excavation 141–2; largest operation 140; longest vertical shaft 140; tunnels 140
Miraflores Lake 257
Mississippi flood protection scheme 143, 144, 154, 197, *241*, 242–4, 252; canal 189; channel improvement 242; floodways 241; levees 242; tributary works 242–4
Missouri River 243; Bridge, Glasgow 24; dams *71*
Mitchell Power Station, Cresap, USA 88
Mittelwork Underground Factory, Germany 139
Moawhango Tongariro Tunnel, NZ 114
Moffat Tunnel, USA 113

Moh's scale of hardness 105, 106
Mohenjo-Daro, Indus River 204, 219
Mohne Dam, Germany 84
Moir 126
Mont Blanc Tunnel, France–Italy 113
Montreal Olympic Stadium *100*
Morandi 228
Morganza Barrage, USA 194, *240*
Mormon Tabernacle, Salt Lake City *88*
Morris 202
Morris, William 36
Moscow 87, 216, 217
Moti-Talav Dam, India 75
motorway bridges 36
Mont Cenis Tunnel, France–Italy 112, 118, 122, 126, 132
Mont Cenis–Villarodin Tunnel, France 115
Mover of Men and Mountains 149–50
Mratinje Dam, Yugoslavia 65
Mudduck Masur Dam, India 83
Mulberry harbours 179
Müller-Breslau 229
Mundford, Norfolk 164–5
Munich Railway Station 104
Municipal Bridge, St Louis 47
Murrumbidgee Tunnel, Australia 114
Murti, Shri N G K 78, 79
Musashi 262
Musto, A A 206
Mycenae 103
mystery towers *259*

Nadyn–Moscow natural gas pipeline 212
Nagarjunasagar Dam, India *63*, 73, 80; Canal 191, 210; spillway 193
Nahrawan irrigation canal 204
Nan Madol city ruins, Caroline Islands 137
Nanking Bridge 57
Nanko Bridge, Osaka, Japan 39, 262
Naples Viaduct 60
Narita No. 1 Hangar, Tokyo 96
Narni Bridge, Italy 30
Naryn River Dam (USSR) 152
Nassiet well, France 153
National Westminster Bank HQ Tower, London 86–7
Navajo irrigation scheme, USA 133; No. 1 Tunnel 133
Navier 229
navvys 272
Needham, Dr Joseph 25, 189, 232, 234, 236, 237, 238, 239
Nervi, Pier Luigi 231
Netzahualcoyotl Dam, Mexico 76, 81
New Cornelia mine tailings dam, USA 143
New Croton Dam, USA 83
New Orleans, 241, 252; Greater New Orleans Bridge 39
New River, London 202; aqueduct 211
New River Gorge Bridge, USA 32, *52*
New Street Station train shed, Birmingham 96
New York 106; port 177
Newcomen's heat engine 202
Niagara: Niagara–Clifton Bridge 29, 34; power stations 19; railway bridge *19*, 42
nicknames for constructions 274–5
Nieuwe Waterweg Canal, Rotterdam 190, 192
Nigg Bay dry-dock, 184, 188, 261
Nikitin, N V 87
Nile River 238; Valley 204
Ninian Field oil platform 178, 262, 268
Niteroi Bridge, Brazil 44, 56, 59, *163*, 275
Nochistongo Tunnel, Mexico 118
Noirieu Tunnel, France 118
Norddalen Iptovtn Tunnel, 114
North Kiangsu Canal, China 143, 151, 154, 191
North Road Tunnel, London 134
North Sea Canal, Amsterdam 168, 178, 190, 192
North Sea oil and gas pipelines 213
North Wirral pipeline 212
Northern Line Tunnel 111
Norwood Tunnel 118
Notre Dame, Paris 89
Nourek Dam, USSR 61, 70, 83, 193
Numrood Dam, Iraq 84

Oahe Dam, USA 70, *129*, 132
obelisk, largest known 266
Oberhasli hydro-electric scheme, Switzerland 128
Ocean Builder I, 262
Oeland Bridge, Sweden 58
office building, largest 99
offshore structures 259–62
Ogoshima Island, Japan 154, 162, 175
Ohio River Bridge, USA 47
Ohnaruto Bridge, Japan 51
Oilfields of the world 153
Old River, USA 241, 244
Old Southwark Bridge 33
Oléron Bridge *55*
One Shell Plaza, Houston 86, 90, 92, 231
Ontario, Lake 67
Ontario Hydro 267
Oospat Tunnel, Bulgaria 115
Orange–Fish Tunnel, USA 107, 109, 272
Orco Tunnel, Italy 128
organisations, largest that designs and constructs 269
Oroville Dam, California 63, 70, 207
Osaka, Japan 178; Expo 1968 Building 102
Oshima Ohasi Bridge *262*
Osiglietta Dam, Italy *65*, 68
Oso Tunnel, USA *131*, 133
Osseo, USA 266
Ostankino TV tower, Moscow 87
Otis, William 145
output of construction industry 270
Ove Arup Partnership 269
Owen Falls Dam, Uganda 67
Owens River aqueduct, USA 210
Owyhee Dam, USA 83

Padawiya Dam, Sri Lanka 81
Paddy's wigwam 273
Paine, Thomas 33
Paisley filter-bed 219
Pakistan irrigation scheme 154, 191, 194, 206, 207, 215, 253
Palace of Fine Arts, Mexico 165
Palasport, Milan *95*, 102
Palladio 16
Panama Canal 147, 154, 183, 190, 192, *257*–9; locks 186
Panda Wewa Dam, Sri Lanka 75
Panshet Dam, India 78–9
Parana Guazu Bridge, Argentina 161
Parana River 200
Paris 217, 220–1; sewers *220*, 221
Park Lane Towers, Denver, USA 94
Parker Dam, USA 209
Parque Flamingo war memorial, Rio de Janeiro 275
Pascagoula floating dock, USA 181
Paschberg Bridge, Austria 13
Pasco Kennewick Bridge, USA 45, 59, 232
Pausilippo Tunnel, Italy 116, 118
Pavia Bridge 234
Peak National Park Bridge 46, 59
Pearson, Weetman (1st Viscount Cowdray) 164, 229
Pedro Miguel, Panama 257
Pennsylvania Canal, USA 228
Pentagon, USA 99
Perronet, Jean Rodolphe 226
Peru tunnel system 136
Peterson 100
Pharos Lighthouse, Alexandria 89
Philadelphia 202
Philbrook Dam, USA 150
Pia Maria Bridge, Oporto *22*, 33

Picardy–Monbauron pipeline 214
Picasso Bridge, Germany *16*
Pien Canal, China 239
Pier 57, Grace Line, New York *159*
Pierhead Limited 268
Pierre Laporte Bridge, Quebec 28
Pieve di Cadore Dam, Italy 65, 68, *79*, 231
piling: bored piles 159, *161–2*, greatest diameter 161, longest world 161, British 162; British load-carrying record 161, *162*; greatest number 162, 163; historical facts 164; longest 159–60; steel cylinders 159, greatest diameter, load 160, longest 160
pipes: cast-iron 214–15; highest head 214; largest inverted siphons 215; world's largest *214*
pipelines: longest crude oil 212; longest sea outfalls and pipeline crossings 212; submarine 213–14, longest 212, deepest 214; world's longest 212
Piper pipeline, North Sea 213
Pirttikoski Tunnel, Finland 115
Piscataqua River Bridge, USA 31
Pittsburgh 86
Pizarro 136
Place Ville Marie, Toronto 86
Plauen Bridge, Germany *11*, 34
pleasure pier, longest 180
Pliny 151
Plougastel concrete arches 225
Plover Cove Dam, Hong Kong 82
Plumstead 219
Poatina Tunnel, Australia 133
Poetsch 172
Point Tupper 174
Poisson 229
Polcevera Bridge, Genoa 45
Pont de la Concorde, Paris 226
Pont de Neuilly, Paris 226
Pont y Cysyllte Aqueduct 225
Ponte Presidente Costa e Silva, Brazil; *see* Niteroi Bridge
Ponte Vecchio, Florence 234
Pontypridd Bridge 31
Porjus Power Station, Sweden 139
Port Colborne lock, Welland Canal, Canada 186, 192
Port de Martorell Bridge, Spain 31
Port Mann Bridge, Canada 32
Port Rashid dry-lock, Persian Gulf 184, 185
Port Said, Egypt 180
Port Talbot 171
Portage Railway Viaduct, USA *17*
Portsdown Hill building, Portsmouth 94
Portillon scheme, France 214
Portland cement 99, 269
ports: largest 177; largest impounded 178
Posey, Mr 120
Posey Tunnel, USA 120
Poughkeepsie Bridge, USA 157
power stations: excavation for nuclear *172*; first underground 139; largest underground, Britain, Europe 139, world 138–9; world's largest hydro-electric 195, 201
Prandtl 230
pregnant oyster (building) *275*
Priest Rapids spillway, USA 191
Priestley mole 129
Prince George Well, Canada 217
Prins der Nederlanden 142–3
production platforms: concrete 261–2; largest for North Sea 261; steel 261
Project Gnome, USA 139
Project Plowshare, USA 139
Property Service Agency (PSA) 269
Prudhoe Bay, Alaska 213
public works, largest project 270
Puentes Dam, Spain: I *77*, 83; II 83
Puget Sound dry-dock, USA 184
Pulaski Skyway elevated road, USA 58
pumping stations: first in modern sense 202; first steam-powered 202; world's largest 203
Putzeling Dam, China 151
Pyramid of Cheops 89, 151
Pyramid of Chephren 151
pyramids: tallest 89; quantity of material 151

Qadirabad Barrage, Pakistan 194
Quebec Bridge 26, 29, 36, 39, 52, 263
Queensboro Bridge, New York *34*, 36, 39

Rafael Urdaneta Bridge, Venezuela *15*, 45, 57
railway bridges, longest 57, 58
Rajasthan Canal, India 191, 210
Rance Dam, France 82, 166, *201*
Rand-McNally Building, Chicago 93
Randfontein Estates Gold Mine, South Africa 140
Raytheon Company 269
Red River, USA 244
Redcar Ore Terminal (BSC) 175, 177, 271; Quay 169
Rendalen Tunnel, Norway 113
Rendesberg Tunnel, Germany 121
Rennie, John 33, 41
reservoirs: underground oil and gas 139; world's largest 67
Reservoirs (Safety Provisions) Act 78
Revin Bridge 60
Reynolds, Osborn 230
Reza Shah Kabir Dam, Iran 66
Reza Shah Pahlavi Dam, Iran 66
Rhine Delta 248
Rhône Valley 255
ribbon bridges, stressed-46; first British 46; slenderest 59
Riga Dam, USSR 81
Rio Colorado Bridge, Costa Rica 46
Riquet, Pierre Paul de 226
river-valley irrigation schemes 252–6
roads, longest urban elevated 58
Robbins mole 129
Robertstown Bridge, Aberdare 22 fn
Rochester Bridge 125, 157
Rockville Bridge, Pennsylvania *38*, 58
Roebling, John August 29, 42, 49, 228
Roebling, Washington 49, 228
Rogun Dam, USSR 61
Rokko Tunnel, Japan 112
Romanian irrigation 206
Rome 210, 219, 221; Roman walls and roads 233
Ronquières shiplift, Belgium 187
roofs: first concrete shell 104, 231; largest aluminium *91*; longest span 96–8; saddle-shaped 101–2; steel space-frame grids 102–3
rope bridges 25, 31
Rosebud County water well, Montana 153
Roseires Dam, Sudan 81
Roseland Dam, France *72*, 73
Rosenkratz crane 263
Rotherhithe Tunnel 120, 122–3
Rotterdam Europoort 145, 150, 154, 175, 177, 180, 183, 184, 185, 212, 248
Rouen Cathedral 89
Royal Pavilion, Brighton 274
Royal Seaforth Dock, Liverpool 169
Ruhr dams, bombing of 84
Runcorn Bridge 32
Rungie Aqueduct, France 217
Rustenburg platinum mines, South Africa 140

Sacsayhuaman, Peru 266
Sadd-el-Kafara Dam, Egypt 10, 75, 84
Sadovia–Corabia irrigation scheme, Romania 206
St Bernard Road Tunnel, Switzerland *136*

St Charles Air-Line Bridge, Chicago 53
St Clair Tunnel, Canada–USA 108
Saint-Denis Bridge, France 22
St Fergus–Samlesbury gas pipeline 213
St Francis Dam, USA *78*
St Francis River, USA 243
St Gotthard Rail Tunnel, Switzerland, 112, 118; Road Tunnel 112
St James's Park Footbridge *24*, 25, 59
St Lawrence Power Station, Canada 267; house moving *265*
St Lawrence Seaway 26, 42, 154, 189, 190, 191, 192, 270
St Louis Bridge 24, 60
St Nazaire Bridge *42*, 44; dry-dock 184
St Nicholas Church, Hamburg 89
Saint-Ouen Dock warehouse, Paris 93
St Paul's Cathedral 89
St Peter's, Rome 229
St Petersburg (now Leningrad), 159–60
Saint-Pierre-de-Beauvais 89
St Sophia, Istanbul 103
Salazar Bridge, Lisbon 27, 157
Salcano Bridge, Italy 34
Saltash Bridge 21, 157, 263
Samos Tunnel, Greece 119, 225
San Fiorano hydro-electric scheme, Italy 214
San Francisco 86
San Francisco Bay 207
San Francisco–Oakland Bridge 57, 157, 166
San Jacinto Tunnel, USA 109, 210
San Joaquin Valley, USA 207
San Juan–Chama irrigation scheme, USA 131
San Luis Dam, USA 70; Canal 215
San Luis Obispo County, USA 207
San Mateo–Hayward Bridge, USA 56, 58
San-Quentin Canal, France 122
Sando Bridge, Sweden 33
Sandström, G E 225, 273
Santa Barbara Channel, USA 261
Santa Barbara County, USA 207
Santa Massenza Power Station, Italy 139
Santo and São Vicente sewage outfall 212
São Paolo elevated expressway, Brazil 275; exhibition hall roof *91*
Sapperton Tunnel 118
Saratov Bridge, USSR 46, 60
Saskatchewan River Dam Tunnel, Canada 133
Saskatoon potash mine, USA 172
Saturn Vertical Assembly Building *90*, 188
Saulte Ste Marie Bridge, USA–Canada 53
Sava Bridge, Belgrade 44
Save River Bridge, Mozambique 46
Sayany Dam, USSR *62*, 63, 66, 68, 80, 195; power station 201
Scapa Flow Dam 82
Schaffhausen Bridge, Switzerland 31, 266
Scherzer rolling lift bridges 53; world's longest spans 54
Schliemann, H 103
Schrah Dam, Switzerland 83
Schwarmindingen building, Switzerland 94
Science and Civilisation in China 232
Sciotoville Bridge, USA 35
Scorpe Dam, Germany 84
Sears Roebuck Company 85
Sears Tower, Chicago 85, 87, 93, 94, 99, 231
Seikan Tunnel, Japan 109, 119, *127*, *130*
Sembawang dry-dock, Singapore 184
Semenza, Carlo 230–1
Sennacherib 75
Serre Ponçon Dam, France 171
Serrell 29
Seven Mile Bridge, USA 56
Severins Bridge, Cologne 35
Severn Bridge 26, 27, 49, *50*, 51, 59, 232; Tunnel 108, 119
sewing machine *273*
Seyrig, T 33
Shandaken Tunnel, USA 109, 117, 132
Shasta Dam, USA 80
Shatt-el-Arab Bridge, Basra 25
Shatt-el-Hai irrigation canal 204
Sheerness boat store 92
Sheridan, R W 167
Sherwood Colliery, Mansfield 133
Shin-Kanmon Tunnel, Japan 111, *130*
Shirley Gut Siphon Tunnel 120
Shoreham Harbour towers *259*
Sienna Town Hall Tower 92, 94
Simplon 1 Tunnel, Switzerland 111, 117, 128, 129, 132
Simplon 2 Tunnel, Italy 111, 117
Singapore 101, 275; Development Bank 161
single-point moorings 178
Sisters opencast mine, Co. Durham 148
Sixtus V, Pope 211
60 Wall Tower, 70 Pine Street, New York 86
Skidmore, Owings and Merrill 231
skyscraper, first of reinforced concrete 90
slave labour 272–3
Smeaton, John 9, 125, *226*
Smith, Dr 126
Smith, Sir Hubert Shirley 159
Smith, William Sooy 24
Snefru North Stone Pyramid, Egypt 89
Snoqualme Falls Power Station, USA 139
Snowy Mountains hydro-electric scheme, Australia 133, *255*
Solzhenitsyn, Alexander 272
Sommeiller 128
Sone Barrage, India 194
'Soo' Canal lock, Canada 186
South Bisanseto Bridge, Japan 157
South Capitol Street Bridge, Washington DC 53
South Fork Dam, USA 84
South Halstead Street Bridge, Chicago 54
South State Mall, Albany, USA 99
Southend Pier 180
space structures 101–4
Spencer Works, Newport 270
Sri Lanka dams and reservoirs 206
Stalingrad power station (now 22nd Congress) 230
Standard Oil Building, Chicago 86
Standedge Tunnel 118
Starrucca Viaduct, USA *37*, 40, 42
Starvation Tunnel, USA 133
steam-navvies 145, 147
steam-shovel, first 145
steel bridges: Bessemer process 22, 23; first 23; first all-steel claim 24; longest 58; longest steel-plate spans 44; shallowest arches 60
steel structures, notable 88
Steinman 39
Stephenson, George 227
Stephenson, Robert *2*, 20, 21, 40, 227
Stevenson's Creek Dam, USA 68
stone, largest quarried and moved 263
Stonehenge 263
Stornoforrs Tunnel, Sweden *7*, 115, 128, 133, 138
Straub 229
Stretto di Rande Bridge, Spain 44
stripping shovels 148
structural continuity 60
structural frame: first building 92; first fully framed all-steel 93
Sturrock dry-dock, Cape Town 185
Subiaco Dams, Italy 75, 84
submersible bridges 25
Sudbury, Canada 87
Suez Canal 154, 190, 227, 256–7, 259

Suisun Bay Bridge, USA 157
Sukian Dam, Pakistan 70
Sunderland Bridge 33
Sunshine Skyway Bridge, USA 56
suspension bridges, 11, 13–14, 25, 31, 43, 49–52; first British to carry heavy traffic 30; first wire 30; longest: continuous length 27, chain span 51, *236*, self-anchored span 51, with steel cables and prestressed concrete decks 46; slenderest 59; ugliest 35; iron-chain 234–6
Sutlej River, Pakistan *152*, 215
Sutton sewerage works 222
Sway Tower, Hampshire ('Peterson's Folly') 100
swing bridges: only one to carry a canal over a canal 54; world's longest spans 53–4
Sydney Harbour Bridge 26, 32
Sydney Opera House 231, *251*, 274

Tacoma Narrows Bridge, USA 28, 49, 52, 274
Talow Tunnel, Iran 113
Tama Tunnel, Tokyo 117
Tancarville Bridge, Le Havre 28
Tappan Zee Bridge, New York 39, 57, 156, *158*, 160, 162
Tarbela Dam, Pakistan *69*, 70, 140, 143, 144, 145, 154, 171, 195, 253, 267, 269
Tarbela Joint Venture (TJV) 269
Tay Bridge, first 52, 58
Techapi Mountains, USA 204
Tecolote Tunnel, USA 113
Telford 228
Telford, Thomas 18, 43, 224–5, 228, 234, *271*
Tennessee River Bridge 53
Tennessee Valley Authority 253, 255
Tepula Aqueduct, Rome 210
Terzaghi, Karl 225, 229
Tete River Bridge, Mozambique 46
Texas Stadium, Irving 96
Thames Barrier 167, 188, 197, *199*, 270
Thames–Lee Tunnel 109
Thames River 217
Thames Tunnel 99, 122, 123, 227
Thames Water Authority 216–17, 219, *220*
Thatcher Bridge, Panama 32
Thijsse 232
Third water tunnel, New York 111
Thirlmere Aqueduct 210
Thistle Field, North Sea 261
Thornton Cleveleys Archimedean screw pumps *202*
Three ugly sisters *273*
Tiahuanaco, Peru 266
Tibetan bridges 236
Tibi Dam, Spain *64*, *76*, 83
Tiel Bridge, Holland, 45
timber bridges 15–17; longest 58; longest span 31
timber building, world's largest *102*
Timoshenko 230
Tinsley Viaduct 60
Tiratsoo, E N 153
Todt organisation, Germany 272
Toktogoul Dam, USSR 65–6
Tolla Dam, Corsica 68
Tomakomai, Japan 178
tools: bulldozers 149–50; chain-bucket excavators 148–9; draglines 147–8; most productive muck-shifting: land 143, sea 142–3; steam-shovels 145, 147; stripping shovels 148
Toulouse building 94
Tourtemagne Dam, Switzerland 73
Tower Bridge 36, 53
Tower Bridge Tunnel 124
Townline tunnel, Canada *187*
Trajana Aqueduct, Rome 211
Trajan's Bridge, River Danube 31
Trans Canada gas pipeline 212
Transamerica Pyramid, San Francisco 86
Transcontinental railroad, USA; joining the rails *227*
transporter bridges 53; longest span 54
Tré-la-tête Dam 69
Tredgold 273
Trent and Mersey Canal Tunnel 122
trestle bridges 13, *17*; world's longest bridge 58
Trevithick 122
Trezzo Bridge, Italy 30, 234
Tronquoy Tunnel, France 122
trussed bridges, 12, 13, 16, 22, 23, 24, 25, 35; longest continuous three-span claim 39; longest prestressed concrete spans 46; only surviving Bollman *21*; world's longest simply supported spans 47
Tubarao Bulk-loading Terminal 145, *179*, *182*; slewing bridge ship loader *179*
tubular bridges 21, 24
tunnels and underground workings: caisson and under-the-roof tunnelling *156*; canal tunnels: first 236, first major British 118, 122; compressed-air techniques 124–6, 128; cut and cover 106; drills *128*; driven 106; early 122; hard ground record rates 131–3, British 133; highest pressure, tunnels and shafts 140; holes, big underground 138–9; immersed tube 106, 117, *122*, 123, largest 120–1; kanats 137; largest cross-section 7, 115–16; longest 106–7, 109, 110–13, 117–19, road 113, hydro-electric 113–15; mining 140; moles *125*, 129, *131*, *134*; oldest story 136–7; pipe-jacking 134–5; power stations, reservoirs and underground factories 139; rigs *126*, *128*; rock 105, 126–30, longest 128; Roman 272; shields, Brunel's first *123*–4, *124*; shutter *124*, soft ground 105, 122–5, record rates 134; subaqueous 107–8, longest 119; tragedy 126; volumes 137–8
Tuticorin, India 180
Tyrrell 17, 30, 34, 42

Union Bridge, Berwick 30, 43, 47
United Californian Bank Building, Los Angeles 86
United Engineers and Constructors Incorporated, Philadelphia 269
University of Georgia Coliseum *103*
University of Illinois Assembly Hall, Champaign 95
Upper Sone Bridge, India 57
Urado Bridge, Japan 45
Uruguay River Bridge 45
US Interstate highway system 271
US Steel Corporation Building, Pittsburgh 86
Uskmouth Power Station 158
USSR irrigation 255–6
Ust–Ilim Dam, USSR 67, 80; power station 201
Utah natural arch, longest span of stone 11
Utzon, Jorn 251–2

Vado Hondo Dam, Argentina 70
Vajont Dam, Italy *61*, 62, 68, 79, 81, 83, 170, 231
Valtellina Tunnel, Italy 114
Vanbrugh, Sir John 41, 273
Vanne Aqueduct, France 217
Vauban, Sébastien le Prestre 164, 226
vaults, earliest 103
Vazie 122
Veen, Dr van 151, 232, 244, 245
Veeranam Dam, India 81
Veerse Gat Dam, Holland 82, *248*
Velsen Tunnel, Holland 167, *168*
Vermuyden, Sir Cornelius 154, 197, 198
Verrazano Narrows Bridge, USA 26, 27, 29, *34*, 42, *48*, 51, *160*

Versailles 215
Vertical Assembly Building, Merrit Island, USA 99
vertical-lift bridges: first 54, world's longest spans 54
V I Lenin Dam, USSR 67
viaduct bridges 13, 17
Victoria, Lake 67
Victoria Bridge, Montreal 40–2, 51
Victoria Underground Railway *129*
Vierendeel bridges 52
Vierlingh, Andries 245
Vinstra Tunnel, Norway 114
Virgo Aqueduct, Rome 211
Volga–Baltic Canal (V I Lenin), USSR 189, 191
Volga–Don Canal (V I Lenin), USSR 154, 191
Volga River 230
Volgograd Bridge 54
Volta, Lake 67
Volta River Bridge 42
von Mitis 23
Voulzie Aqueduct, France 217
Vyrnwy Dam 76

WAC Bennett Dam, Canada 67, 70
Waddell 36
Wadi Kuf Bridge, Libya 45, 228
Wadi Tharthar Dam, Iraq 67, 81; flood-protection scheme *198*, 200
Walcheren dikes, Holland 232
Walt Disney World, Florida 94
Walt Whitman Bridge, Philadelphia 28
Wanapum spillway, USA 191
Wapping Tunnel 119
Warszawa radio mast, Poland 87, 89
Washington Memorial, USA 89, 92
Washita County wildcat well, USA 153
water-closet, world's most important invention 219
water supply: largest authority in England and Wales 216–17; world's largest system 216
Water Tower Place, Chicago 86
water treatment: earliest 219; first successful filter-bed plants 219; largest plant, world, Britain 219
Watergrove Dam 170
Waterloo Bridge 15
Wedgwood, Josiah 122
weight: heaviest ever dropped 263; heaviest lifted 264
well: deepest Europe, Middle East 153; deepest productive 153; deepest for water, world 153, 217, Britain 217
Welland Ship Canal, Canada 186, *187*, 190, 192, 215
Werwag 228
West Bay Bridge, San Francisco 28, 266
West Delaware Tunnel, USA 109
West Gate Bridge, Melbourne 44, 52
West Side Highway elevated road, New York 58
West Side sewage plant, New York 160
Western Avenue Extension elevated road, London 58, 60
Western Deep Levels gold-mine, Carltonville, South Africa 140
Westfield Mine, Fife 155
Westminster Bridge 164
Westminster Hall 104
Whakapapa Tawhitsikuri Tunnel, NZ 114
Wheeling Bridge, USA 29, *43*
Whipple 229
Wieringen Island, Holland 246
Wieringermeer Polder, Holland 246–7
Wilkinson, W B 99–100
Willamette River Bridge, USA 53
William of Orange 189, 192
Williams, Sir Leader 54
Williamsburg Bridge, New York *35*
Wimpey and Company 269
Winch Bridge, Middleton 18
Windscale pipeline 212
Winningen Bridge, Germany *12*
Winslow, E M 211
wire-winding, biggest 268
wisdom tooth (building) *275*
Wittengen Bridge, Baden *17*, 31
Wolstanton Colliery 140
Woolley, Sir Leonard 103
Woolwich Tunnel 125
Woolworth Building, New York 88, *90*
works, largest in Britain 270
World Register of Dams 72
World Trade Center, New York 85, 94, 99
world's great bridges, The 159
Wright, Benjamin 226
Wuhan flood-prevention, China 151 fn

Yangtze River bridge *236*
Yellow River 150; length of irrigation canals 189
Yewbarrow Tunnel 132
Yokohama 178

Zaire River 200
Zarate–Brazo Largo Bridge, Argentina 44, 56, 58
Zdakov, Czechoslovakia 32
Zeiss 104
Zermatt inverted siphon, Switzerland 215
Zeya Dam, USSR 67, 69, 80, 196
Zola, Emile 76
Zola Dam 76
Zoo Bridge, Cologne 44
Zuider Zee Dam, Holland 81, 82, 260
Zuider Zee reclamation plan 228, 245, 246–7
Zuk Hydroproject Institute, USSR 269
Zuk, Sergei Yakovlevich *230*

Index compiled by Gordon Robinson